Foundations of Real Estate Financial Modelling

Foundations of Real Estate Financial Modelling is specifically designed to provide an overview of pro forma modelling for real estate projects. The book introduces students and professionals to the basics of real estate finance theory before providing a step-by-step guide for financial model construction using Excel. The idea that real estate is an asset with unique characteristics which can be transformed, both physically and financially, forms the basis of discussion.

Individual chapters are separated by functional unit and build upon themselves to include information on:

- Amortization
- Single-Family Unit
- Multifamily Unit
- Development/Construction Addition(s)
- Waterfall (Equity Bifurcation)
- Accounting Statements
- Additional Asset Classes.

Further chapters are dedicated to risk quantification and include scenario, stochastic, and Monte Carlo simulations, waterfalls, and securitized products. This book is the ideal companion to core real estate finance textbooks and will boost students' Excel modelling skills before they enter the workplace. The book provides individuals with step-by-step instruction on how to construct a real estate financial model that is both scalable and modular.

A companion website provides the pro forma models to give readers a basic financial model for each asset class as well as methods to quantify performance and understand how and why each model is constructed and the best practices for repositioning these assets.

Roger Staiger (FRICS) is Managing Director for Stage Capital, LLC, a global advisory firm in real estate financial modelling, portfolio management and asset repositioning. He holds faculty positions at George Washington University, Georgetown University and Johns Hopkins University. He has held many senior positions including Managing Director for Constellation Energy's Retail Commodity Division and CFO for America's Best Builder 2006.

www.routledge.com/cw/staiger

Foundations of Real Estate Financial Modelling provides the reader with a clear path towards appropriate methods and processes used to properly model various types of real estate investments. In addition, the book provides the "why" behind the models by introducing practical explanations of valuation, capital stack formation and assessment of risk. The book also defines key terms, which are especially important in today's real estate industry where jargon and undefined "rule of thumb" benchmarks proliferate.

What is noteworthy is Professor Staiger's use of visual enhancements (graphs and actual pro-formas) throughout the book that bring clarity and meaning to the content. This is effective because many adult learners are visual in their comprehension of new material.

In summary, students will appreciate the thoughtfulness of the book that provides a roadmap to a clear and concise explanation of the process and practical application of real estate models for financial forecasting and risk assessment.

Bob Rajewski, Adjunct Professor, John Hopkins University

The book is aptly named. The author does an excellent job of providing the context in which real estate modelling should be viewed in. Rather than being a step-by-step instruction manual, Staiger reviews the appropriate supporting theory that must be understood before diving headlong into real estate financial modelling. As real estate is not a commodity, there is no "one size fits all" financial model. Given the extreme variations in assumptions, risk profiles, and deal structures, real estate financial models are as diverse as the assets they attempt to quantify. Failure to understand these concepts along with the industry standard metrics will almost guarantee the construction of flawed models.

In general terms, pro forma modelling is fairly straightforward with a basic understanding of finance. However, many seasoned analysts struggle with modelling waterfalls. As the waterfall is designed to separate risk and associated returns by tranches, it therefore requires the adjustment of payment streams. Done incorrectly the outcome can be disastrous causing potential investors to lose faith in the integrity of the model and possibly the entire project. The author devotes a substantial amount of time covering this topic and incorporates a variety of diagrams to simplify the concepts.

I recommend this book to anyone attempting to build a defensible real estate financial model. This book will inspire confidence to both beginner and advanced model builders.

Keith A. Hopkins, MBA, Managing Director, Stage Capital Group, LLC

Foundations of Real Estate Financial Modelling was created as part of Professor Staiger's curriculum within the School of Real Estate at Georgetown University's School of Continuing Studies. It is a superb step-by-step process of how to build various models which are both scalable and modular. The textbook will be an excellent resource for students, as well as non-finance faculty, to understand the mechanics of building a real estate pro forma.

Roger's disciplined and purposeful approach to pro forma modelling has been so well received by the students that we regularly offer workshops, free to students, to ensure the skill is gained and focused upon in the program. This textbook is an excellent resource for the students to supplement the workshops and explain the basic theory of real estate finance coupled with technical skills.

We are proud that we will be incorporating this textbook within our curriculum at Georgetown University. This textbook could well be required reading for all students.

William H. Hudnut III, Executive Director, Georgetown University

Foundations of Real Estate Financial Modelling

Roger Staiger

LONDON AND NEW YORK

First published 2015
by Routledge
2 Park Square, Milton Park, Abingdon, Oxon OX14 4RN

and by Routledge
711 Third Avenue, New York, NY 10017

Routledge is an imprint of the Taylor & Francis Group, an informa business

British Library Cataloguing-in-Publication Data
A catalogue record for this book is available from the British Library

Library of Congress Cataloging-in-Publication Data
Staiger, Roger P., III.
Foundations of real estate financial modelling / Roger Staiger.
pages cm
Includes bibliographical references and index.
1. Real estate investment—Mathematical models. 2. Real estate investment—Finance—Mathematical models. 3. Real estate development—Finance—Mathematical models. 4. Real property—Finance—Mathematical models. I. Title.
HD1382.5.S72 2015
332.63′24015118—dc23
2014042657

ISBN: 978-1-138-02516-5 (hbk)
ISBN: 978-1-138-02517-2 (pbk)
ISBN: 978-1-315-77532-6 (ebk)

Typeset in Times New Roman
by Apex CoVantage, LLC

Contents

Acknowledgements

I would like to express my sincere gratitude to the many people who helped me author this book over the past two tumultuous years of my life – a period which saw my father die under suspicious circumstances and resulted in a will caveat. This ended with a unanimous jury verdict finding that his will was forged and not duly executed. (This will be the subject of a different book.)

On the darkest of days, this book was my lifeline to sanity and progression. The support over the past two years this book and I, in particular, have received has been tremendous. Bill Hudnut, at Georgetown, absolved me of all administrative responsibilities. Mike Anikeeff, at Johns Hopkins, ran interference with the administration, who always seems to trip up on themselves. Bob Rajewski, my respected friend and colleague, stood in whenever court appearances disallowed me to cover a specific class (I will never be able to thank you enough). Ed Harding, of Team13System, the smartest accountant/finance person I have ever known, edited the book, providing countless hours of improvement.

Two students authored chapters: Reagan Mosley and Michael Cardman, Pro Forma Portfolio Modelling and Structured Products, respectively. Your tireless hours and superb efforts are appreciated and noted. Reagan, I promise to never tease you in class about Tinder. Michael, your efforts as TA made it possible to write my book in the evenings. To both, thank you.

Finally, Jennifer, my secret weapon and the smartest person I have ever met, thank you. At the beginning of the court saga regarding my father, I put my hand in the air and said, "You always said you would help me if I needed it. Well, I need 'help,' please help me." The tireless hours you spent focused on my court case which brought justice for my family will always be remembered. To all reading this, while Jennifer would never say anything because she is soft-spoken and polite, she attended Harvard '91 and Yale Law '95. For all your love and assistance, I promise to never question your degree in women's studies at Harvard (a dual major actually with economics) or ask why you cannot iron a shirt when that should have been covered by the end of your first year at Harvard!

Although I have mentioned a few people here, the actual list is long and unending. From students encouraging me to finally publish to peers anxious to see the long-awaited book, I thank you for your consistent and constant encouragement. While my name may be on the cover (which I love, by the way), I fully recognize it has been a long journey, and please know I recognize everyone's efforts and readily admit this was far from a singular effort.

Thank you!

Preface

Foundations of Real Estate Financial Modelling is a book designed to assist individuals in developing real estate pro forma models beginning with a fresh spreadsheet. Although the first three chapters cover theory and provide background for real estate, the book is actually a "how-to" book on pro forma construction.

In the first three chapters readers review the basics of real estate, valuation, and cash flow distinctions between asset types (i.e. hotel, multifamily, retail, etc.). In addition, readers are introduced to a new metric, P(Loss), developed by the author. P(Loss) quantifies the probability of not returning 100.0% of invested capital, that is, the most important question for an investment. It is innovative in its simplicity and requirement of risk quantification.

In later chapters, beginning with Chapter 4, readers follow the chapters to construct ever-more complex real estate financial models. The finished versions can be found on the companion website for use after the reader has personally completed the model. What the book does not do is analyze the use of these models, though real case studies are included on the companion website. The book focuses on construction rather than use. This book is best used as a classroom tool or by the novice to intermediate financial analyst who has a background in real estate.

Later books will include more analysis of case studies using the base models constructed within this text. For now the focus is on model construction rather than case study analysis.

1 What is real estate?

When considering the question "What is real estate?" more questions are produced than answers. Real estate is considered to be real property. This translates the question to what is "property"? For the purposes of this book, real estate is defined as investment(s) in physical buildings and land, in both combined, or in financial products backed by physical buildings and land, such as mortgage-backed securities (MBS). Therefore, the mental image of real estate is generally a building, home, or farm – i.e. a physical structure residing on land. However, the picture of real estate should actually be a return distribution which includes both return and risk. Real estate is tangible, it can be seen, touched, felt, and at times of distress, kicked. Real estate is also intangible as it is a financial instrument which can be transformed physically and financially. Finally, the value of real estate, in reality, is the cash flow it produces, which is a direct result of the tenants and leases and the risks associated with these cash flows. Conceptually, the physical real estate is an expense item while the tenants produce the cash flows for a real estate project. The characteristics of these cash flows depend on the type of real estate (i.e. asset class) and the current economic environment.

To consider real estate in a broader context, it is an asset class which attracts investment funds. In fact, real estate is the second largest asset class in the US and was actually the largest asset class in 2006, prior to the bubble bursting. To provide some context, in 2010 the US GDP was approximately $14,500bn, with total US asset values (aggregation of Fixed Income, Equity, and Real Estate) of approximately $69,000bn. In 2010, the asset split was Fixed Income ($32,500bn), Equity ($18,000bn), and Real Estate ($18,500bn). Therefore, real estate, in 2010, was approximately 27% greater than US GDP and the same percentage of US asset value.

Real estate in the context as an asset class must be bifurcated further into residential and commercial components. Residential is four times the size of the commercial asset class and is actually a leading indicator by 7 months (based on historic analysis of the Case-Shiller index versus RCA's commercial index in 2014). Residential real estate is therefore approximately $14,800bn while commercial real estate is $3,700bn (note: commercial real estate in 2007 was $6,000bn).

Although the differences between residential and commercial may appear to be obvious, it is important to reclassify each to best describe their financial characteristics – i.e. residential is non-income producing and commercial is income producing. Classifying commercial as income producing is the conventional methodology. However, the rhetorical question remains, "Are both real estate asset classes not both non-income producing?" The difference is that the commercial asset class has historically been purchased for investment purposes (i.e. an investment-type asset), while traditionally residential real estate purchased as a primary home is considered for personal consumption (i.e. residential home). To further extrapolate, residential is also quasi-income producing when one considers the rental expense that would be incurred if individuals were not in their own homes. MBAs call this deference of rental expense "opportunity cost" and model it as if it were an expense to determine price.

At the core, all real estate has the same cash flow diagram (Figure 1.1).

Specifically, real estate is unique as an asset class because, unlike equity and fixed-income securities, the purchase is rewarded not with income (e.g. dividends and/or coupon payments) but rather with invoices. These invoices are maintenance, in the case of a physical structure, property taxes, etc. Basically the 'rewards' for owning real estate, in its purest form, are expenses which are both timely and constant. Of course the owner also has use of the property and owner's rights to the property, but it is the tenants that provide the income for the asset. Further, these expenses are known and can be accurately forecast given historic depreciation tables and tax forecasts. Therefore, the value of real estate, in purity, is negative.

Then why does real estate trade with positive pricing? The answer: the ability of real estate to attract cash flows, realized or not. Of course there are other reasons as well – for example the ability to absorb large amounts of capital in singular transactions, community investment and presence, and so forth – but in summary, it is the ability to actively manage the asset to produce positive cash flows (i.e. income over maintenance, taxes, and debt service). Therefore, the value of real estate is not the physical structure itself but rather the rents, the leases, the asset can attract.

This is the subject of Chapter 12, but the value of real estate can be modelled using Portfolio Theory. In the case of an office project where multiple tenants occupy the asset, the value is determined as the portfolio of leases rather than as a single project, or office asset. This addresses repositioning given that a single-tenanted building provides no diversification of industry and risk while a multi-tenanted building provides diversification, provided the tenant's industries are not highly correlated. Extrapolating further, a multi-tenanted real estate asset can be modelled as and considered a portfolio of forward rate agreements (FRAs) rather than a single asset. Again, this is the subject of later chapters.

An actively managed real estate investment with tenants will have the more traditional cash flow diagram associated with equity and fixed-income securities (Figure 1.2).

Note that the previous diagram assumes a terminal value at the end of the period. There are two methods of valuing a real estate asset: (1) assume cash flows continue into perpetuity and (2) assume cash flows end upon sale, or terminal value. For the purposes of this text, a terminal value will be assumed in calculating yield; however, it should be understood that

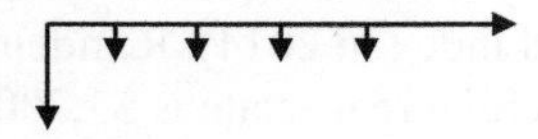

Figure 1.1 Real Estate Cash Flow Diagram

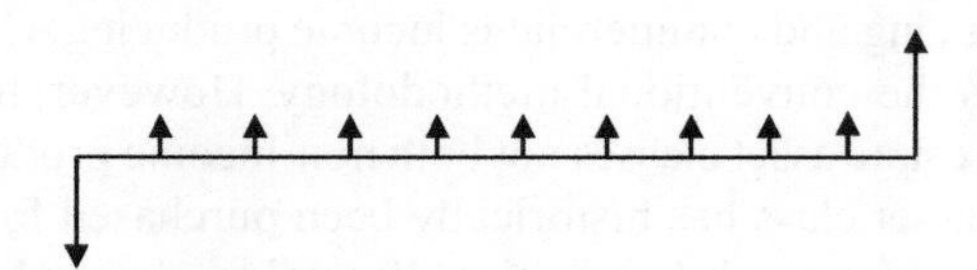

Figure 1.2 Real Estate Cash Flow Diagram with Tenants, i.e. Leased

$$\text{Sales Price}_n = \frac{\text{NOI}_{n+1}}{\text{Capitalization Rate}}$$

Figure 1.3 Sales Price/Capitalization Rate Equation

the terminal value can be removed if the property is considered held (i.e. valued) into perpetuity (note: terminal value in this text will be quantified using the traditional capitalization rate method). See Figure 1.3.

Note that sales price is calculated as a function of net operating income (NOI) rather than net income. Net income is NOI less debt service, that is, cost of financing. This is consistent with Modigliani and Miller's propositions from corporate finance, which state the value of an entity is independent of financing concerns.

Yield

Yield is a valuation method used for Western finance – i.e. Europe, North and South America. It is the solved discount rate for the series of cash flows when setting the Net Present Value (NPV) equal to zero, in other words, the Internal Rate of Return (IRR). This is more simply viewed in Figure 1.4 and Figure 1.5.

The NPV equation is a polynomial. Therefore, it is important, to avoid multiple IRR possibilities, that there be only a single cash flow change – i.e. cash flows change from negative to positive once. If there are multiple changes in cash flow signs, as in the following example, there is the possibility of multiple IRR solutions.

Note: there are other pitfalls when completing an IRR; however, they are not all covered in this text. Some of these pitfalls include type of project consideration (i.e. financing versus investment) and using IRR to compare multiple projects with different initial capital and varying time lines.

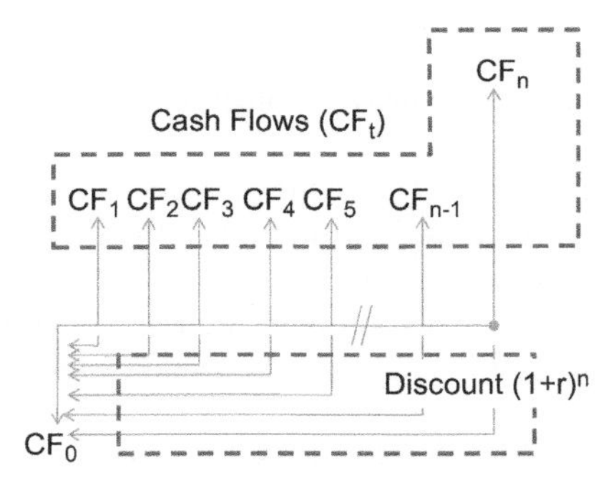

Figure 1.4 Cash Flow Diagram

$$NPV = CF_0 + \sum_{t=1}^{n} \frac{CF_t}{(1+r)^t}$$

Figure 1.5 Net Present Value Equation

Net Present Value (NPV)/Internal Rate of Return (IRR) in detail

Net Present Value (NPV) is the Present Value (PV) of all future cash flows, discounted at the appropriate market rate or the rate of alternative investments, minus the initial cash outlay. The Net Present Value rule is that an investment is worth considering when the NPV is a positive. It is calculated using the formula in Figure 1.6:

$$NPV = C_o + \sum_{i=1}^{N} \frac{C_t}{(1+rate)^i}$$

Figure 1.6 Net Present Value Equation (Rearranged)

where
C_o = – Initial Cash Outflow
i = period or timing of cash flow
rate = discount rate used to evaluate the cash flows

Without exception, NPV can be used to determine value of projects to the equity holders. Accepting projects with a positive NPV always benefits the equity holders, assuming an appropriate discount rate has been applied.

Internal Rate of Return is the rate calculated by setting the NPV equation to zero for a series of cash flows. It is the premier method for evaluating capital expenditures. It provides an intrinsic value of a project, expressing the realized rate of return for future cash flows. However, unlike NPV, IRR has pitfalls and does not always determine which cash flows offer positive value to equity holders unless it is completely understood.

The following is a list of some, but not all, of the pitfalls to using IRR for cash flow evaluation.

- Multiple IRRs
- Investment versus financing
- Scaling (specific to mutually exclusive projects).

The first, multiple IRRs, is important since the NPV formula is a polynomial when solving for the rate. Therefore, for specific cash flows, those flipping from negative to positive or positive to negative more than once, multiple IRRs may be found. The following example illustrates a series of cash flows, with six flips in sign, in which two IRRs exist. Neither is correct and neither is incorrect. It is impossible to evaluate these cash flows using the IRR method. See Figure 1.7.

When graphing the NPV versus rate, two IRRs are located at each point of an X-intercept (3% and 58%). Both are correct and neither offers a true representation of the worth of the cash flows to equity holders. See Figure 1.8.

Date	1/1/90	1/1/91	1/1/92	1/1/93	1/1/94	1/1/95	1/1/96	1/1/97	1/1/98	1/1/99
Cash Flow	–4,500	4,000	3,000	4,000	–3,000	4,000	–1,000	3,000	–4,000	–6,500

Figure 1.7 Cash Flow Example

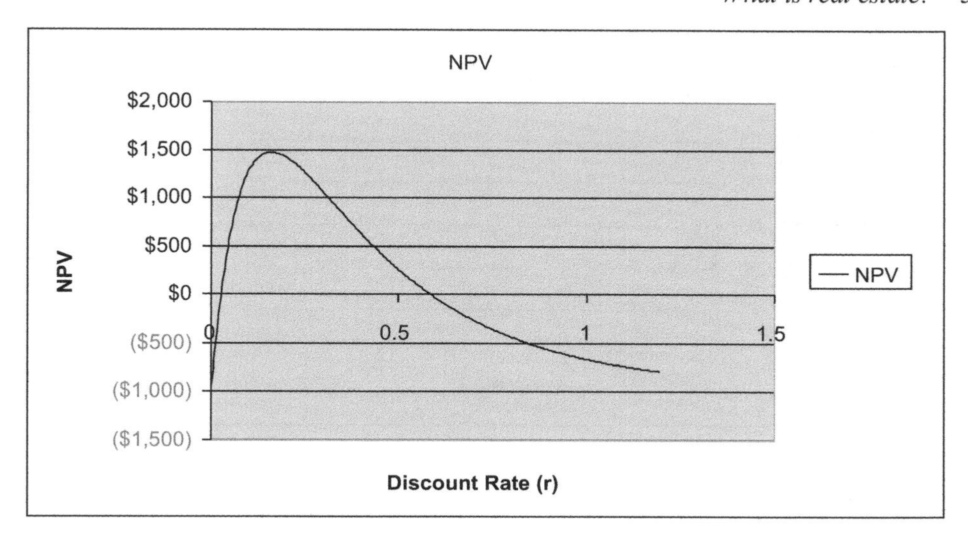

Figure 1.8 Two IRR Examples

The Net Present Value, discounted at 10%, yields a positive value of $1,408. Since the NPV is positive, the cash flows yield value to the equity holders. This would not be the conclusion if the IRR approach were used since it could yield a value of 3%, which is below the discount rate of 10%. Therefore, an NPV analysis yields the correct conclusion.

The second problem with IRR analysis is the altering valuation criteria for investment versus financing projects. Consider the following two projects, A and B, and their respective cash flows:

Project	A	B
Initial Cash	–$100	$100
Cash @ T = 1	$130	–$130

The Internal Rate of Return for each project is 30%; however, the NPVs for projects A and B are positive ($18.20) and negative (–$18.20), respectively, at a 10% discount rate. The IRR analysis would seem to state that both projects are of equal value but the NPV analysis clearly demonstrates that only project A adds value to the equity holders.

The difference is that project A is an investment project whereas project B is a financing project. An example of a financing project is a seminar where the money is received months in advance of the outlays, which occur at the time of the seminar. The rule for acceptance under the IRR criteria is to accept if the IRR is greater than the discount rate for investment-type projects and accept if the IRR is less than the discount rate for financing-type projects. See the graphs in Figures 1.9 and 1.10. This rule change for type of project is confusing and nonexistent in the NPV analysis.

A third pitfall for IRR analysis concerns mutually exclusive projects. Scaling is a critical issue when two opportunities are available but only one can be executed. An example is the

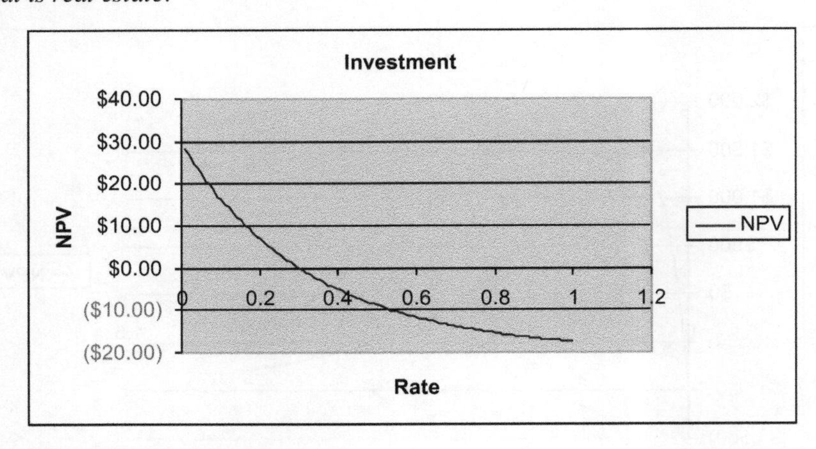

Figure 1.9 Investment-Type Project

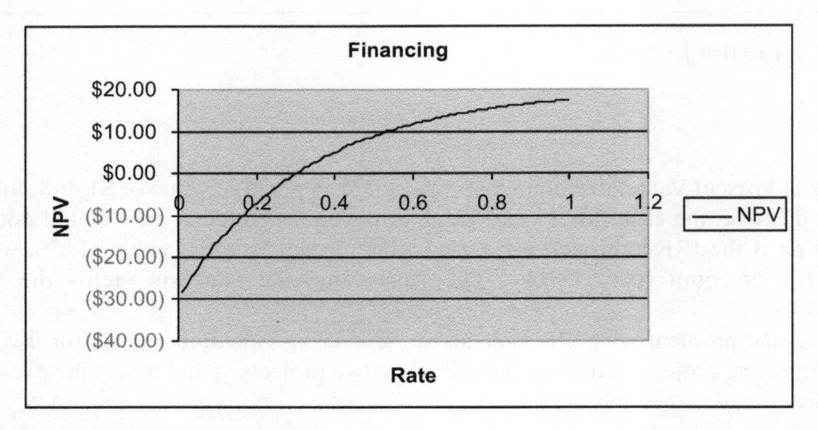

Figure 1.10 Financing-Type Project

ability to earn $50 on one investment and $100 on another. Consider the following two projects, A and B, and their respective cash flows:

Project	A	B
Initial Cash	−$50	−$1000
Cash @ T = 1	$100	$1100

For this example, assume the period for the investment to be short, virtually instantaneous, so that timing effects can be ignored. The NPV for project A is $50 and for project B is $100. The IRR for project A is 100% and for project B is 10%. Using IRR analysis project A should be pursued since it yields the highest return to equity holders; however, NPV analysis shows that project B yields the highest value to equity holders. The IRR analysis is flawed because it neglects to account for the scaling issue that $100 is worth more to the equity holders than $50. This can be adjusted using an incremental IRR analysis but that is beyond the scope of this book.

The three pitfalls mentioned here demonstrate the limitations of using the IRR analysis to evaluate cash flows to equity holders. IRR is beneficial in stating the intrinsic value of future cash flows but it must be completely understood. NPV analysis always yields correct conclusions but does not state the intrinsic value. Both methods should be used together to ensure accurate valuation of cash flows.

Modified Internal Rate of Return (MIRR)

Modified Internal Rate of Return (MIRR) adjusts for the pitfall of traditional IRR analysis which assumes all cash flows are reinvested at the calculated IRR rate. Therefore, traditional IRR analysis may misstate the implicit return for a project by failing to quantify the effect that earned cash flows during the project are not reinvested at a project's IRR but rather at a corporate reinvestment rate (estimated at the corporate weighted average cost of capital, or WACC).

Modified IRR corrects this misstatement by converting a project's cash flows to a zero coupon security. The project's future cash flows are compounded to the final period at the reinvestment (WACC) rate. Using the initial cash outflow, the yield on the zero coupon security is then calculated (MIRR). The equation for MIRR is in Figure 1.11.

$$MIRR = \left[\frac{\sum_{s=1}^{n} (CF_S)(1+r)^{n-S}}{CF_0} \right]^{1/n} - 1$$

where

CF_S = Cash Flows in period S

CF_0 = Initial Cash Flow (Cost)

n = Number of periods

r = Reinvestment rate

S = Current period

Figure 1.11 Modified Internal Rate of Return

An example of the Modified Internal Rate of Return for a series of cash flows follows in Figure 1.12:

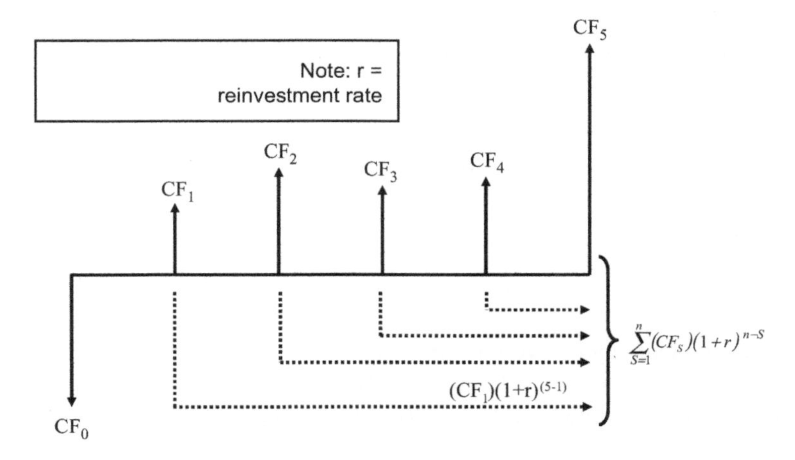

Figure 1.12 Future Valuing Cash Flows for MIRR

The compounding of the cash flows creates a zero coupon security maturing at time period *n* (Figure 1.13).

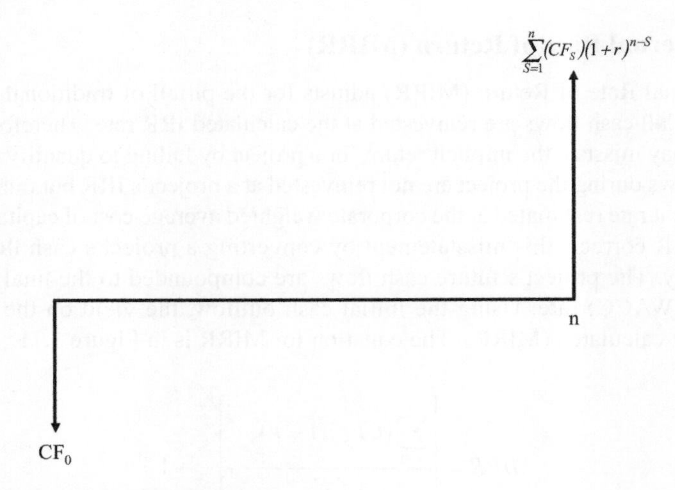

Figure 1.13 MIRR Zero Coupon Bond

Note the Price (CF_0) for the zero coupon security is as follows in Figure 1.14:

$$P = \frac{M}{(1+yield)^n}$$

$$where$$

$$M = \sum_{S=1}^{n}(CF_S)(1+r)^{n-S}$$

$$yield = \text{MIRR}$$

$$n = \text{Periods}$$

Figure 1.14 Price for MIRR

Solving for MIRR yields the equation in Figure 1.15:

$$MIRR = \left[\frac{M}{P}\right]^{1/n} - 1$$

Figure 1.15 MIRR Equation

Risk

'Risk' is understood largely by everyone but the true definition seems to be elusive. Before a discussion of risk commences, risk must be bifurcated into project risk (i.e. single entity) and portfolio risk (i.e. multiple assets held together). Single entity risk, which will be discussed here, quantifies deviation of an expected return for a single project; for example project A had an expected return of 18% with a standard deviation of 5%. Single entity risk assumes a project is held in a vacuum and does not consider additional assets held together. Project risk assumes assets are held together and viewed in their entirety. Portfolio risk therefore not only considers individual project return and risk but also the correlation (i.e. linear association of projects), as well as the respective weights of each asset held in the portfolio. For discussion purposes, project risk will be discussed early in this book whereas portfolio risk will be discussed in later chapters. Therefore, the discussion that follows is for single entities only.

The actual definition of risk is simple: the deviation or variation from an expected outcome. Basically risk is the range of outcomes from the expected value. If one assumes the distribution type is normal, risk, in its most basic form, is the deviation (standard deviation) of a normal curve (Figure 1.16).

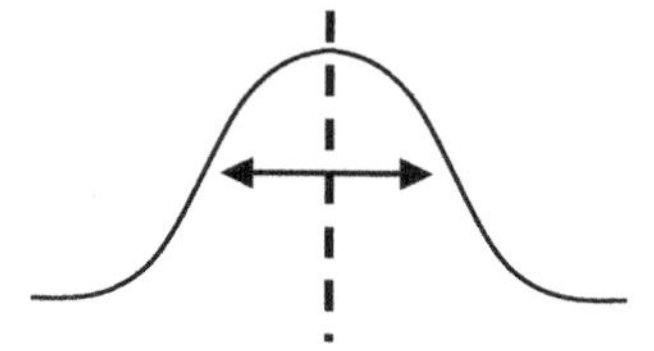

Figure 1.16 Normal Distribution Graphic

It is essential to understand that while this drawing of risk is a representation, it does *not* depict *all* representations of risk. Fundamentally, risk is simply deviation from an expected return. This deviation is often represented by a normal distribution, as humanistic and nature data are best represented as normal distributions. However, risk can be modelled using a myriad of distribution types to include hypergeometric, uniform, triangular, Poisson, etc. The family of distributions includes those both continuous and discrete. The actual characteristics of each should be understood; if they are not, please reference a basic statistics text. For the purpose of this text about asset reposition, the normal distribution, which is a continuous distribution, will be used as a proxy for risk deviation and quantification. Note: there are two general types of distributions, continuous and discrete. Continuous distributions have no probability associated with a singular point but only within a range of points, such as between two points.

There are three main methods to quantify a project's risk: (1) variance, (2) standard deviation, and (3) range. Understanding that the three are all related is essential. Further, each can be used as an approximation for another quantitative measure through the use and understanding of the relationship in Figure 1.17:

$$\sigma = \sqrt{\sigma^2} \approx \frac{Range}{6}$$

Figure 1.17 Risk Relationships

The approximation assumes the distribution is best described by a normal distribution and assumes no outliers in the distribution (Figure 1.18); that is, the empirical rule states +/− 3σ represents approximately 99.7% of the data. Outliers are defined as data values falling outside this range.

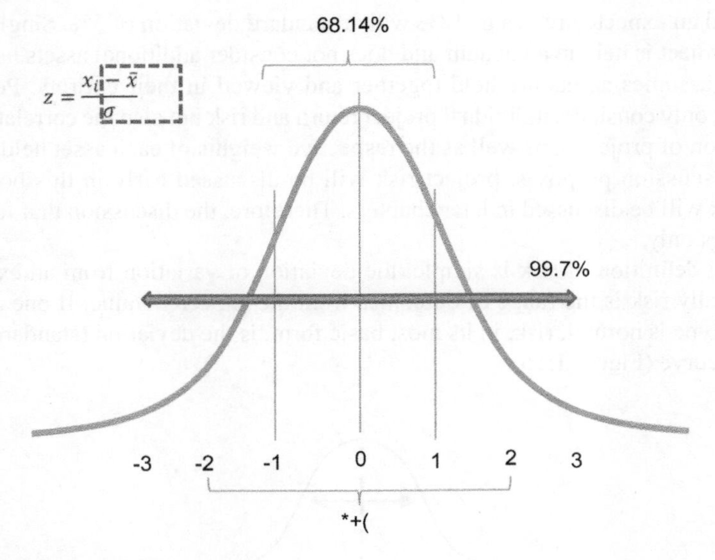

Figure 1.18 Normal Distribution

Understanding the approximation is critical for a simple calculation of risk for projects. Note that a more detailed quantification for risk must use a stochastic approach (such as Monte Carlo) and is beyond the discussion for this text. The simple calculation for risk of a project uses the best-case IRR (yield) and the worst-case IRR. The range is calculated by taking the delta (Figure 1.19):

$$IRR_{\text{Best Case}} - IRR_{\text{Worst Case}} = Range$$

Figure 1.19 Range

The best case will be the highest IRR possible for the project with all 'bull' projections: highest rents, lowest vacancies, etc. The worst case is the polar opposite for the project with all 'bear' projections: lowest rents, highest vacancies, etc. For example, if the best- and worst-case IRR were 50% and −10%, respectively, then the range would be the delta between them. For example, 50% − (10%) = 60% (note: parentheses indicate negative value). Dividing the range by 6 yields a result of 10%. Therefore, the standard deviation – risk – of this project, assuming distribution is normal, is 10% (i.e. $\sigma \approx 10\%$); see Figure 1.20.

$$\frac{Range}{6} \approx Risk(\sigma)$$

Figure 1.20 Risk Approximation

Scenario risk quantification

Although range divided by 6 is a simple calculation, it can be misleading because it assumes the underlying distribution is normal. For real estate this is often not the case. While still a relatively simplistic risk quantification, scenario analysis is superior as it allows for a nonsymmetrical distribution to be approximated. The basic equation for scenario risk (variance) quantification is in Figure 1.21.

This equation may appear new to many, but it is actually the broader definition of variance than what was provided in lower-level education. The two equations most often demonstrated are sample and population variance. The difference is P_i, which allows for non-equal probabilities for each event, R_i.

For example, the two equations for sample and population variance follow in Figure 1.22.

In both cases, population and sample, the probability of each event R_i is equally probabilistic. The reason for the difference between population and sample (i.e. the minus 1) is that the sample uses x-bar to estimate the population parameter μ. This approximation requires a degree of freedom loss adjustment in the denominate (i.e. $n-1$).

The scenario risk (variance) equation is therefore most applicable to generic cases where the probability of each event is not equal. As a demonstration, assume a real estate project has the following returns and probabilities of events (Figure 1.23). Note that event probabilities are difficult to quantify and often are utilized from past experiences on similar projects or projections based on known events going forward. The return calculations for the real estate project were estimated using Internal Rate of Return (IRR); however, the return could have been calculated utilizing numerous methods: MIRR, average Cash-on-Cash, etc.

$$\sigma^2 = \sum_{i=1}^{n} \left[E(r_i) - E(r)\right]^2 p_i$$

Figure 1.21 Scenario Risk Equation

$$\sigma^2_{\text{Population}} = \sum_{i=1}^{n} \frac{(R_i - \mu)^2}{n} \text{ and } \sigma^2_{\text{Sample}} = \sum_{i=1}^{n} \frac{(R_i - \bar{x})^2}{n-1}$$

where

$$\mathrm{P}_{i_{\text{Population}}} = \frac{1}{n} \text{ and } \mathrm{P}_{i_{\text{Sample}}} = \frac{1}{n-1}$$

Figure 1.22 Population and Sample Standard Deviation Equations

	E(R$_i$)	**P(i)**
1)	40.0%	10%
2)	10.0%	60%
3)	–20.0%	30%

$$E(R) = (0.4\times0.1) + (0.1\times0.6) + (-0.2\times0.3) = 0.04 + 0.06 - 0.06 = 0.04$$

$$\sigma^2 = (0.4-0.04)^2(0.1) + (0.1-0.04)^2(0.6) + (-0.2-0.04)^2(0.3) = 0.013 + 0.0022 + 0.0173$$

$$\sigma^2 = 0.0173 \therefore \sigma = \sqrt{0.0173} = 0.1315$$

Figure 1.23 Project Return and Risk Example

The project parameters are therefore:

(4.00%, 13.5%)

Note that had each event been assumed equally probabilistic, the expected return, E(R), that is, the central location point, would have been 10% and the risk 10%. This would result in significant overestimation of expected return and underestimation of risk.

Further note: if the assumption remains that the distribution is normal, the empirical rule can be used to determine the distribution's characteristics. If a normal distribution is not appropriate, Chebychev's formula must be employed.

The important point here is to recognize projects have both an expected return (mean) and risk (standard deviation). Without the understanding of both for a project the most basic question of investing cannot be addressed, that is, what is the probability one will lose money.

Distribution shape(s)

Returning to the definition of real estate with an understanding of return and risk, a visual best describes the asset class. When one considers real estate in the context of both equities and fixed-income securities, the unique characteristics are exemplified. For instance, a direct investment in physical real estate usually involves leverage (i.e. borrowed capital). A typical million-dollar purchase will have the traditional 80% debt and 20% equity capital stack.

The 20% equity is the 'down payment' and represents capital injected by the owner. The 80% is borrowed funds from an external source, generally a lender. Further, the 80% debt may require an external guarantee from the owner. Finally, the ownership of the physical real estate also exposes an owner to the liabilities associated with the real estate, for example cleaning toxic, environmentally unfriendly soils. Therefore, a physical real estate purchase may expose the owner to considerably more financial risk than the initial equity commitment or even the entire capital stack. The total loss in a physical real estate investment may be unlimited. For example, a $10m property is leveraged 80%. If the property just disappears (not realistic but this is theoretical), the $8m debt is still owed. Now, if the property has disappeared *and* the soil is toxic and must be remediated, this is the responsibility of the owner and therefore poses risk greater than the asset's original price.

For the moment, as a point of contrast, the return distribution for equity and fixed-income security (Figure 1.24) limits the loss of investment to 100.0%. In statistics parlance, this is similar to a lognormal distribution where the loss is anchored at 100.0% and the gain trails to positive infinity. With the rare exception of purchasing an equity below par (the details of this are beyond the scope of this text), the maximum loss for an equity purchase is the total invested capital. The same is true for a fixed-income security. The maximum loss for a fixed-income purchase is the total invested capital as well.

Returning to real estate, as discussed earlier, the maximum loss, in theory, is negative infinity (Figure 1.25). This is largely due to the unlimited liability associated with the

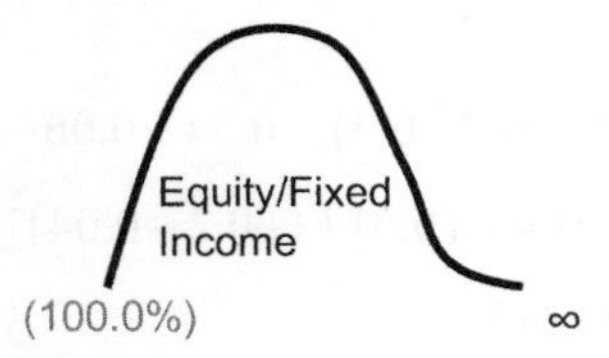

Figure 1.24 Equity/Fixed-Income Return Distribution

Figure 1.25 Real Estate Income Return Distribution

purchase of an asset. While this can be mitigated through legal maneuvering (i.e. purchasing the asset in an LLC, bankruptcy, etc.), the risk of an enormous loss remains real. This risk is real in real estate, as the 2007–08 Great Recession demonstrated. Using the Case-Shiller pricing index as a guide to US residential real estate, peak-to-trough differences were over 35%. Therefore, a $500,000 home with 80% leverage, purchased at the peak of the market and sold at the trough, using simple math, required a $75,000 check to be written at time of sale to satisfy the lender (ignoring principal pay down and the time period between peak and trough, which was about 2 years).

Efficiency/probability of loss (P(Loss))

How is a project best described? Most often a project is described by the yield alone. This characterization is not only immature but also dangerous. Yield fails to answer the most basic question of investment: will I get my money back? Stated differently, yield fails to answer the question pertaining to Return *of* Capital. The question that yield does address is how much money will the project earn (i.e. yield) if all assumptions and forecasts prove to be correct. Yield is therefore, fundamentally, Return *on* Capital.

To summarize, the three fundamental goals of investing are as follows, in order of importance.

1 Stay out of jail
2 Return *of* Capital
3 Return *on* Capital.

For the purposes of this text, only numbers 2 and 3 will be covered. Legal structures and issues with real estate ownership are left to texts which specialize and focus on legal issues. However, it is important to Question: note that earning money is actually the tertiary goal of investing and the least important.

To answer both the second and third goal for investing in real estate, a project's characteristics must summarize both return and risk, that is, $(\bar{x}, \sigma)$, where $\bar{x}$ is the project expected return. It is important to extrapolate both of these metrics and value projects, even though they are singular, as a portfolio consisting of one asset. Therefore, from Portfolio Theory, it is understood that the goal is *not* to maximize return but rather to maximize the efficiency. The equivalent is true for an individual asset held in isolation: the goal is to maximize the efficiency of the project.

What is efficiency? Loosely defined, efficiency is the point of maximum return for any given level of risk, in other words, the ratio of risk and return. A complete discussion of efficiency utilizing Portfolio Theory is beyond the scope and purpose of this text but for the knowledge-seeking reader, the Markowitz Portfolio Theory should be referenced for a greater understanding.

How is efficiency quantified? There are three main methods to quantify efficiency: (1) Coefficient of Variation, (2) Sharpe Ratio, and (3) Treynor Ratio. This text will focus exclusively on the Coefficient of Variation metric. The Coefficient of Variation (CV) quantifies efficiency as in Figure 1.26.

Note that risk is in the numerator and return is in the denominator. Though it may at first be counterintuitive, maximizing efficiency is therefore minimizing the Coefficient of Variation.

Next, how does the CV respond to the goal for investing of Return *of* Capital? Again, we start with the assumption that project yield follows a normal distribution. Further, we remember from our Statistics 101 classes that a normal distribution is defined by two parameters: (1) central location and (2) standard deviation. Both values presumably have been quantified for the project.

It must also be understood that to quantify the Return *of* Capital, a project which has an expected return and standard deviation (risk) not equal to 0 and 1, respectively, must be translated to the standard normal distribution. A z-score is a method to translate a nonstandard normal distribution to the standard normal distribution, which has a central location (expected return) of 0 and a standard deviation of 1. The z-score equation is seen in Figure 1.27.

For the purposes of analysis, the z-score (which in Statistics 101 was an abstract formula to be used on an examination as a means to an 'A' grade) is now a major contributor to the understanding of risk for the sophisticated and the layperson. When using the z-score to translate a project's performance, x-bar is the E(R) and σ the project's risk. X_i is the data value whose location is being transformed from a nonstandard normal curve via the z-score to the standard normal curve.

When analyzing $X_i = 0$, it is at this point that the project actually loses capital, that is, returns less than the initial invested capital. Therefore, a z-score using $X_i = 0$ represents the point at which a project loses money; stated differently, the area to the left of the z-score is the P(Loss) of the project. Demonstrated more simply, setting $X_i = 0$ in the z-score simplifies to the image in Figure 1.28.

$$CV = \frac{\sigma}{\bar{x}}$$

Figure 1.26 Coefficient of Variation Equation

$$\text{Z} - \text{score} = \frac{-\overline{\text{X}}}{\sigma}$$

where

$$\text{x}_\text{i} = 0$$

Figure 1.27 Z-score

$$\text{Z} - \text{score} = \frac{-\overline{\text{X}}}{\sigma}$$

where

$$\text{x}_\text{i} = 0$$

Figure 1.28 Z-score (cont'd)

The next step is recognizing the relationship between this special case of z-score and the Coefficient of Variation (CV). Notice that the z-score is the negative reciprocal of the CV. Therefore, the area to the left of the z-score as calculated by the negative reciprocal of the CV is P(Loss). Remember that P(Loss) addresses the first basic question in investments (aside from legality), Return *of* Capital. (Note: P(Gain) = 1 − P(Loss).) See Figure 1.29.

An example of this translation is the 15-year historical return of the Case-Shiller index, composite-10 MSA. The historical year-over-year average return as calculated on a monthly basis is 5.38%, and the risk (i.e. standard deviation of returns) is 10.28% for the data ending July 2010. The results are summarized as such:

$\bar{x} = 5.38\%$
$\sigma = 10.28\%$

Although the data does respond to the concern of Return *on* Capital in the value 5.38%, it does not, untransformed, address the Return *of* Capital. To do this the Coefficient of Variation, efficiency, must be calculated. Once the CV is quantified, it can be transformed through the use of the z-score to calculate P(Loss), as in Figure 1.30.

$$CV = \frac{\sigma}{\bar{x}} \therefore -\frac{1}{CV} = -\frac{1}{\sigma/\bar{x}} = -\frac{\bar{x}}{\sigma} = z - score$$

Figure 1.29 Coefficient of Variation

$$CV = \frac{\sigma}{\bar{x}} = \frac{0.1028}{0.0538} = 1.9108$$

Figure 1.30 CV Calculation

Transform to the z-score as in Figure 1.31:

$$z - score = \frac{x_i - \bar{x}}{\sigma} \Rightarrow z - score = -\frac{\bar{x}}{\sigma} = -\frac{1}{CV} = -\frac{1}{1.9108} = -0.5233$$

Figure 1.31 Z-score Quantification

Therefore, the area to the left of −0.5233 on a standard normal curve is the probability of loss for this investment as described by the composite-10, Case-Shiller index. Using the "= normsdist(z)" function in Microsoft Excel to calculate the value, we determine P(Loss) = 30.04%. Stated differently, there is a 69.94% chance that an investment in a project with the characteristics as quantified by the historic 15-year performance of the composite-10 will return the initial capital invested or more. Therefore, it is the efficiency measure that, when transformed utilizing a z-score and with the assumption of the underlying distribution being normal, provides the first and most important answer to the investment concern Return *of* Capital.

Asset class consideration(s)

In an example of asset class distributions, the three main asset classes are equity, fixed income, and real estate. To demonstrate an understanding of the differences, indices will be used to approximate each: equity (S&P 500), fixed income (VBMFX), and real estate (Case-Shiller composite-10). (See Figure 1.32.) For the period ending July 2010, the following return/risk characteristics were calculated on a historic 5- and 10-year basis.

What is striking is the difference when comparing the probability of loss for each asset class. Fixed income, as expected, has a significantly lower risk/return ratio. Equity and residential real estate have significantly higher probability of losses. For both periods of this analysis, the risk for equity is higher than real estate and the return, while higher in the 5-year case, may not warrant the risk of investment.

Although the returns can vary and the relationships between the asset classes adjust depending upon the historical period, it is important to view the investment, any investment, in any asset class as a distribution – i.e. with return *and* risk – rather than as simple return. Decisions made on asset investment without risk quantification fails to address the most simple investment premise, not losing money.

Further, an understanding of the return/risk dynamic is essential for an asset reposition. A slight change in either can drastically change asset performance and appearance. It should generally be a rule that when risk can be reduced at a greater rate than the return, the asset becomes more efficient.

A qualitative real estate example would be swapping out a single tenant in a building with another single tenant for the same lease terms. The difference is that the initial tenant could be a B-grade tenant while the new tenant is ExxonMobil, that is, AAA-rated. Because the lease terms have not changed, the expected return for the asset (building) is unchanged; however, the risk profile has significantly changed (i.e. risk has been reduced). This certainly would justify a change in tenant and may even justify a slight price-per-square-foot adjustment downward for ExxonMobil as acknowledgement of the higher-quality tenant. As described, this is a very basic, very successful asset reposition at the tenant level.

5-Year	S&P500	VBMFX	Res RE
Avg Return	0.75%	5.19%	-2.95%
Std Deviation	21.37%	2.92%	12.12%
CV	28.64	0.56	(4.11)
Z-Score	(0.03)	(1.78)	0.24
Prob of Loss	48.61%	3.77%	59.62%

10-Year*	S&P500	VBMFX	Res RE
Avg Return	0.63%	5.32%	4.15%
Std Deviation	20.01%	2.72%	12.91%
CV	31.56	0.51	3.11
Z-Score	(0.03)	(1.95)	(0.32)
Prob of Loss	48.74%	2.53%	37.41%
* Apr10' - Dec01'			

Figure 1.32 Asset Class Comparison

2 Asset valuation defined

Repositioning is simply the transformation of project characteristics, that is, the changing of return and risk of a project, through physical changes in the operating characteristics and/or financial adjustment of cash flow distributions. It is generally done for an asset in distress or an underperforming asset though it can also be done as a measure to optimize an existing portfolio. The two main methods of transformation are considered to be physical and financial.

Physical repositioning is the method to transform an existing project's characteristics to best support portfolio consideration or, at the extreme, transform a losing project into a winner. This is completely 'physically' (i.e. through a physical transformation of the building's façade). A simple example of physical transformation is an office to apartment conversion. An existing office building is repositioned physically as multifamily apartment units, drastically changing the risk/return profile of the asset. Essentially, the transformation achieves what is depicted in Figure 2.1.

To understand this transformation process, it is critical to digest a project at its core, in its cash flow characteristics. It is the cash flow characteristics that lead to the quantification of return and risk. The change in the return and risk for a project is the transformation (i.e. repositioning).

The initial cash outflow for the project is the purchase price, CF_0. The initial cash outflow, CF_0, may also include development and construction costs incurred up to commercial operating date (COD), the completion date (Figure 2.2). The additional negative cash flows up until the large grey dot are expenses incurred, if any, prior to the operations date when the property becomes income producing. In the case of a turnkey project which is fully leased and completed prior to purchase, there may not be any capital outflows prior to operations; however, this would be a special case. It is rare that a project requiring repositioning will not incur negative cash flows at the beginning after purchase and prior to lease-up. Cash flows $1 - n-1$ are net positive cash flows from lease revenue less operating costs (a more complete discussion of these is reserved for the individual asset class sections), and CF_n is the terminal income, that is, net sales proceeds.

What is essential is the understanding and realization that a real estate asset is not the physical but rather the financial cash flows that the physical will attract. Viewing a real estate project not as a physical asset but rather as an asset utilized to achieve cash flows is critical. In truth, the physical structures are actually expense items used in the process of attracting cash flows. The true value of a real estate project is from the leases (i.e. paper), whereas the physical structure is necessary to attract the tenants, resulting in leases.

In fact, a property manager is responsible for the physical whereas an asset manager is responsible for the leases (financial). Of course, in smaller companies with limited project sizes – below $5.0m – these roles are often combined.

$$\mathrm{Pr}\,oj_{\alpha}(\bar{x},\sigma) \Rightarrow tranformed \Rightarrow \mathrm{Pr}\,oj_{\alpha^{*}}(\bar{x}^{*},\sigma^{*})$$

Figure 2.1 Project Transformation

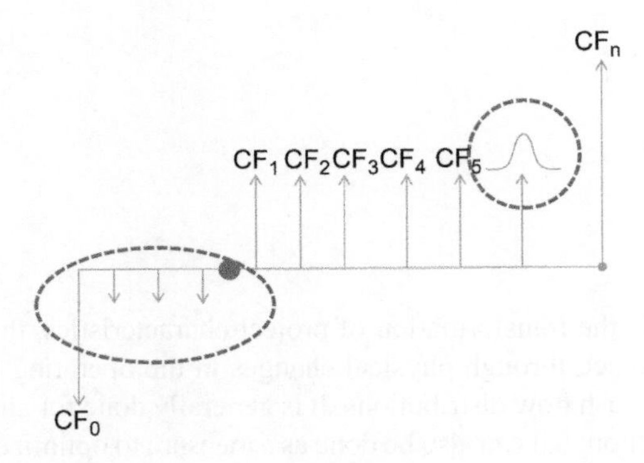

Figure 2.2 Cash Flow Diagram

Physical methods to reposition

A physical repositioning is the act of changing some active element of a project's characteristic, such as the return or risk of the project. These methods, while not mutually exclusive, are in Figure 2.3.

1 Cost Basis
2 Cash Flow Magnitude
3 Cash Flow Timing
4 Cash Flow Riskiness.

Changing any of these is accomplished by physically changing some characteristic of the project – e.g. debt/equity ratio, lease terms, tenant type, office to apartment, etc. It is this type of physical transformation of a project which differentiates real estate investment from equity and fixed-income investment. More accurately, an investor cannot change the characteristics of an equity and fixed-income investment because these are passive investment strategies.

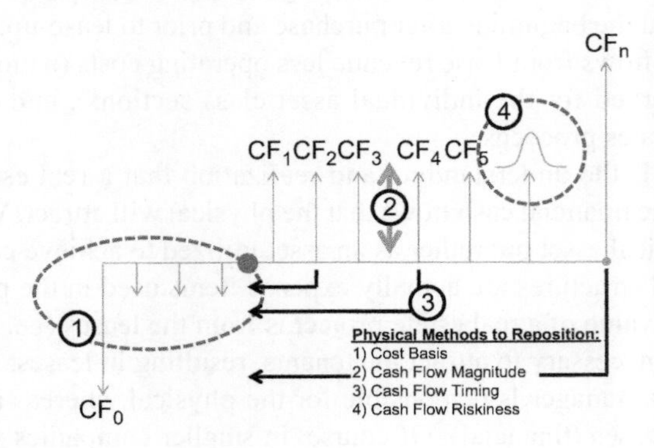

Figure 2.3 Value Creation

However, the characteristics of a real estate investment, which is an active investment strategy, can be adjusted through implementation of different strategies and tactics. An active investment is basically 'hands-on' whereas a passive investment is 'hands-off' (i.e. invest and forget). In a passive investment, the investor has no decision-making authority, only the buy/sell authority of the asset in its entirety.

As the numbers correspond to the diagram in Figure 2.3, the effects of changes in project characteristics will be reflected by corresponding changes in the project's financial attributes, return, and risk. Note the differences in real estate investment, as depicted by the diagram, compared to equity and fixed income. In both equity and fixed income, the only adjustment available to the owner is the cost basis (i.e. purchasing at a lower price). However, with real estate, as this is an active investment strategy with the necessity to manage all aspects of the project, more adjustments are possible to the cash flow of the project.

Single-family detached home example

For the purpose of this introductory chapter, the example which will demonstrate asset repositioning is a single-family detached home example. This asset type is used for simplicity because it has a singular tenant (i.e. single source of revenues), a shorter expense listing, and a capital structure and loan product (30-year constant payment mortgage) which is familiar to most individuals. Further, the model structure will be similar, albeit on a smaller scale, than the other real estate asset classifications discussed later in this text – i.e. retail, office, multifamily/apartment, and hotel.

A brief introductory description of the basic pro forma, which will be used for the entirety of this text, is important. A basic pro forma consists of three sheets, or pages: input/summary, debt amortization, and pro forma. The input/summary page is both the front- and back-end of the pro forma. This sheet captures the inputs (i.e. assumptions) that drive the results, and the lower half summarizes the output and quantifies the summary output in a valuation section. The debt page is a basic amortization for a constant payment mortgage (CPM), in the case of the single-family detached model, a 30-year CPM. Finally, the pro forma page quantifies all cash flows, aggregates the debt payments, and provides net cash flow for the project. Although it is possible to consolidate the model further, to reduce the number of sheets, the three sheets are chosen to allow users to see and understand the scalability of this structure for much larger real estate projects.

A graphical depiction of the model is shown in Figure 2.4. The steps are for modelling purposes and are recommended to be followed by individuals interested in reconstructing the model. (Note: the completed model is provided to purchasers of this book at www.oxfordmodels.com.) However, a short summary of the steps are included for the adventurous individual who intends to reconstruct the model.

1. Construct the inputs of the model, e.g. sources/uses, operating characteristics, financial assumptions, exit considerations, and valuation parameters.
2. Develop the pro forma page (capital inflows and outflows, i.e. basic income statement) to net operating income (NOI).
3. Develop the amortization table. Note, this single-family detached model uses a horizontal construction for simplicity. Later, more complex and larger models will use a vertical structure.
4. Link the debt output (i.e. costs) to the pro forma page.

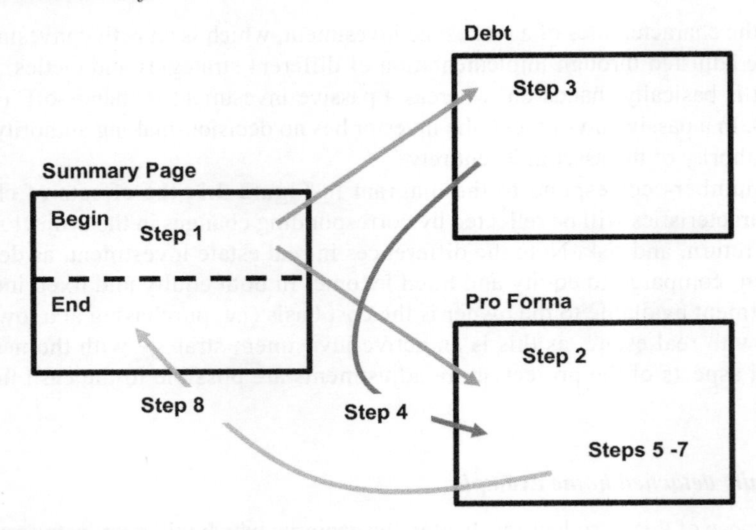

Figure 2.4 Pro Forma Steps

5 Complete Net Income calculation, e.g. NOI less debt payment.
6 Complete sales summary, the exit strategy. Quantify net sales proceeds, i.e. gross sale, less sale expense and principal repayment.
7 Complete net free cash flow – e.g. net income plus (provided net sales is a positive value) net sales proceeds.
8 Summarize pro forma and net cash flow on summary page (bottom). Quantify return and risk calculation to determine project characteristics.

The input/summary page for the single-family detached model is in Figure 2.5.

As stated, the top half, above the grey line, is the input section. This includes the initial capital structure (i.e. Sources/Uses) and the remaining operating, valuation, and sales considerations. The bottom half, below the grey line, is the summary output section. This includes the annual summary of cash flows (i.e. revenues and expenses), debt services (principal and interest as this is a before-tax analysis), and net sales to free cash flow. The free cash flow is bifurcated between operating cash flow and reversion to differentiate the proportion of value coming from each. This bifurcation is another indicator of risk – i.e. the higher the percentage from reversion the more risk in the project.

The debt page is self-evident; it is simply a horizontal amortization schedule. The schedule separates interest and principal payments, the sum of which is the total payment per period. The structure appears in Figure 2.6.

The pro forma page is a scalable summary of all the inflows and outflows from the model. It is basically an income statement without any accounting adjusted values (depreciation, amortization, taxes, etc.). It is this page which is summarized on the bottom half of the input/summary page. See Figure 2.7.

Note that while the analysis states a final sales price in period 10, the pro forma models 11 periods. Recall that sales price in period n is quantified as the NOI period $n + 1$ divided by the terminal capitalization rate; in other words, a 10-year analysis requires an 11-year forecast of net operating income.

Sources	325,294
Equity	75,061
Investor 1	75,061
Investor 2	
Debt	250,233
Tranche 1	250,233
Tranche 2	
Uses	325,294
Purchase Price	312,791
Closing Cost	10,000
Finance Cost	2,502
Surplus/(Deficit)	-

Operations	
Rental/mo	3,200
Rental Escl	2%
Vacancy	5%
Expense Escl	2%
Maintenance/mo	100
HOA	150
Property Tax/yr	5000
Insurance	250

Finance	
Loan-to-Value	80%
Principal	250,233
Term (years)	30
Interest	7%

Valuation	
Cap Rate	8%
Sales Expense	4%
Discount Rate	10%

Valuation: (Taken from Pro Forma tab)

		1	2	3	4	5	6	7	8	9	10
Revenue											
Rental		38,400	39,168	39,951	40,750	41,565	42,397	43,245	44,110	44,992	45,892
Vacancy		1,920	1,958	1,998	2,038	2,078	2,120	2,162	2,205	2,250	2,295
Total Revenue		**36,480**	**37,210**	**37,954**	**38,713**	**39,487**	**40,277**	**41,082**	**41,904**	**42,742**	**43,597**
Expense											
Maintenance/mo		1,200	1,224	1,248	1,273	1,299	1,325	1,351	1,378	1,406	1,434
HOA		1,800	1,836	1,873	1,910	1,948	1,987	2,027	2,068	2,109	2,151
Property Tax/yr		5,000	5,100	5,202	5,306	5,412	5,520	5,631	5,743	5,858	5,975
Insurance		3,000	3,060	3,121	3,184	3,247	3,312	3,378	3,446	3,515	3,585
Total Expenses		**11,000**	**11,220**	**11,444**	**11,673**	**11,907**	**12,145**	**12,388**	**12,636**	**12,888**	**13,146**
Net Operating Income		**25,480**	**25,990**	**26,509**	**27,040**	**27,580**	**28,132**	**28,695**	**29,269**	**29,854**	**30,451**
Debt		**20,165**	**20,165**	**20,165**	**20,165**	**20,165**	**20,165**	**20,165**	**20,165**	**20,165**	**20,165**
Net Sales Proceeds											**159,087**
Reversion FCF		-	-	-	-	-	-	-	-	-	**159,087**
Operating FCF		**5,315**	**5,824**	**6,344**	**6,874**	**7,415**	**7,967**	**8,529**	**9,103**	**9,689**	**10,286**
Total FCF	**(75,061)**	**5,315**	**5,824**	**6,344**	**6,874**	**7,415**	**7,967**	**8,529**	**9,103**	**9,689**	**169,373**
Cash-on-Cash	**9.93%**	**7.08%**	**7.76%**	**8.45%**	**9.16%**	**9.88%**	**10.61%**	**11.36%**	**12.13%**	**12.91%**	**225.65%**

PV Reversion	61,335	57.73%
PV Operating	44,905	42.27%
Total	106,240	
CFO	75,061	
NPV	31,180	
IRR	15%	

Figure 2.5 Pro Forma Example

Model assumptions and base case summary

For simplicity, six inputs will be varied within the model, holding all else constant (note: a seventh will be varied when discussing the timing of cash flows). The four methods of repositioning an asset will be discussed in context with changing these six inputs only. As a start, the baseline for the model must be determined. To determine the baseline, three cases for each assumption must be defined: best, worst, and most likely. The three cases are required because a range of possible outputs is necessary to quantify risk. These cases, in statistics parlance, are states of nature. A table summary of the cases appears in Figure 2.8.

The range of each input must be utilized to determine the quantification of risk using the approximation of range/6. Again, this assumption assumes a normal distribution of the forecast cell (i.e. IRR). For the base case of this single-family detached real estate asset, the characteristics are summarized as return = 13.27% and σ = 7.00%. Therefore efficiency – as measured by the Coefficient of Variation – is 0.53 and the probability of loss, assuming normal distribution, follows as 2.91%. In lay terms this means there is a 97.09% chance that the project will return at least the invested capital.

Principal 250,233 (Summary Page)
Interest 7% (Summary Page)
Term 30 (Summary Page)

	1	2	3	4	5	6	7	8	9	10	11
Beginning of Period	250,233	247,584	244,749	241,717	238,471	234,999	231,283	227,308	223,054	218,503	213,632
Interest	17,516	17,331	17,132	16,920	16,693	16,450	16,190	15,912	15,614	15,295	14,954
Principal	2,649	2,835	3,033	3,245	3,472	3,715	3,976	4,254	4,552	4,870	5,211
End of Period	247,584	244,749	241,717	238,471	234,999	231,283	227,308	223,054	218,503	213,632	208,421
Payment	20,165	20,165	20,165	20,165	20,165	20,165	20,165	20,165	20,165	20,165	20,165
Check	-	-	-	-	-	-	-	-	-	-	-

Figure 2.6 Debt Page Example

Period		1	2	3	4	5	6	7	8	9	10	11
Revenue												
Rental Income		38,400	39,168	39,951	40,750	41,565	42,397	43,245	44,110	44,992	45,892	46,809
Vacancy		1,920	1,958	1,998	2,038	2,078	2,120	2,162	2,205	2,250	2,295	2,340
Total Revenue		**36,480**	**37,210**	**37,954**	**38,713**	**39,487**	**40,277**	**41,082**	**41,904**	**42,742**	**43,597**	**44,469**
Expenses												
Maintenance/mo		1,200	1,224	1,248	1,273	1,299	1,325	1,351	1,378	1,406	1,434	1,463
HOA		1,800	1,836	1,873	1,910	1,948	1,987	2,027	2,068	2,109	2,151	2,194
Property Tax/yr		5,000	5,100	5,202	5,306	5,412	5,520	5,631	5,743	5,858	5,975	6,095
Insurance		3,000	3,060	3,121	3,184	3,247	3,312	3,378	3,446	3,515	3,585	3,657
Total Expenses		**11,000**	**11,220**	**11,444**	**11,673**	**11,907**	**12,145**	**12,388**	**12,636**	**12,888**	**13,146**	**13,409**
Net Operating Income		**25,480**	**25,990**	**26,509**	**27,040**	**27,580**	**28,132**	**28,695**	**29,269**	**29,854**	**30,451**	**31,060**
Debt		**20,165**	**20,165**	**20,165**	**20,165**	**20,165**	**20,165**	**20,165**	**20,165**	**20,165**	**20,165**	**20,165**
										Sales Price	**388,250**	
										Sales Expense	**15,530**	
										Principal Repay	**213,632**	
										Sales Proceeds	**159,087**	
Reversion FCF		0	0	0	0	0	0	0	0	0	**159,087**	
Operating FCF		**5,315**	**5,824**	**6,344**	**6,874**	**7,415**	**7,967**	**8,529**	**9,103**	**9,689**	**10,286**	
Total FCF	**(75,061)**	**5,315**	**5,824**	**6,344**	**6,874**	**7,415**	**7,967**	**8,529**	**9,103**	**9,689**	**169,373**	

Valuation:		
PV Reversion	61,335	58%
PV Operating	44,905	42%
	106,240	
CFO	75,061	
NPV	31,180	
IRR	15%	

Figure 2.7 Summary

Input	Worst	Best	Most Likely
Purchase Price	325,000	300,000	310,000
Loan-to-Value	75%	90%	80%
Rental Rate	2,900	3,500	3,000
Maintenance	250	50	100
Vacancy Rate	10.00%	2.50%	5.00%
IRR	–9.42%	32.60%	13.27%
Range	42.02%		
Risk	7.00%		
CV	0.53		
P(Loss)	2.91%		

Figure 2.8 Risk Quantification

$$\sigma = \sqrt{\sum_{i=1}^{n} \left[R_i - E(r) \right]^2 P_i}$$

Figure 2.9 Scenario Standard Deviation Equation

Note that a more accurate method of risk quantification is the scenario risk equation, which requires an estimate of the probability of occurrence for each state of nature (Figure 2.9).

For simplicity, range/6 was used as an approximation for standard deviation. Later calculations in this text will utilize the scenario risk equation.

Cost basis

The cost basis is the amount of capital invested in a project. Initially it is the invested capital (i.e. equity contribution) at closing, CF_0. Graphically depicted, the cost basis is shown in Figure 2.10. Just like a fixed-income security bond, the yield, project IRR, is inversely related to the magnitude. If the cash flows are fixed then the only change to yield is created by changing the initial cash inflow, CF_0.

In a real estate asset repositioning, the initial cash flow can be adjusted using several methods. The first, and easiest, is simply to purchase the asset at a lower basis. This can be accomplished by superb negotiating skills upfront or by contractual agreement that the seller shares in the upside of the project later. If the upside is not achieved, the cost basis remains low.

As changing the cost basis does not affect the cash flows, these will remain constant; only the yield is changed (assuming changing the cost basis does not affect construction quality or scheduling). This is precisely what occurs with a US Treasury note: the only variable is purchase price as the coupons and terminal value are fixed. For the single-family detached case, prices have been varied between \$275,000 and \$350,000, producing the graphic in Figure 2.11 relating price to yield.

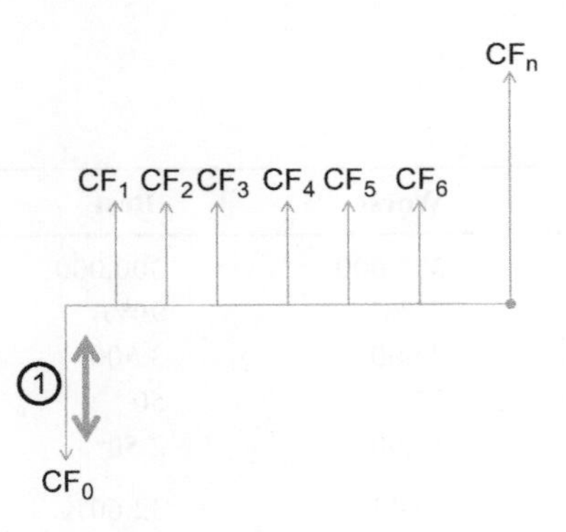

Figure 2.10 Cost Basis

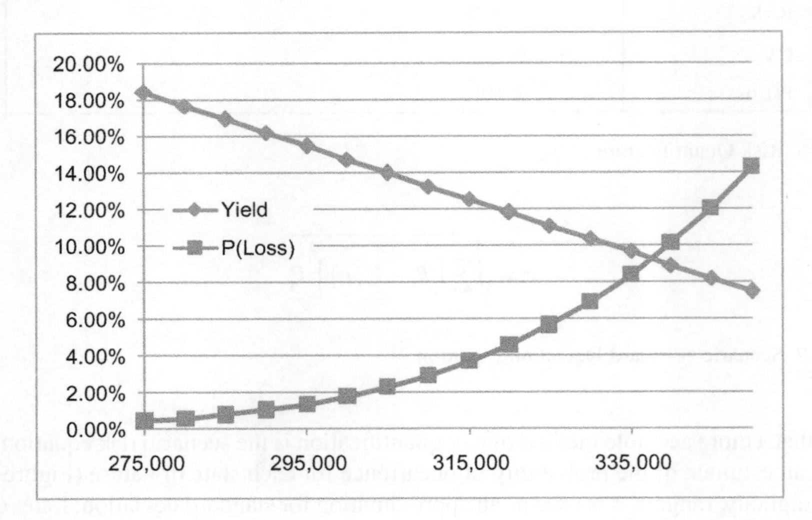

Figure 2.11 Yield v P(Loss)

The relationship between cost and yield is inversely related: $IRR\alpha\frac{1}{CF_0}$ Therefore, in a repositioning, an adjustment to the cost basis is a very important consideration because a reduction translates to an immediate gain in yield without a corresponding effect on the operations of the real estate asset. More important than the consideration for a decreasing yield as the cost basis increases is the exponential growth in the risk as measured by the probability of loss. The riskiness of the project increases at a greater rate than the yield decreases. (Note: this assumes a normal distribution of yield and a constant risk for the project, i.e. cash flows are unchanged by the cost basis.)

Another consideration for a cost basis adjustment of an asset currently owned is the accounting treatment. Whereas a detailed discussion of accounting is beyond the scope of this text, a write-down of equity is possible. This basically reduces the equity asset value for the property on the balance sheet, with a corresponding reduction in equity, while maintaining the debt principal on the property. This would be achieved through lender negotiations and the forgiveness of debt.

Later discussions involving alternate asset classes will offer a third option for reducing the cost basis, reduction in development/construction costs. This is the owner seeking lower-price alternatives without affecting the asset's ability to attract tenants and ultimately not affecting the cash flows earned by the asset.

Cash flow magnitude

The cash flow magnitudes are the net cash flows per period to a project. For a fixed-income security and equity these are considered coupons and dividends, respectively. In real estate parlance, cash flows are considered net income (i.e. NOI less debt payments). Note that this assumes the analysis is completed on a pre-tax basis.

As a matter of background, it is important to understand and recognize the cash flow diagram for a real estate asset. Real estate is unique in that it does not produce positive cash flows directly from purchase. There can be residual value which is positive, provided there is appreciation, but that is not guaranteed. The cash flows, as drawn in Figure 2.12, represent property tax payments, maintenance, and other ongoing expenses for the asset. Further, the cash flows are not equally spaced, as maintenance issues are not evenly spaced, though there can be accruals (i.e. a reserve accrual account utilized to save for unanticipated expenses). Property taxes do occur at regular intervals as does depreciation, which is an accounting recognition of reduction of value of a physical structure over time; however, this text discusses all projects pre-tax and therefore depreciation is not considered in the income

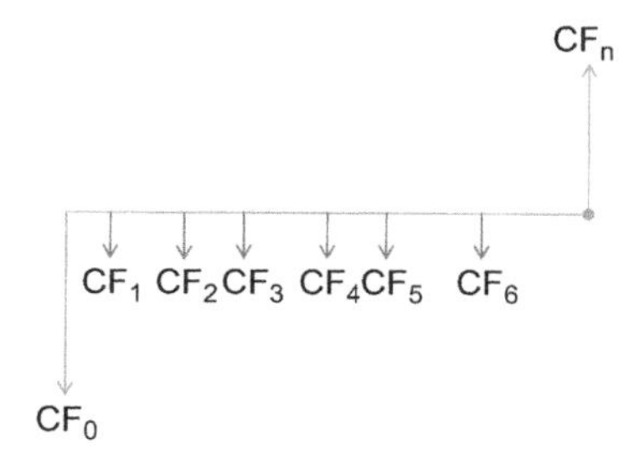

Figure 2.12 Cash Flow Diagram

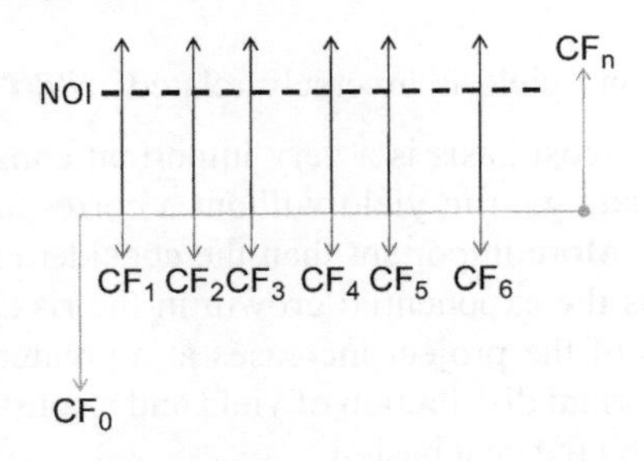

Figure 2.13 Cash Flow Diagram with Leases

statements. Given that the cash flow diagram of a real estate asset demonstrates all negative values, where is the value derived? Leases!

The value in a real estate asset is simply the cash flows that it can attract, that is the leases that will be signed for the project (Figure 2.13). The cash flows are derived from active management of the project and attracting tenants who sign leases and thus produce cash flows. The lease is the contractual obligation to pay certain cash flows at certain times through the period of the lease. The net lease amount received, after expenses, is the net operating income for a project: cash flow before debt service. (While the base case for this single-family detached example uses 80% leverage, the use of leverage in adjusting yield will be discussed in the financial section of this chapter.) The graphical depiction of a real estate asset fully leased with tenants is shown in Figure 2.13. The dotted horizontal line represents the net operating income (NOI) for the projects and is the lease income net expenses for the owner. The lease structure can vastly change the characteristics of a project and must be considered in a financial repositioning.

Understanding that the cash flow to a real estate project is leases net expenses addresses the two methods to reposition a real estate asset: increase lease income and decrease expenses. Either or both will have the same net effect – i.e. an increase in the project's net operating income. Again, note the base case for this example uses 80% leverage. However, in later cases, the base case will not incorporate leverage. Further, the lease will also affect the riskiness of the cash flows (i.e. higher-quality tenants have lower risk of achieving cash flows).

To exemplify the effects on a project's yield for an asset repositioning, two considerations will be made: changes to the magnitude of the rental stream and changes to the maintenance which represents expense adjustments. In later sections it will be discussed how active management and asset transformation can be accomplished to achieve both of these goals (raising rent and decreasing expenses). However, for this section these can be explained by internet marketing to attract more potential renters, thus attracting higher rents (the assumption is that internet market has zero cost), and expenses are reduced by performing all or part of expenses by the owner and not contracting externally.

In the first case, rental income is varied in increments of $100/month from $2,500 to $4,000 (Figure 2.14). In the second case, expenses are adjusted in $50/month increments from $0 to $500 (Figure 2.15). As expected, the results are inversely related. As rental rates increase, yield increases and the probability of loss decreases. However, as the maintenance expense increases, yield decreases and the probability of loss increases. The rate of change of the slope of each yield curve is indicative of the sensitivity of yield to the changing variable (e.g. rental rate), which has a greater rate of change reflective in the yield than maintenance expense. A repositioning of this property, or any property, must include a complete analysis

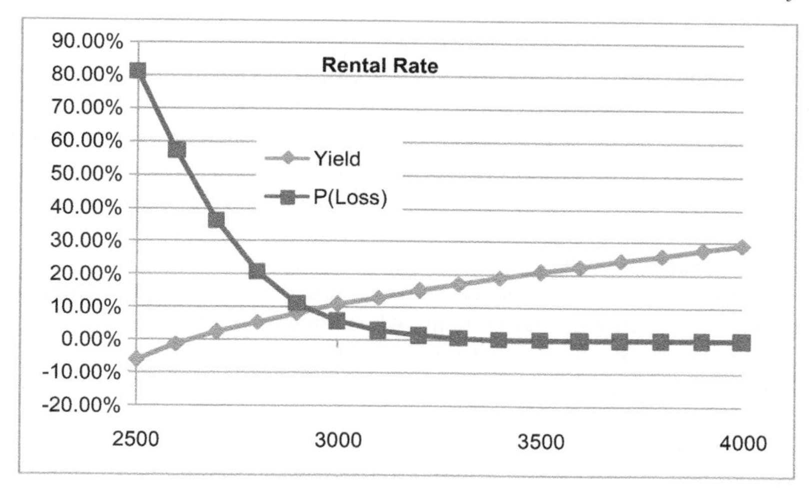

Figure 2.14 Rental Rate

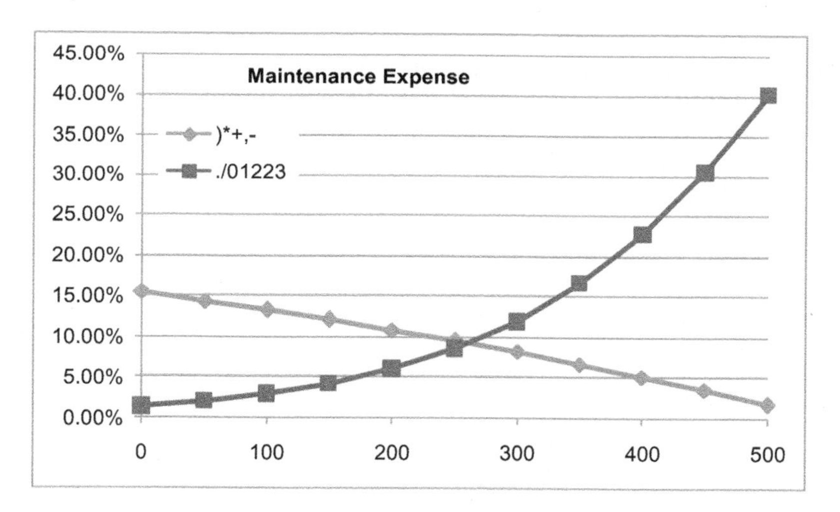

Figure 2.15 Maintenance Expense

of all rental income (leases), as well as a market study to determine potential rental income with associated costs. Further, a complete assessment of expenses is critical to maximizing net operating income. Again note that risk is held constant for this analysis. This allows for P(Loss) to be quantified more easily and without secondary effects of other variables.

Methods to attract higher rents for a single-family detached residence may include an increase in amenities, such as washer/dryer units, pool (could be a liability and actually a reduction in rent), included trash and water, or covered parking (if not the normal for the neighbourhood). Methods to reduce expenses could be home warranty service purchase, whereby a third-party contractor warrants home maintenance, or agree to self-perform services (if skills and time exist for this as an option), or simply passing the expenses on to the tenant as part of the negotiated and executed lease.

Cash flow timing

The cash flow timing reflects the time period when cash flows are received. For fixed-income securities and equities, these are fixed and the owner of the assets does not have control over timing. However, real estate is an actively managed asset, thus providing the owner some control over when cash flows are received. Of course, the limiting factor is market conditions, but even within a given market, there exists flexibility determined by the lease structure and tenant type.

The timing of cash flows is basically moving receivables forward for a project. For example, it is receiving rental payments on the 15th and 30th of the month rather than in total on the 30th (Figure 2.16). To a greater extreme, when considering weekly rentals, this would basically be receiving weekly rents on the 7th, 14th, 21st, and 28th. In simple form, the diagram shows the cash flows for each period shifted to the left (i.e. they are received at an earlier date than first indicated). The solid arrow indicates the initial timing of the cash flow and the dotted arrow indicates the revised, sooner, receipt date.

Prior to providing the real estate example for the single-family detached case, a more simple example has been chosen. Assume that an investor is offered four different timing scenarios for a $100 investment which will return $120 over the course of one year. The four scenarios are as follows:

1 End-of-Year ($120 is received at the end of the year)
2 Semiannual ($60 is received at the end of each 6-month period)
3 Quarterly ($30 is received at the end of each 3-month period)
4 Monthly ($10 is received at the end of each month).

A summary of the cash flows with dates appears in Figure 2.17.

The results of the analysis are shown in both tabular and graphical form (Figure 2.18).

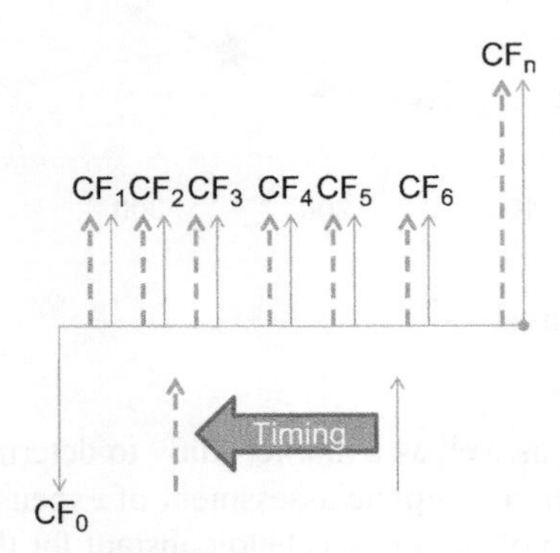

Figure 2.16 Cash Flow Timing

CF Timing	1-Jan-10	31-Jan-10	28-Feb-10	31-Mar-10	30-Apr-10	31-May-10	30-Jun-10	31-Jul-10	31-Aug-10	30-Sep-10	31-Oct-10	30-Nov-10	31-Dec-10
Annual	(100)												120
Semi-Annual	(100)						60						60
Quarterly	(100)			30			30			30			30
Monthly	(100)	10	10	10	10	10	10	10	10	10	10	10	10

Figure 2.17 Cash Flow Summary

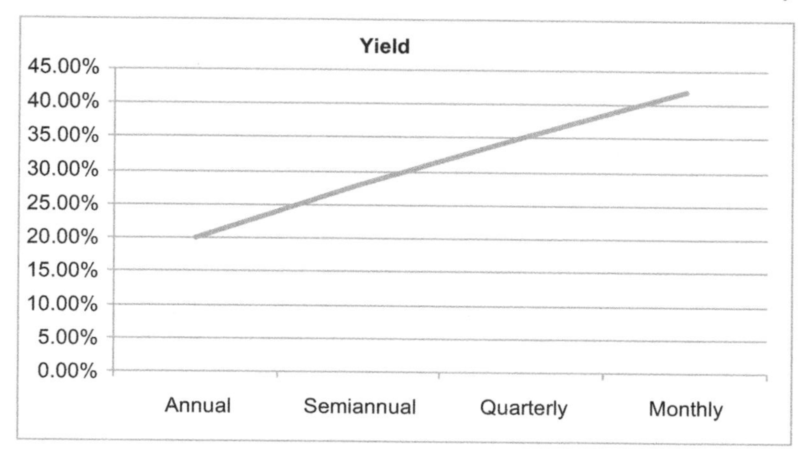

Figure 2.18 Yield

CF Timing	IRR
Annual	20.06%
Semi-Annual	28.05%
Quarterly	34.97%
Monthly	41.78%

Figure 2.19 CF Timing and IRR

Please note that the calculation of yield (Figure 2.19) was performed using the XIRR function in MS Excel rather than the IRR function traditionally used in most analyses. The difference is that the IRR function is a periodic quantification of yield, which assumes equidistant payments in terms of timing. The XIRR function discounts using a daily basis and therefore requires the timing as well as the magnitude of the cash flows as part of the function. See the MS Excel help function for a detailed explanation of the XIRR function.

The graphical representation of yield and cash flow timing indicates a linear relationship. That is, as cash flows are moved to the left (i.e. received sooner) yield increases. Careful consideration, therefore, should be given when negotiating lease terms for the timing of payments.

To provide a shift in cash flows for the single-family case which uses the IRR function in MS Excel, the periods were changed from single, annual annuities using the end-of-period convention to specific dates, which can be adjusted daily using the XIRR function. The adjustment for the case is that initial cash outflow is modelled as occurring on 1 January 2010 and each subsequent cash flow occurs, in the base case, on 31 December of the following year. Note, for the first period, the first cash flow occurs on 31 December 2010. Also, as the cash flows are all annual, adjusting the timing of cash flows will be accomplished by adjusting the dates to monthly increments. For example, 31 December will become 30 November, which will represent a single- (1-) month shift. These shifts will be accomplished from 31 December which represents the most conservative option, end-of-period, to 1 January which represents the most aggressive option, beginning-of-period (Figure 2.20).

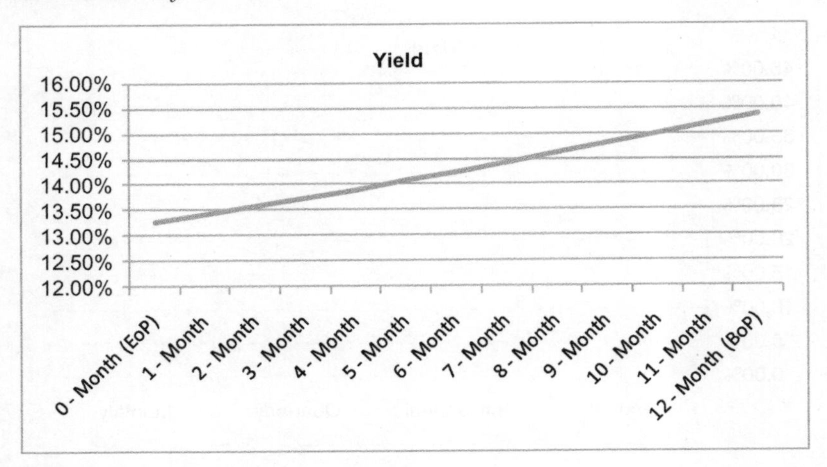

Figure 2.20 Yield

The results provide an almost positive linear relationship between timing of cash flow receipts and yield. The sooner cash receipts are received, the higher the yield. Of course this can only be accomplished through the design of the lease. It also relates to why there are trade credits for early receipt of payment and default interest rates for late payments. Both of these are contractual methods to protect and actually enhance the yield as the tenant delays payment(s).

An international example for adjusting the timing of cash flows is a retail shopping centre in China. As part of the lease, the centre was paid a portion of the gross receipts. Concerned that vendors were underreporting gross receipts, the owners of the centre put cash registers at the entrances and all receipts were centrally collected with vouchers provided to the vendors. This way the centre not only reduced the risk of collection losses but also increased timing of payment from monthly to daily.

Riskiness

The cash flow riskiness is a difficult value to quantify without the use of stochastic processes, for example Monte Carlo simulation. It is complicated as the riskiness is the amalgamation of all the inputs for a model. Therefore, for the purposes of this text, which does not use stochastic processes for risk quantification for an asset reposition, the risk of cash flow will be modelled as the vacancy factor, which inversely varies cash flows per period (Figure 2.21).

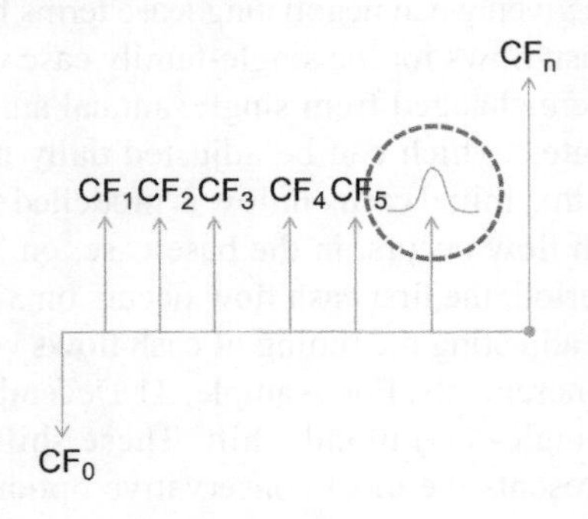

Figure 2.21 Cash Flow Riskiness

As discussed, risk is the deviation for an expected return. For individual cash flows it is therefore the risk of achieving the expected cash flow. As depicted earlier, risk of a cash flow is represented by the normal distribution. The graphic in Figure 2.21 demonstrates for a single cash flow, but again, there is risk associated with achieving each individual cash flow for a project. This is best exemplified in a retail centre which has a base rent and a percentage of gross sale. The reason for the percentage of gross sale as part of the lease is to ensure that the real estate owner and the tenants have an alignment of interests (i.e. it is to the real estate owner's advantage to assist in driving traffic to the facility for the tenant).

Risk, as quantified by vacancy, will vary from 0% (100% certainty of achieving expected cash flow) to 15%. This allows for variation of the cash flow using an input to the overall model.

The result is an inverse relationship between vacancy and yield (Figure 2.22). The critical aspect of the relationship is that the probability of loss steepens as vacancy increases. Basically, things get worse at a worsening rate. Of course, this probability is quantified for the entire project and, once again, assumes a constant risk that is normally distributed.

What is also difficult is quantifying risk for a project's individual cash flows and then modifying the risk as a strategy for asset repositioning. The method to modify the risk of cash flows directly is with financial instruments, that is, derivatives. Detailed derivatives analyses are beyond the scope of this text, and therefore indirect methods to control risk will be evaluated for each asset class (e.g. vacancy rates). Assuming vacancy rates can be directly controlled (in truth vacancy is indirectly controlled through rental rates) for this text, vacancy will be considered a proxy for 'risk.'

The most common risk in a lease and therefore real estate investment is rent rollover, or lease renewal. This must be strongly considered when negotiating and structuring new leases. The risk and control of the space at renewal can be affected by renewal options and the options put on the space. A seasoned asset manager is critical for lease discussions.

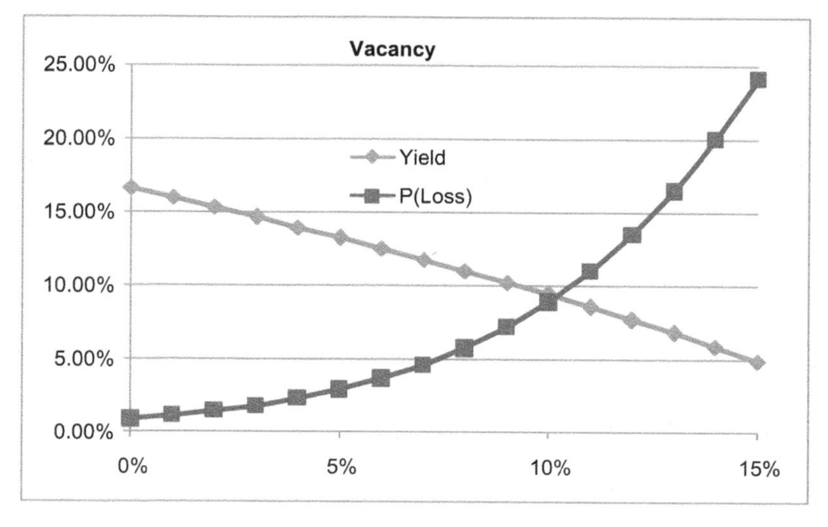

Figure 2.22 Vacancy

Finance method to reposition

A financial repositioning is the act of changing some active element of a project's characteristic (i.e. return or risk of the project). Several methods, while not mutually exclusive, are as follows:

1 Capital Structure, i.e. debt/equity mix
2 Futures/Forwards/Options, i.e. using financial instruments (derivatives) to adjust risk profile
3 Waterfall/Securitization, i.e. transformation of risk through separation of final cash flows into risk tranches.

Changing or using any of these is accomplished by active management of a project and an understanding of future trends and projections, both macro- and microeconomic. With the exception of the capital structure, the purpose is to transform the return and risk characteristics through external strategies entirely, without adjusting the operational characteristics at all. None of these methods adjust the operating characteristics but rather adjust the final return and risk for the project externally – i.e. these methods affect cash flows below NOI whereas physical methods affect those above NOI.

Capital structure

Adjusting the cash flow (i.e. leverage) is an important consideration for any project. Conventional financial theory states maximum value is achieved at the point of minimum weighted average cost of capital (WACC). However, in this single-family model, the cost of equity is not a direct cost. Therefore, a complete WACC analysis cannot be completed without assigning a cost of equity to the source of capital.

In this model, adjusting the Loan-to-Value ratio has two end results: (1) lowering or raising the capital base for which to compute yield, CF_0, and (2) reducing the expense incurred by debt payments, that is, principal and interest payments. When the Loan-to-Value is 100%, the entire purchase price is financed through debt (Figure 2.23). Therefore the equity base is

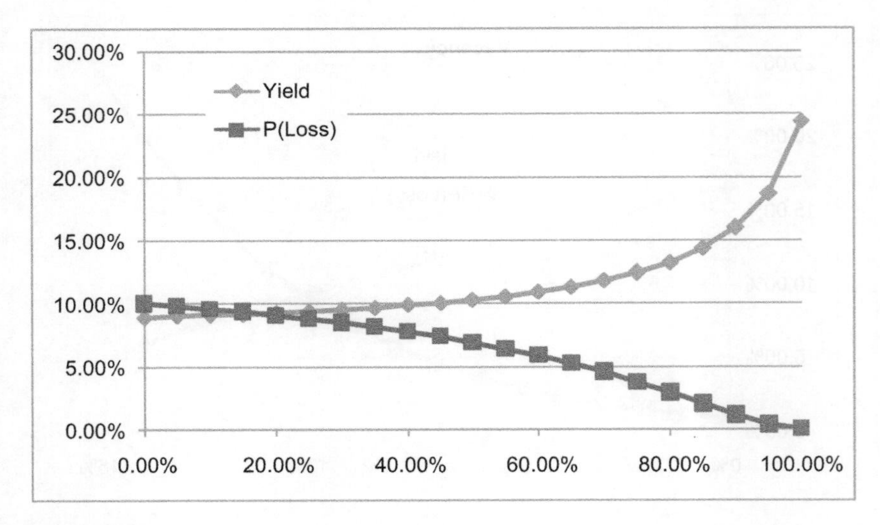

Figure 2.23 Yield v P(Loss)

lowest though the corresponding principal and interest payments are the highest. At the opposite, when Loan-to-Value is 0%, the entire purchase is equity financed and no debt financing is used. Therefore principal and interest payments are $0.00 in each period but the capital base to quantify yield is highest (i.e. includes 100% of purchase price).

By adjusting the Loan-to-Value ratio, the risk structure, as measured by P(Loss), changes dramatically. As less and less debt is used within the capital structure, the chance of loss of a portion of invested capital increases. To some degree, this is counterintuitive, as the elimination of debt negates all financial risk through 100% equity financing. However, the capital base for which yield is computed is maximized and for the purposes of quantifying P(Loss), the risk of the project is constant. Note, this is not realistic; when cost of equity is imputed within a model and considerations are made for the increase in risk due to increased debt, the risk will change.

The important point is that risk does change with a changing capital structure. Thus, a real estate asset can be repositioned through the use of a dynamic capital structure.

Futures/forwards/options

While a complete discussion of financial derivatives and their uses to mitigate risk and adjust the payoff diagram for a real estate project or a portfolio consisting of physical real estate assets is beyond the scope of this text, a lack of mention would be imprudent. Futures/Forwards/Options are financial instruments which trade upon an underlying pricing index. For US real estate, the Case-Shiller index has Futures and Options, exchange-traded products, offered by the CME Group in Chicago. These instruments can be used to bet (speculate) about price movements or hedge (mitigate) against a price movement. This market began May 2006.

For instance, if there is a concern that prices may go down, causing a loss in the future, the adept and knowledgeable chief financial officer may short a futures contract or long a put option. However, if there is a purchase decision in the future and the concern is a price rise, the CF_0 may long a futures contact or long a call option. Both strategies provide financial payoffs if the market (i.e. index) moves in a particular direction.

This strategy of risk mitigation and repositioning – or rather, adjusting – the final cash flows is entirely outside of the operations and physical management of the real estate asset. This is a passive strategy and adjusts the final payoff(s) for the asset but does not affect the physical asset at all.

Waterfall/securitization (bifurcating cash flows)

Waterfall

The waterfall is the method of distributing profits among partners within a transaction. In more complex transactions profits do not follow an even distribution – i.e. profits are not distributed pari passu. (Pari passu is the proportional distribution of profits in accordance with percentage of capital provided for an investment. That is, an investor that provides 80% of capital receives 80% of profit if distribution follows pari passu.)

The simplest form of a waterfall is the income statement for a project. Revenue is received, direct expenses are paid, and net operating income remains. With remaining cash, debt payments are made and the residual is paid, or accrued, to the owners (i.e. equity providers). A graphic depiction is found in Figure 2.24.

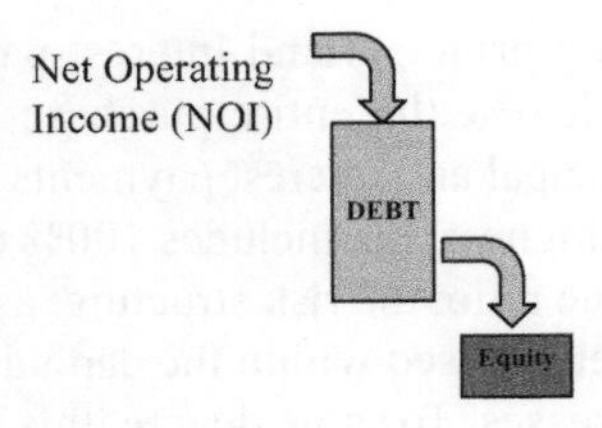

Figure 2.24 Basic Waterfall

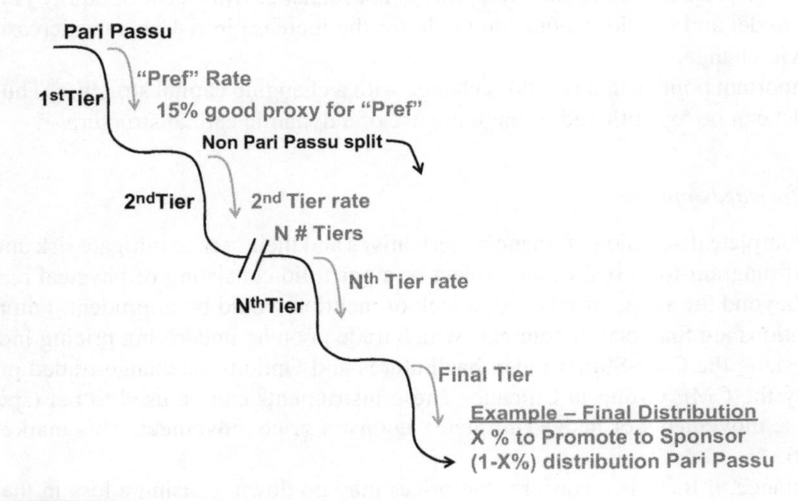

Figure 2.25 Waterfall

A general waterfall structure is shown in Figure 2.25. Please note that rates discussed are internal rates of return (i.e. accrual rates on capital) and not simple interest rates. The distribution of profits by investor can be different at each 'tier' and is pre-negotiated from project onset.

The waterfall is therefore used to separate risk, similar to the purpose of structured products with regard to the selling of tranches. The waterfall allows for a financial repositioning of risk for owners. By adjusting the payment streams contractually, the risk of return is varied greatly depending upon the position of cash flows in the waterfall, which are 'owned.'

The general structure has an investor receiving a preference ('pref') rate. This 'pref' rate is a base rate which is generally received prior to any distributions. Afterwards, distributions are made between partners at stated percentages up to stated rates.

The reason for the term 'waterfall' is obvious; it is about cash flows being distributed according to a partnership agreement. This is similar to securitizations.

In the waterfall structure in Figure 2.25, cash flows are split by an initial amount up to the first tier, or waterfall, of 15.00%. The next tier will vary the distributions of cash flows according to partnership. This may continue for any number of waterfalls. Obviously the larger the project the more likely it will be to have multiple tiers. However, the structure will remain the same.

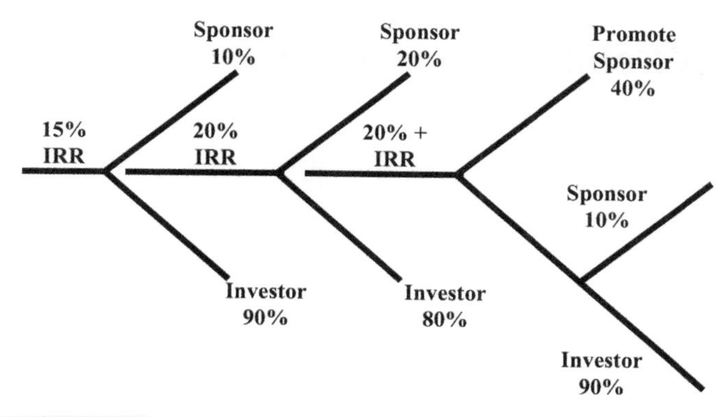

Figure 2.26 Waterfall Alternate

An alternative method of waterfall structure is for a partnership as described in Figure 2.26. There are two parties: Sponsor and Investor. The initial capital percentage by each partner for Sponsor and Investor is 10% and 90%, respectively. For the first waterfall tier, the cash flows are split pari passu for the first 15% IRR. For the next 5% IRR, up to 20%, the cash flows for the Sponsor and Investor are split 20% and 80%, respectively. Following the project returning a 20% IRR, the Sponsor returns a 40% promote (i.e. 40% of all remaining), with the final 60% being split pari passu.

Figure 2.26 shows the waterfall structure linearly. The splits are demonstrated top and bottom while the actual waterfall (i.e. IRR amounts) follow linearly along the base trunk of the diagram. The diagram is intended to be viewed from left to right.

Securitization

Securitization separates cash flows according to an ownership agreement. These separations of cash flows are listed in order a priori for each tranche (i.e. ownership cash flow stream). Generally larger portfolios are securitized, which is essentially the transformation of a portfolio of financial obligations into a liquid form attractive to a myriad of investors. As an example of a securitization, a visualization of the difference between syndication and securitization follows in Figure 2.27. In the most basic form, a securitization is a division of ownership horizontally while a syndication is a division vertically.

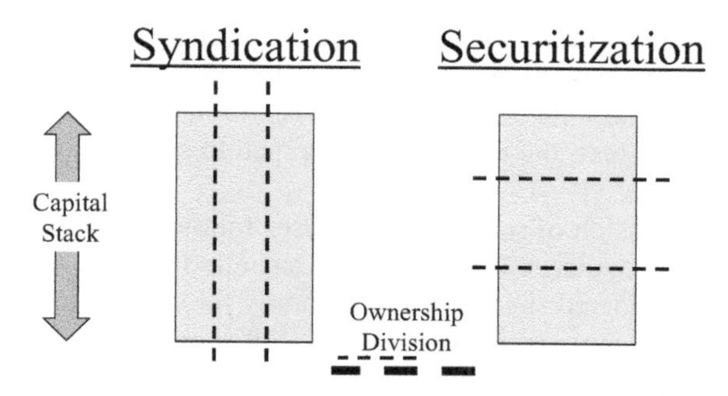

Figure 2.27 Securitization

3 Project characteristics defined

While physical structures can vary greatly in appearance – heights, uses, footprints, etc. – it is really the intended use for a facility that separates the asset classes. This separation, though physical in structure, affects the cash flows the most with regards to the lease structure, terms, and conditions. Offices tend to have the longest tenants as a class, though a retail anchor tenant may be many times longer than the average office tenant. Hotels are just the opposite, with daily leases for clients.

Industry and geographic norms prevail, in general, when leases are considered. At the maximum, leases are the final position in a discussion in the Western world. However, in less developed parts (emerging markets), the lease can only be the beginning of a discussion. Further, length of term, rate, optionality, and so forth are all different by asset class and geographic location. This, more than anything else, is why Real Estate is geographic as well as technical. How leases are structured varies greatly depending upon asset class and physical location for the property. What is appropriate in Washington, DC, is not appropriate in Kuala Lumpur, which is quite different from Paris, France. These differences contribute to the different shapes of the asset cash flow distribution. As a simple example and a general statement, in Western leases the tenant (occupier) has the legal superiority, whereas in emerging markets it is generally the owner of the real estate itself. The legal system plays another huge role in real estate lease structures and asset cash flow distributions.

Rents are another defining factor which are different between asset classes. Generally there is a base rent which is paid monthly (daily for hotels) with annual or periodic step-ups to adjust for expense increases, inflation, and so on. However, retail also has a variable component (percent rent) and, if the lease is triple net, an expense component. There are also different lease structures that allow for risk and cost savings, for example triple net lease versus gross. Further, a common expense shared in Western commercial leases is Common Area Maintenance (CAM) charges. This is generally allocated by square foot leased and is a pass-through of expense to the tenants. The goal of the CAM charge is to ensure that the facility is maintained and thus, for a retail facility, traffic is created for sales to the establishment. Of course there are many additional structures and pass-through expenses, most of which are beyond the scope of this text; the reader is encouraged to seek out texts focusing on lease structures singularly.

A more detailed discussion of several asset classes follows with a graphical depiction of the cash flow diagram. The discussion begins with a single-family residential (for sale property). This could be considered similar to a multifamily for sale produce (e.g. condominium building). The remaining commercial asset classes (office, retail, hotel, multifamily for rent) follow the traditional structure for commercial real estate asset class and are assumed to be turnkey for the point of discussion – i.e. no development/construction/renovation is needed.

Prior to a discussion by asset type, a comparison of the major asset classes is prudent, in other words, the differences between equities, fixed income, and real estate (residential and commercial). A historical comparison of asset class returns compared to inflation from 2001 to 2013 provides an exemplar of performance. Please note that the Vanguard Total Bond Index, VBMFX, has been used as a proxy for fixed income and the S&P 500 index for equities. Real Capital Analytics commercial price indices have been used for commercial real estate and the Case-Shiller index for residential values.

A comparison of returns demonstrates the multifamily asset class (i.e. apartment) as the darling of those compared (Figure 3.1). Over the 12-year period, the geometric growth rate as calculated from the index is 5.72% per year, more than doubling inflation. Surprisingly, over the same period, fixed income returned 5.29%, again more than doubling inflation over the period.

Retail and residential, both of which are most influential on the US economy, were the laggards, returning 2.82% and 3.17%, respectively. However, an analysis of the returns is incomplete without an understanding of the distribution shape (risk) associated with each asset class. For instance, while multifamily may have had the highest geometric return over the 12-year period, a reasonable comparison cannot be made without viewing the asset class efficiency, that is, what risk is accepted for each unit of return.

A similar graphic but with the inclusion of risk is provided to compare the asset classes from an efficiency perspective (Figure 3.2). The only asset class whose return exceeds the

Figure 3.1 Asset Class Returns

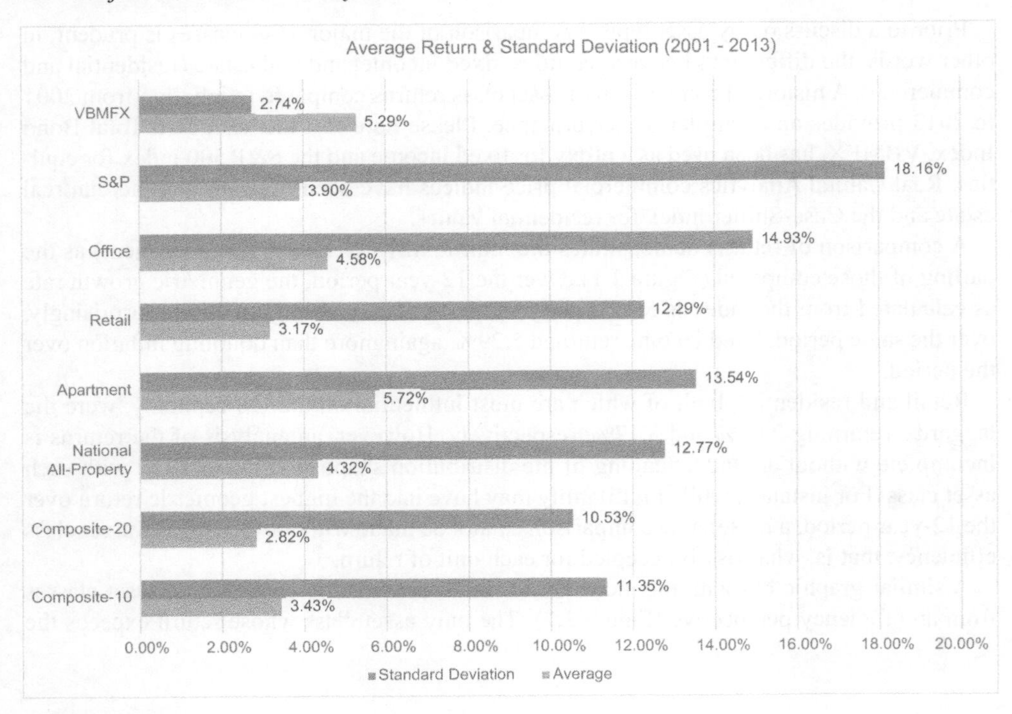

Figure 3.2 Risk/Return

risk is fixed-income securities. The return is 5.29% over the period and the standard deviation is 2.74%. From an efficiency perspective, fixed income CV is 274/529 = 0.52. Simple math transforms this as Z = −1/CV = −2. Therefore, the P(Loss) for fixed-income securities is 2.50%, very small.

Equities proved to have the most risk over the period, with a standard deviation of 18.16%. The return for equities is 3.90% and therefore the CV is 1816/390 = 4.66. The z-score is thus z = −0.21. Using linear extrapolation, the P(Loss) for equities is 43.20%, significantly higher than fixed income over the same period.

For real estate the market is bifurcated. Commercial and residential are viewed as different asset classes (Figure 3.3). The National All-Property RCA index is used to discuss commercial and the Case-Shiller Comp-20 is used to discuss residential. The commercial return over the period is 4.32% and risk is 12.77%. The residential return is 2.82% with 10.53% risk. The Coefficient of Variation for commercial and residential are 2.96 and 3.73, respectively. The z-scores for each are therefore −0.34 and −0.27, respectively. Again assuming linear interpolation for a normal distribution, the P(Loss) for commercial is 38.44% and for residential is 40.82%. While both have returns providing an inflation premium, both are significantly less efficient over the 2001–2013 period than fixed-income securities.

Another method to compare return and risk for the four asset classes is the year-over-year performance from 2001 to 2013. Rather than viewing risk as the standard deviation and using the empirical rule associated with a normal distribution, risk can be simplified to range (i.e. maximum less minimum). A more striking comparison demonstrates the range of returns as a time series for the period. Fixed income, when viewed across time, has little deviation

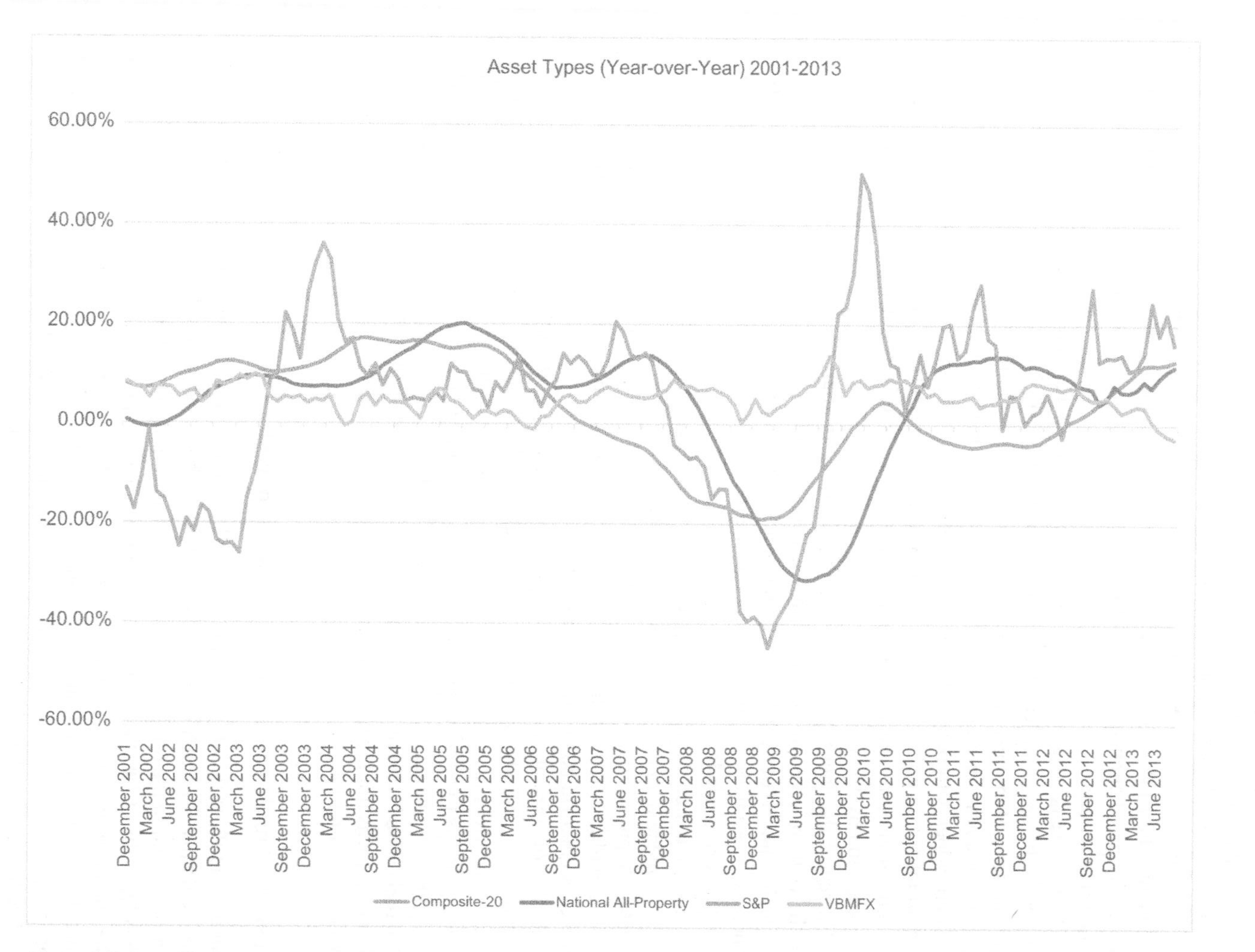

Figure 3.3 Asset Types

(range) when compared to the S&P 500 index. Although the maximum return for the S&P 500 is over 40%, the maximum loss exceeds 40% as well. This is problematic because of the unequal nature of gains and losses. That is, a 50% loss requires a 100% gain to become whole again. For example, a $100,000 investment loses 50% in the first year. At the end of the first year the investment is now worth $50,000. To regain the loss, the gain over the subsequent period must be 100%. The average return for the 2 years appears to be 25% – i.e. (100–50)/2 – but reality has demonstrated the return is 0.00%. Hence volatility cannot be ignored when comparing returns.

The most striking aspect when comparing the commercial and residential year-over-year returns is the apparent lead/lag relationship (Figure 3.4). Notice the trough for residential appears several months prior to that for commercial. Visually this demonstrates a potential leading indicator for commercial pricing changes (i.e. residential leads commercial).

There appears to be a 6-month lead for residential to commercial. Historically, the residential inflection point during the Great Recession was July 2006, whereas for commercial the inflection point was the fourth quarter of 2007. That is, residential led commercial into the recession by 16 months.

Currently, the data supports a 6-month lead by residential for commercial pricing as based on the year-over-year historical analysis. This is significant at a 1.00% confidence level and has an R-squared of 61.90%. The actual equation for the 6-month lead is $F(x) = 0.97X + 0.02$ where X is residential year-over-year and the dependent variable is year-over-year RCA All-Property pricing index.

This discussion compares asset classes utilizing risk/return historical performance only. The one piece of analysis which has been excluded but is vital and which will be discussed in later chapters is the correlation of the asset classes. Since the turn of the century, diversification within portfolios has been more difficult to achieve as assets prices have tended to move together (to be positively correlated).

A table of correlations of year-over-year pricing from 2001 to 2013 is included for discussion (Figure 3.5). The only negative correlation is fixed income with all other asset classes. The high correlation of real estate asset classes demonstrates that adding additional product types or assets to a real estate portfolio may not achieve the desired diversification intended by a portfolio manager. This will be discussed later, but a cursory viewpoint demonstrates that real estate investing is best done singularly rather than in multi-asset classes. Please note that Hotel was not included in this correlation matrix due to the lack of parallels with RCA Analytics Hotel Index.

A single-family residential (for sale property)

Single-family residential communities or Planned Unit Development (PUD) communities are generally for sale property where land is subdivided into individual lots for building/construction. In the traditional structure the land is purchased 'raw', that is, not entitled. The developer then takes the land (this is a North American process) through entitlement, adjusts the zoning from agricultural to residential, and then subdivides the property into individual lots. Traditionally the lots are approximately 10,000 sq. ft. or a quarter of an acre. The developer then puts in curbs, gutters, sewer, roads, and major infrastructure.

The 'developed' lots are then sold in bulk or piecemeal to a residential builder, who builds homes on the lots for final sale to the end customer (Figure 3.6). The lots are generally 'taken down' in increments contractually agreed between the developer and the builder. The greatest risk both the builder and developer have for this process is the speed (i.e. velocity) of lot

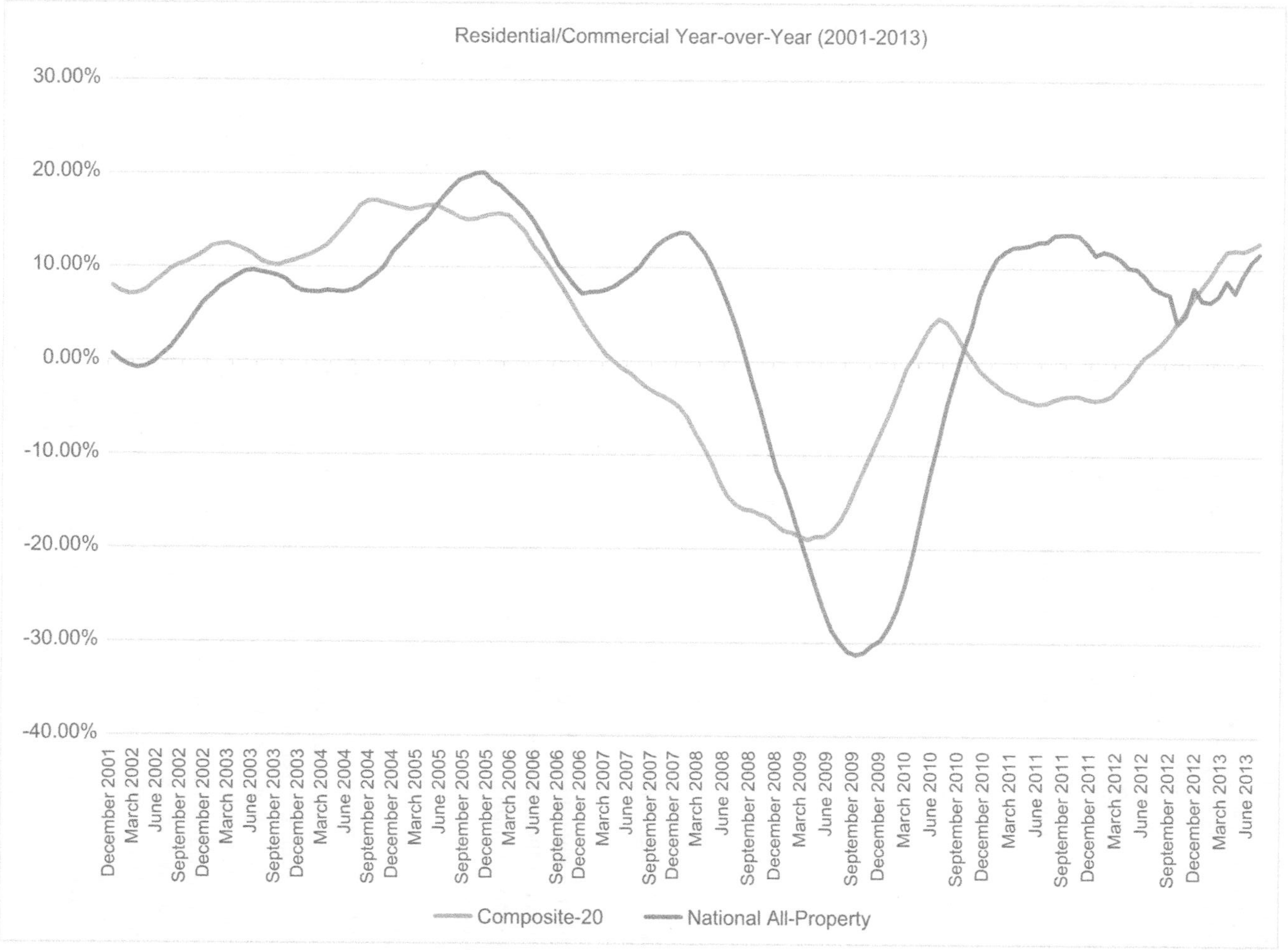

Residential/Commercial Year-over-Year (2001-2013)

30.00%
20.00%
10.00%
0.00%
-10.00%
-20.00%
-30.00%
-40.00%

December 2001
March 2002
June 2002
September 2002
December 2002
March 2003
June 2003
September 2003
December 2003
March 2004
June 2004
September 2004
December 2004
March 2005
June 2005
September 2005
December 2005
March 2006
June 2006
September 2006
December 2006
March 2007
June 2007
September 2007
December 2007
March 2008
June 2008
September 2008
December 2008
March 2009
June 2009
September 2009
December 2009
March 2010
June 2010
September 2010
December 2010
March 2011
June 2011
September 2011
December 2011
March 2012
June 2012
September 2012
December 2012
March 2013
June 2013

Composite-20
National All-Property

Figure 3.4

	Comp-10	*Comp-20*	*Nat'l All Props*	*Apartment*	*Retail*	*Office*	*S&P*	*VBMFX*
Comp-10	100.00%							
Comp-20	99.63%	100.00%						
Nat'l All Props	54.60%	57.54%	100.00%					
Apartment	56.81%	59.39%	96.43%	100.00%				
Retail	70.36%	72.25%	88.62%	85.41%	100.00%			
Office	42.27%	45.43%	97.42%	91.48%	77.39%	100.00%		
S&P	36.54%	38.95%	38.79%	36.59%	30.87%	40.35%	100.00%	
VBMFX	−23.38%	−26.38%	−29.42%	−26.48%	−35.22%	−27.88%	−9.96%	100.00%

Figure 3.5 Residential/Commercial Correlation

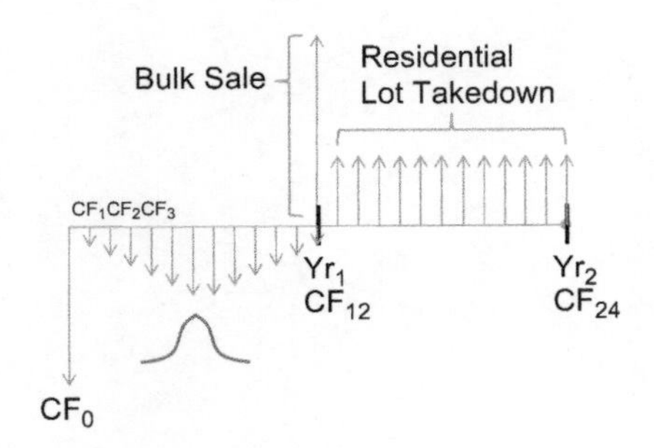

Figure 3.6 Single-Family Residential

takedown. Other than development issues (soils, etc.), the greatest effect on the yield for the developer is the schedule for the lot takedown. This is equivalent to a fixed-income instrument whose coupons are delayed – i.e. yield is compressed and at the extreme a capital loss is incurred.

Retail

Retail Real Estate is generally considered sites of commerce, that is, where items are purchased and sold. The traditional image of retail is the shops along highways where households purchase goods and services. Using this image (and reality) as a base, retail, even while one single structure, that in inline (i.e. containing multiple shops) must be bifurcated into two types of tenants: (1) anchor tenants and (2) inline stores. The distinction is necessary because both have considerably different lease terms, cash flow streams, and rollover risk for the tenants.

Anchor tenants, affectionately called 'big box stores,' will have leases exceeding 25 to 30 years. These are critical to the entire centre's economics as many times they drive the traffic (customers) to the site. Customers will be driven to the anchor, for example a grocery store, and in turn complete additional shopping at the smaller tenants. This is the main reason why the anchor tenants have long lease terms: the commitment to the area is long lived.

Inline stores will have significantly shorter leases at 5 years with an equivalent extension option, as negotiated. Examples of inline stores are services centres and small clothiers. Typically these are not large chains, though some may occupy this space, such as a chain coffee shop or sandwich shop. The credit quality can be less for the inline stores because they generally do not drive the economics of the individual retail site.

Further, the retail asset class differs in the method for which cash flows (i.e. lease terms) are quantified and considered. Retail leases will have a base component and a percent of sales component. The base rent is usually fixed with regular step-ups and the percent of sales is a portion of the sales (performance) of the retailer. There is a third portion if the lease is triple net – i.e. Common Area Maintenance (CAM), insurance, taxes – however, this is ultimately a net zero. The triple net components are paid by the tenants but netted against the expenses by the owner and therefore considered zero. (Note: triple net components are not depicted in Figure 3.7.)

This lease structure is to best align the interests of the owner and the tenant. The owner is incentivized to drive traffic to the site, as a portion of this is returned through a percent of sales provision in the lease. The owner will then ensure rapid snow removal, attractive landscaping, and clean and safe parking for patrons. Interest alignment is critical for the landlord and the tenant.

The cash flow diagram for retail follows. Note the risk in percentage rents, which are collected monthly, and the larger risk associated with rent rollover for an inline tenant at year 5. The rollover for an anchor tenant is not diagrammed because of the overall length. What is not demonstrated but still a significant risk is the anchor turnover at the end of the term. However, given the length of anchor tenants' terms, the cash flow is considered perpetual and the rollover risk, in the early years, as having insignificant risk (Figure 3.7).

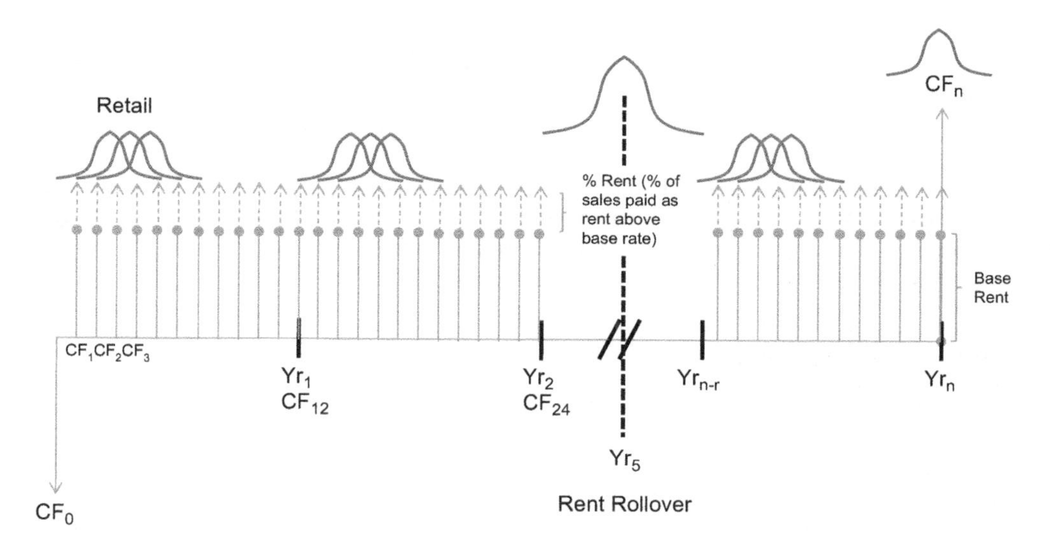

Figure 3.7 Retail

Office

Office is considered space for white- and blue-collar administrative uses. Every company and corporation uses office space to house workers and, ultimately, produce product.

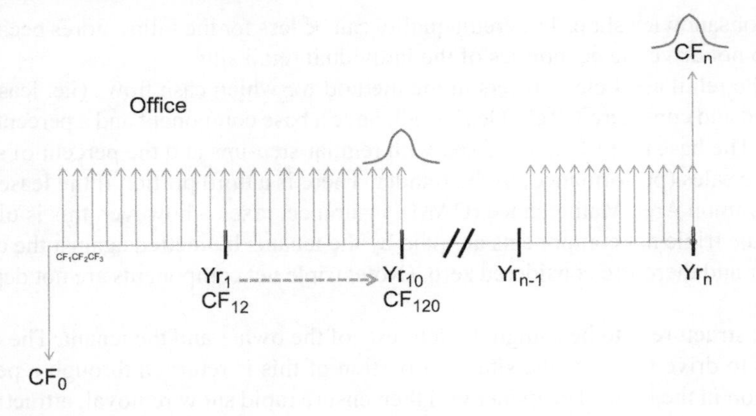

Figure 3.8 Office

Generally the space is not used to produce an end-product, as would manufacturing or warehouse space, but rather for intellectual purposes and back office support. (Industrial properties would be used for actual production of corporate 'widgets.')

Office real estate asset class when visualized is a large tower, like in many of the world's financial districts: Singapore, London, New York, etc. Most visualize the physical but as discussed in this text, the reality is that office real estate has a distribution shape to its cash flows. If one views a large tower that is multi-tenanted, it can be considered a portfolio of smaller leases. If we extrapolate further, these leases may be considered forward rate agreements (FRAs) and the office real estate is really a portfolio of FRAs. This concept will be explored later in the text.

An office cash flow diagram (Figure 3.8) depicts the monthly cash flows from tenants and the rollover risk which occurs. Office tenants typically will sign a 10-year lease with extension options. For this text, rollover is considered the main risk and tenant default is largely left to texts focusing entirely on the asset class individually.

Multifamily/apartment

Multifamily/apartment is basically for rent rental properties. These are generally apartment complexes, though they can be garden style or even individual units. The asset type will vary depending upon amenities offered, structure type and size, proximity to key employment centres, and demographic target.

The attribute of the asset class that is fairly consistent across multifamily are lease tenor, for instance 1 year. Of course there does exist executive housing which can be leased on a monthly or weekly basis. However, this text will consider the traditional annual lease. Rent rollover risk exists at the end of each annual lease, as the risk of nonrenewal and vacancy and improvement costs. Generally in a large multifamily facility, the leases will be staggered, thereby having expiration of some leases every month (i.e. the use of the portfolio concept in multifamily leasing). Therefore, rent rollover risk is omnipresent for this asset class. Figure 3.9 depicts the risks associated with the multifamily/apartment asset class.

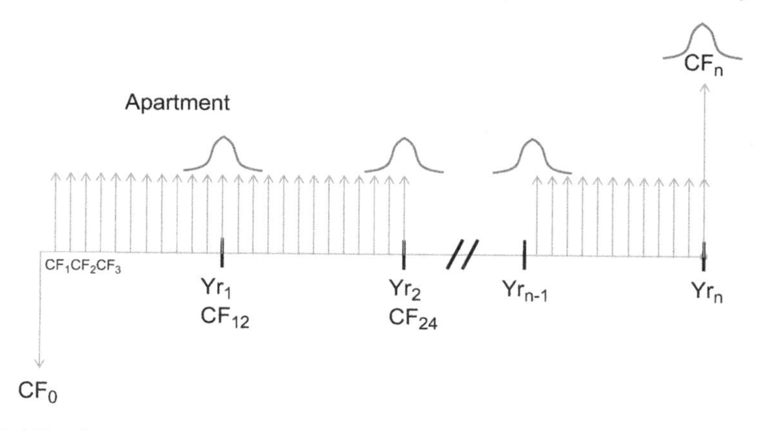

Figure 3.9 Multifamily/Apartment

Hotel

Hotel is the most unique asset class given the tenor of leases (single days). Hotels are the most correlated to economic events and the most responsive to changes in pricing, attitudes, and preferences. Further, 'rents' are highly variable depending upon market conditions and can be changed virtually instantaneously. As the terms of the 'lease' are daily, risk is omnipresent on a daily basis given that leases can end for the facility any day.

Modelling the hotel real estate asset class requires daily calculation of cash flows because this asset class can re-price its lease on a daily basis. As such, this asset class also offers the portfolio manager that holds multiple real estate asset classes the greatest inflation hedge possible. The downside risk is that hotels are first into a recession but also first out. Hotels are a leading indicator of economic events, in general, for real estate portfolios as they react the most quickly to economic change.

Types of hotels can vary by class, size, location, amenities, and whether it is 'flagged' (i.e. part of a network or chain) or unflagged. An example of a flag is a Marriott or Hilton hotel where there is centralized registration. An unflagged hotel might be a boutique or individual hotel without affiliation with a particular hotel manager. The cash flow diagram for a hotel follows in Figure 3.10. Note the daily risk factor, which is reflective of the daily 'lease' term.

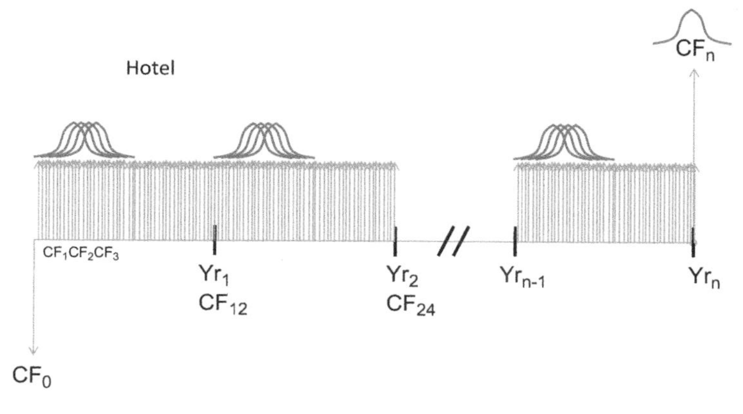

Figure 3.10 Hotel

While not discussed in great detail, hotel as a real estate asset attempts to combine various income sources within a single property. Examples are meeting spaces, ballrooms, and signage. These are meant as core income and used to offset, usually partially, the risk of the daily lease term for the overnight guest.

4 Amortization schedule

Three types of amortizations will be discussed in the context of real estate pro forma modelling. However, additional methods exists (e.g. constant amortization mortgages) but are significantly less common. The three covered will be constant payment mortgage (CPM), interest only (IO), and custom amortization. The CPM has been widely used for decades in the post-Depression era for both home and car loans. The CPM provides level payments throughout the term of the mortgage and generally is fully amortized (i.e. no principal balance remains at the end) on most traditional schedules. Interest only, though prevalent prior to this century, came to the forefront of American borrowing in the early to mid-2000s with exotic loans. It is a method to reduce monthly payments for the borrowing but does not amortize (pay down principal) throughout the life of the mortgage. Custom amortization schedules are used for larger projects and typically have zero amortization at the beginning when cash flows for a commercial project are low or negative (as in the start-up phase) and ramp up quickly depending upon the schedule.

Constant payment mortgage

As discussed, this is the traditional loan structure utilized by most home and car loans in the US. A typical structure for a home loan will be 30-year amortization, 4.5% constant mortgage paid monthly, and zero balance due at the end of the 30 years (i.e. fully amortized). As with most valuations in real estate the method of calculation follows the traditional present value/ future value relationship.

$$PV = \sum_{t=1}^{n} \frac{PMT}{(1+i)^t}$$

As mortgage payments are paid monthly, interest must be adjusted (divided by 12) to reflect this payment. Generally, however, the mortgage interest is quoted as annual. The relationship adjusts to the following to account for monthly payments:

$$PV = \sum_{t=1}^{n*12} \left[\frac{PMT}{\left(1+\frac{i}{12}\right)} \right]^t$$

where
i: annual interest
PV: loan principal

PMT: constant payment
n: number of payments

As an example, assume a $1,000 loan which is fully amortizing and payable at the end of the next 12 months at 6.00%. The present value equation is solved for payment.

$$1000 = \sum_{t=1}^{12}\left[\frac{PMT}{\left(1+\frac{0.06}{12}\right)}\right]^t = \sum_{t=1}^{12}\left[\frac{PMT}{(1.005)}\right]^t$$

Using MS Excel's Solver function we set up the cash flow series in Figure 4.1. Therefore, in the previous example, the present value formula simplifies as follows:

$$1000 = \sum_{t=1}^{12}\left[\frac{86.07}{(1.005)}\right]^t$$

That is, $86.07 is the constant payment necessary per month to fully amortize the $1,000 loan over 12 months. The final row of the Excel example uses the Excel function "PMT()" to calculate the constant payment over the 12 months. While it has additional dependents, the three utilized are interest (on a monthly basis, i.e. divided by 12), periods (12 months for 1 year), and principal (−1000). The negative principal was used to provide a positive result for the formula. Note that typically in finance a negative is utilized to indicate direction of cash flow (i.e. inflow or outflow), rather than meaning a negative is 'bad'.

Continuing with this example, an amortization table is developed to demonstrate principal and interest paid per period. The recommended structure for an amortization table is vertically constructed with the column headers in Figure 4.2.

Period is the payment number. BoP is Beginning of Period principal balance. Princ is the principal amount paid in the current period. Int is the interest amount paid in the current period. EoP is the End of Period principal balance, the principal remaining to be paid at the end of the period.

The periods, for ease, are started in cell A2, beneath the header, as a static value '1' for the first period. Each period beneath references the one above with a relative reference and adds one. This allows the developer to copy and drag the formula down to as many periods as desired for the analysis; for example, for a 30-year amortization it would be 360 periods.

Principal	1000											
Payment	86.07											
Discount	1.005											
Period	1	2	3	4	5	6	7	8	9	10	11	12
1,000.00	85.64	85.21	84.79	84.37	83.95	83.53	83.11	82.70	82.29	81.88	81.47	81.07
$86.07	=PMT(6%/12,12,-1000)											

Figure 4.1 Cash Flow

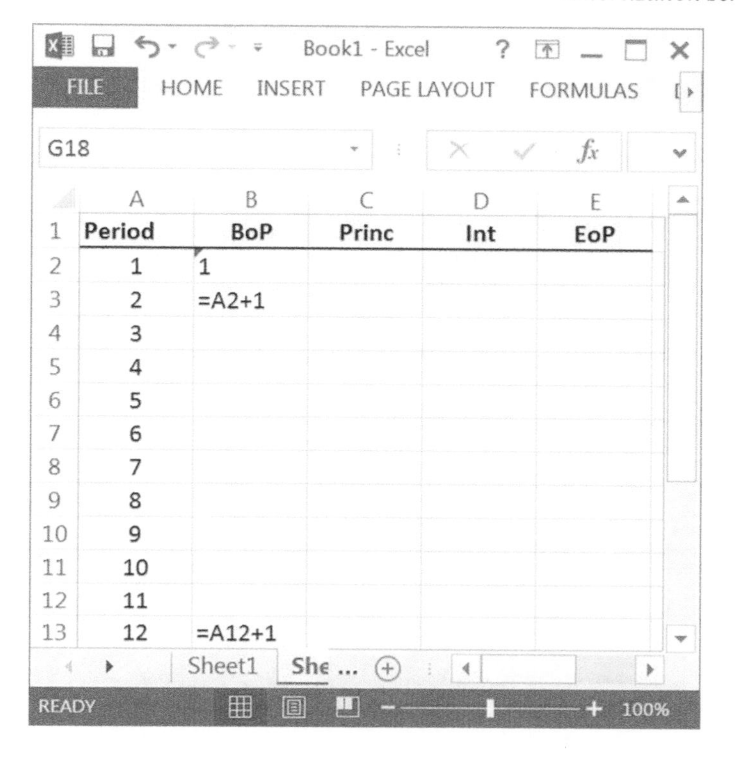

Figure 4.2

Column B has the contents of Column A as an example. As A3 references A2 with a relative reference, it can simply be copied or dragged to the final period.

Cell B2 references the initial principal amount. In the example it is 1,000. Cell C2 is a calculation of the principal paid per period. For this cell the function in Excel is "=PPMT()" which calculates the 'Principal' payment per period. The formula PPMT utilizes four dependents in order: Interest Rate, Period, Number of Periods, and Initial Principal. The formula used in C2 is therefore "=PPMT(6%/12,A2,12,-$B$2)". There is an intermixing of relative and absolute references. This allows for the formula to be copied vertically without the need to rekey.

While there is an Interest per Period formula in Excel – i.e. "=IPMT()" – this author finds it most simple to calculate directly. Therefore, interest per period is simply the product of outstanding principal per period and periodic interest. In the case of the example, cell D2 is "=B2*6%/12". Column E has EoP (End of Period) and this is calculated as the Beginning of Period balance less principal (amortized amount) paid down. The formula in E2 is therefore "=B2-C2". Finally, cell B3, the BoP principal amount for period 2 is equal to the end of period balance for period 1. Therefore cell B3 is "=E2". Currently the amortization table has the start of the second period completed (Figure 4.3).

At this point cells C2:E2 should be copied directly (dragged) to complete cells C3:C4. Once this is done the analyst can copy B3:E3 to row 13 or the 12th period to complete the amortization table (Figure 4.4).

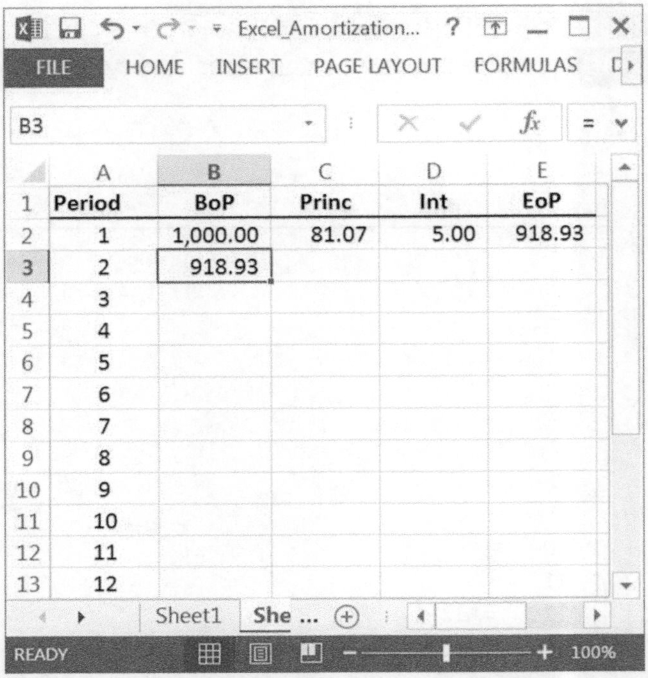

Excel_Amortization...

FILE | HOME | INSERT | PAGE LAYOUT | FORMULAS

B3

	A	B	C	D	E
1	**Period**	**BoP**	**Princ**	**Int**	**EoP**
2	1	1,000.00	81.07	5.00	918.93
3	2	918.93			
4	3				
5	4				
6	5				
7	6				
8	7				
9	8				
10	9				
11	10				
12	11				
13	12				

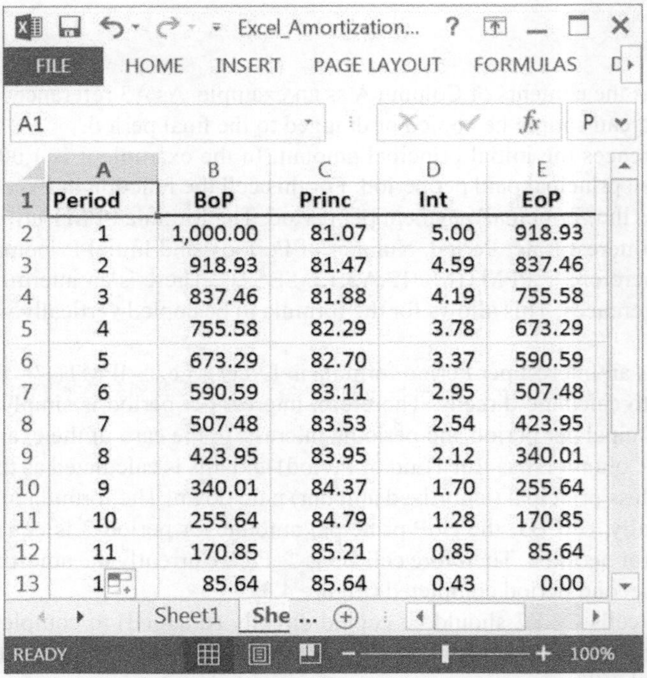

Sheet1 | She ...

READY 100%

Figure 4.3

Excel_Amortization...

FILE | HOME | INSERT | PAGE LAYOUT | FORMULAS

A1

	A	B	C	D	E
1	**Period**	**BoP**	**Princ**	**Int**	**EoP**
2	1	1,000.00	81.07	5.00	918.93
3	2	918.93	81.47	4.59	837.46
4	3	837.46	81.88	4.19	755.58
5	4	755.58	82.29	3.78	673.29
6	5	673.29	82.70	3.37	590.59
7	6	590.59	83.11	2.95	507.48
8	7	507.48	83.53	2.54	423.95
9	8	423.95	83.95	2.12	340.01
10	9	340.01	84.37	1.70	255.64
11	10	255.64	84.79	1.28	170.85
12	11	170.85	85.21	0.85	85.64
13	1	85.64	85.64	0.43	0.00

Sheet1 | She ...

READY 100%

Figure 4.4

Note that cell E13 has an End of Period balance of 0.00. This demonstrates that the loan is fully amortized and that no principal balance remains.

Interest only

An interest only (IO) loan product provides zero amortization (principal paydown) throughout the life of the loan. It is characterized as a constant balance and constant interest payment (assuming the interest rate is fixed) throughout the term of the loan.

For setting up the loan amortization schedule, the structure remains the same, thus the heading is equivalent (see Figure 4.5). For this example the 6.00% interest rate will remain fixed. The first period's balance remains 1,000. However, as this is interest only, cells C2:C13 (principal) are all zero. The interest expense in D2 is the product of BoP balance (B2) and 6.00%/12 (monthly interest rate). The formula remains the same as before for D2:D13, for example, for D2 (=B2*6%/12).

Note that the EoP balance in cell E13 remains 1,000. There is no amortization for this type of loan payment. However, the loan payment per period is significantly less. In the IO structure only $5.00 is paid while in the constant payment structure $86.07 is paid. This represents 5.81% of the CPM structure. It provides a significant benefit for affordability but absolutely no 'forced' savings through amortization of the product.

Interest only loans are an 'exotic' product utilized in commercial loan products and some residential mortgages. For a cash-flowing property (i.e. commercial), the product reduces finance payments and forced equity buildup in the product while maintaining a higher debt

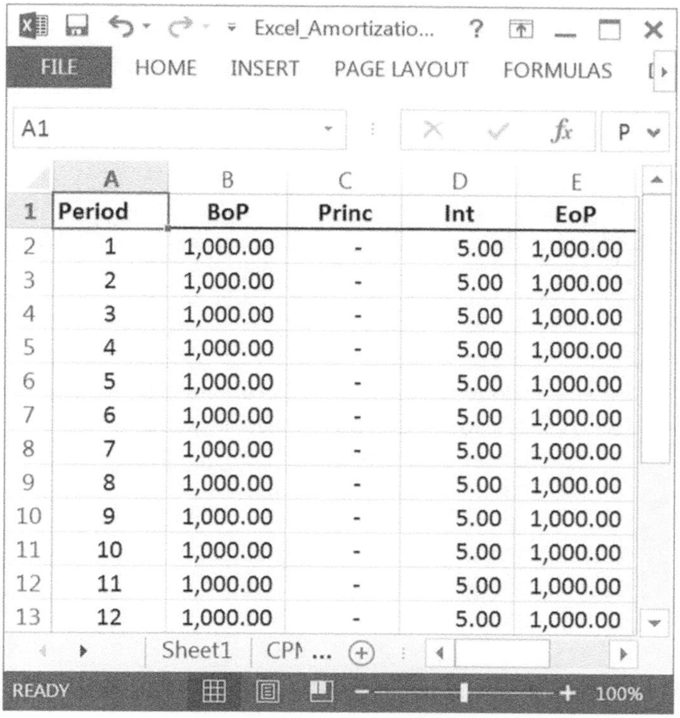

Excel_Amortizatio...

FILE HOME INSERT PAGE LAYOUT FORMULAS

A1

	A	B	C	D	E
1	Period	BoP	Princ	Int	EoP
2	1	1,000.00	-	5.00	1,000.00
3	2	1,000.00	-	5.00	1,000.00
4	3	1,000.00	-	5.00	1,000.00
5	4	1,000.00	-	5.00	1,000.00
6	5	1,000.00	-	5.00	1,000.00
7	6	1,000.00	-	5.00	1,000.00
8	7	1,000.00	-	5.00	1,000.00
9	8	1,000.00	-	5.00	1,000.00
10	9	1,000.00	-	5.00	1,000.00
11	10	1,000.00	-	5.00	1,000.00
12	11	1,000.00	-	5.00	1,000.00
13	12	1,000.00	-	5.00	1,000.00

Sheet1 CPM ...

READY 100%

Figure 4.5

service coverage ratio, that is, NOI/debt payment. In a residential mortgage, the interest only structure also lowers the payment but provides no benefit of forced savings (equity pay-down). Provided the price of the residential home increases, equity is gained through market forces. However, during the real estate crisis of 2007–2010, when home prices were decreasing, the interest only product led to many loans increasing over the value of the residence itself.

Custom amortization

Custom amortization, as the name indicates, is a customized amortization schedule. Generally full amortization is achieved at the end of the term (Figure 4.6), though some more complicated structures may have a balloon payment at the end allowing for a restructuring. In this text full amortization at loan term will be utilized.

The structure headings for the amortization table remain the same, but an additional column will be added (column F) to indicate the percentage of amortization in the period. Cell F1 is therefore titled 'Amort%'. Column F will then state the percentage of amortization attributable to an individual period. Column F should be formatted in blue to indicate input cells. Note that for ease of explanation, the loan amount, constant percentage, and so forth have been input directly into the cell. This will be made dynamic later. Also note that the summation of the values in column F is 100.00%. This further indicates full amortization.

Column C, Princ, therefore must have the formula adjusted to indicate the amount of principal paid per period. Cell C2 is adjusted to be the product of the original principal amount and the amortization percentage. For C2 the formula is adjusted to "=$B$2*F2". The

Excel_Amortization_Chap4 - Excel

FILE HOME INSERT PAGE LAYOUT FORMULAS DATA REV

C2 =$B$2*F2

	A	B	C	D	E	F
1	Period	BoP	Princ	Int	EoP	Amort%
2	1	1,000.00	-	5.00	1,000.00	0.00%
3	2	1,000.00	-	5.00	1,000.00	0.00%
4	3	1,000.00	50.00	5.00	950.00	5.00%
5	4	950.00	50.00	4.75	900.00	5.00%
6	5	900.00	50.00	4.50	850.00	5.00%
7	6	850.00	50.00	4.25	800.00	5.00%
8	7	800.00	100.00	4.00	700.00	10.00%
9	8	700.00	100.00	3.50	600.00	10.00%
10	9	600.00	150.00	3.00	450.00	15.00%
11	10	450.00	150.00	2.25	300.00	15.00%
12	11	300.00	150.00	1.50	150.00	15.00%
13	12	150.00	150.00	0.75	-	15.00%

Sheet1 CPM IO ...

READY 100%

Figure 4.6

$B$2 is an absolute reference to the original loan balance, $1,000, while F2 is the dynamic reference to the custom amortization percentage in the respective period.

In this structure, also note that the highest payment period is period 9. Principal is $150.00 while interest is $3.00 for a total payment of $153.00. This customization allows for commercial products to better align finance charges with forecasted cash flows. In theory, this will smooth the debt service coverage ratio (DSCR). Also, in the early periods with no amortization percentage, start-up is not burdened with heavy finance charges.

Amortization schedules (the combination)

There are three general structures for amortization schedules: (1) constant payment mortgage, (2) interest only, and (3) custom amortization. How are each of these combined into a single, dynamic spreadsheet providing for the transition between each and allowing the analyst to rapidly select the best structure for a real estate project? The answer is to combine each methodology with a 'switch,' allowing easy comparison of the three within the same structure.

As a start the original spreadsheet will be expanded, delineating the hard inputs (Figure 4.7). Five rows are added atop the original structure to provide the necessary inputs to the amortization table. These inputs will originally be linked (i.e. referenced) from an input

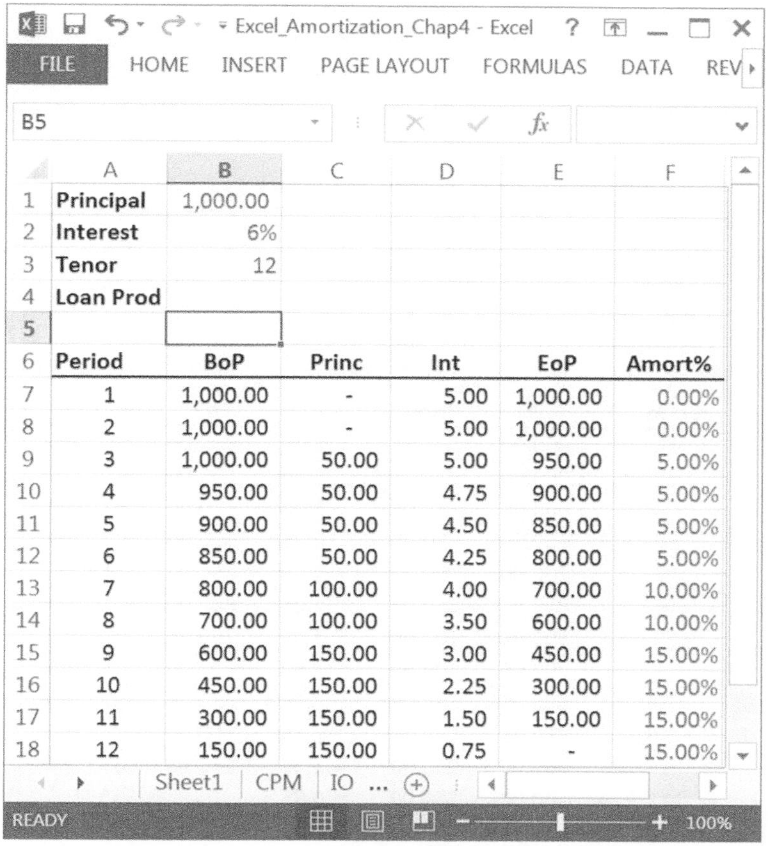

Excel_Amortization_Chap4 - Excel

FILE HOME INSERT PAGE LAYOUT FORMULAS DATA REV

B5

	A	B	C	D	E	F
1	Principal	1,000.00				
2	Interest	6%				
3	Tenor	12				
4	Loan Prod					
5						
6	Period	BoP	Princ	Int	EoP	Amort%
7	1	1,000.00	-	5.00	1,000.00	0.00%
8	2	1,000.00	-	5.00	1,000.00	0.00%
9	3	1,000.00	50.00	5.00	950.00	5.00%
10	4	950.00	50.00	4.75	900.00	5.00%
11	5	900.00	50.00	4.50	850.00	5.00%
12	6	850.00	50.00	4.25	800.00	5.00%
13	7	800.00	100.00	4.00	700.00	10.00%
14	8	700.00	100.00	3.50	600.00	10.00%
15	9	600.00	150.00	3.00	450.00	15.00%
16	10	450.00	150.00	2.25	300.00	15.00%
17	11	300.00	150.00	1.50	150.00	15.00%
18	12	150.00	150.00	0.75	-	15.00%

Sheet1 CPM IO ...

READY 100%

Figure 4.7

page. However, this is the general structure that will result. The structure therefore adjusts, using the custom amortization example described in the previous section, as shown.

Each of the input cells B1:B4 will be named as follows: Principal_1, Interest_1, Tenor_1, and Loan_Prod_1, respectively (Figure 4.8). Naming the cells provides absolute references and provides ease of troubleshooting later. Please note that naming can be done manually by overwriting the name in the reference box that has the respective location. In the example in Figure 4.7, "B5" is the location. However, when naming cell B1, this reference will be "B1". The names have been placed in cells C1:C4 for ease of explanation (see Figure 4.8). Notice that cell B4 has been left blank. As this is the 'switch' between the loan products, a drop-down box will be created to provide the user only three static choices for the cell: 1–CPM, 2–IO, and 3–Custom Amortization. This is done by selecting the cell, C4, and under the data title on the ribbon select 'Data Validation' (Figure 4.9). Three drop-downs will result: Data Validation, Circle Invalid Data, and Clear Validation Circles. Select the first drop-down, 'Data Validation'. A new input box will appear titled "Data Validation". It is here where the analyst must set up the inputs for the loan product input (i.e. cell B4). Under the Settings tab there is 'Validation Criteria' (Figure 4.10). Select the 'Allow:' drop-down box and a list of options will be provided. 'List' is the selection that is appropriate. In the 'Source' section the analyst must write "1,2,3" (Figure 4.11). This is a list of alternatives for the loan products to be selected. The number 1 will designate Constant Payment Mortgage, 2 will designate

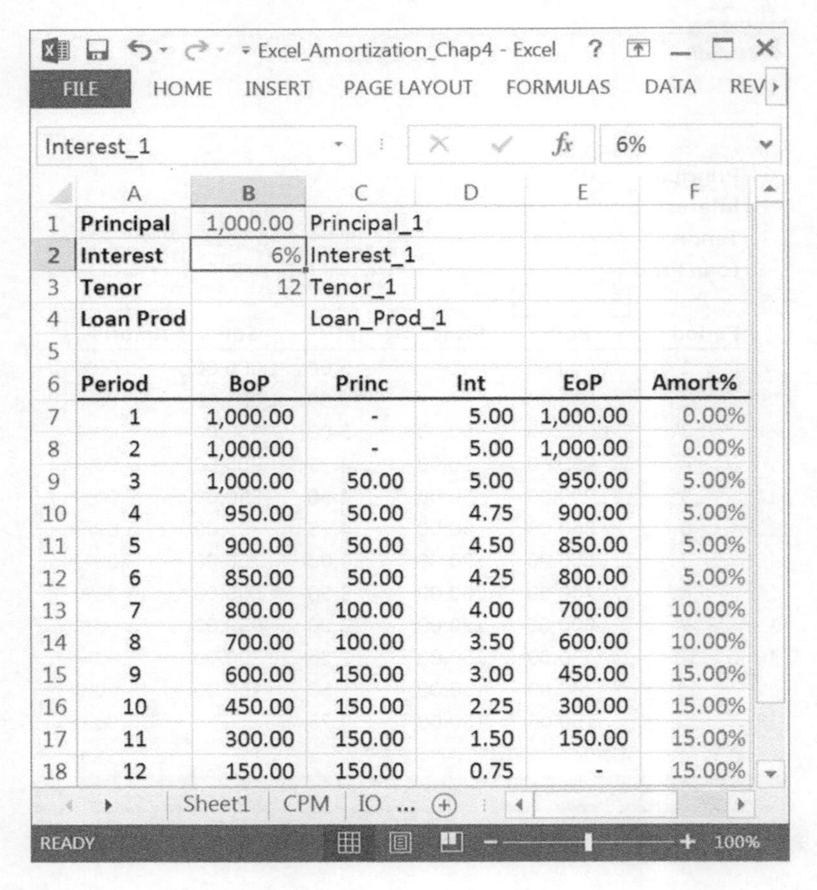

	A	B	C	D	E	F
1	Principal	1,000.00	Principal_1			
2	Interest	6%	Interest_1			
3	Tenor	12	Tenor_1			
4	Loan Prod		Loan_Prod_1			
5						
6	Period	BoP	Princ	Int	EoP	Amort%
7	1	1,000.00	-	5.00	1,000.00	0.00%
8	2	1,000.00	-	5.00	1,000.00	0.00%
9	3	1,000.00	50.00	5.00	950.00	5.00%
10	4	950.00	50.00	4.75	900.00	5.00%
11	5	900.00	50.00	4.50	850.00	5.00%
12	6	850.00	50.00	4.25	800.00	5.00%
13	7	800.00	100.00	4.00	700.00	10.00%
14	8	700.00	100.00	3.50	600.00	10.00%
15	9	600.00	150.00	3.00	450.00	15.00%
16	10	450.00	150.00	2.25	300.00	15.00%
17	11	300.00	150.00	1.50	150.00	15.00%
18	12	150.00	150.00	0.75	-	15.00%

Figure 4.8

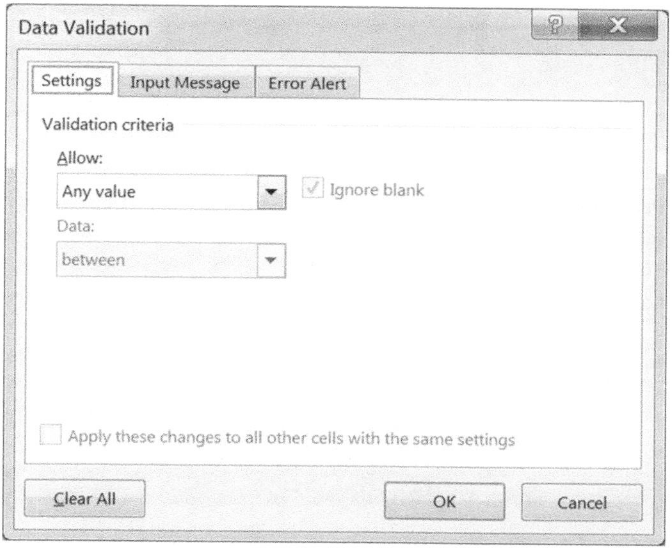

Data Validation

Settings | Input Message | Error Alert

Validation criteria

Allow:

Any value ☑ Ignore blank

Data:

between

☐ Apply these changes to all other cells with the same settings

Clear All | OK | Cancel

Figure 4.9

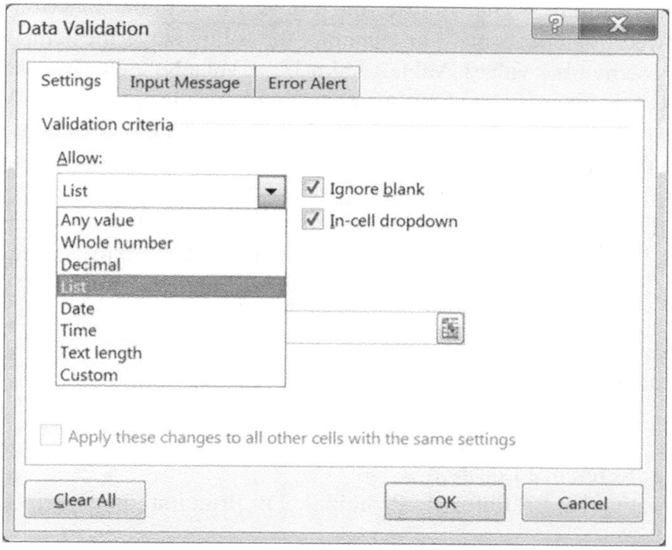

Data Validation

Settings | Input Message | Error Alert

Validation criteria

Allow:

List ☑ Ignore blank

☑ In-cell dropdown

Any value
Whole number
Decimal
List
Date
Time
Text length
Custom

☐ Apply these changes to all other cells with the same settings

Clear All | OK | Cancel

Figure 4.10

Interest Only, and 3 will designate Custom Amortization. After this is done and 'OK' is selected, the input for B4 will be a drop-down with the selection 1, 2, 3. These inputs will then be utilized in a "=Choose()" function to select the type of loan product.

Now adjustments must be made to that amortization table to accept the inputs and ensure the table is dynamic. The first is to set cell B7, the BoP balance for period 1, equal to the initial principal. Therefore cell B7 becomes "=Principal_1" which is an absolute reference to cell B1 (i.e. principal amount). This does not utilize the Choose function and is absolute as it is the beginning principal balance.

The next cell in the amortization table, Principal, will be dynamic and is dependent upon the structure – i.e. Principal = F(CPM, IO, Custom). Much as the function states, the Choose

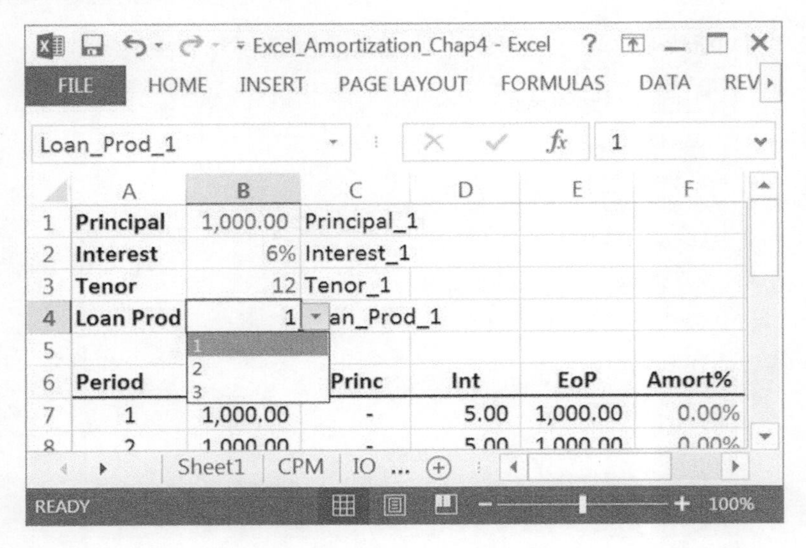

Figure 4.11

function operates similarly. The Choose function is a directional function with an initial argument followed by dependents. For example, the following is the setup for the Choose function: "=choose(number,Value1, Value2, Value3, . . ., ValueN)". Whatever natural number – i.e. {1,2,3,. . .N} – is placed in the first location, 'number', the appropriate 'ValueN' will be selected. For instance, if "2" is placed in the first location of the Choose function, then Value2 will result from the function. If "4" is placed in the first location of the Choose function, then Value4 will result, and so on.

The Principal cell, cell C7, will then be adjusted to account for each structure. Again, the three structures, in order, are constant payment mortgage, interest only, and custom amortization. Therefore, cell C7 utilizing the Choose function becomes as follows:

=CHOOSE(Loan_Prod_1,PPMT(Interest_1/12,A7,Tenor_1*12,-Principal_1), 0,F7*Principal_1)

where

Loan_Prod_1: Reference Location

PPMT(Interest_1/12,A7,Tenor_1,-Principal_1): Principal per period for Constant Payment Mortgage

0: Principal paid for Interest Only Mortgage

F7*Principal_1: Principal per period for Custom Amortization Product

The interest and EoP cells remain the same, as they are not dependent upon structure. Also note that the units for 'Tenor' are months whereas for 'Interest' it is annual. This was done to provide examples to the analyst that different units can be selected and must be monitored closely and understood carefully.

To select the constant payment mortgage structure, the analyst/developer must simply select "1" from the drop-down box in cell B4. See Figure 4.12. For interest only see Figure 4.13. For custom amortization see Figure 4.14.

Excel_Amortization_Chap4 - Excel

E2

	A	B	C	D	E	F
1	Principal	1,000.00	Principal_1		Example	One (1)
2	Interest	6%	Interest_1			
3	Tenor	12	Tenor_1			
4	Loan Prod	1	Loan_Prod_1			
5						
6	Period	BoP	Princ	Int	EoP	Amort%
7	1	1,000.00	81.07	5.00	918.93	0.00%
8	2	918.93	81.47	4.59	837.46	0.00%
9	3	837.46	81.88	4.19	755.58	5.00%
10	4	755.58	82.29	3.78	673.29	5.00%
11	5	673.29	82.70	3.37	590.59	5.00%
12	6	590.59	83.11	2.95	507.48	5.00%
13	7	507.48	83.53	2.54	423.95	10.00%
14	8	423.95	83.95	2.12	340.01	10.00%
15	9	340.01	84.37	1.70	255.64	15.00%
16	10	255.64	84.79	1.28	170.85	15.00%
17	11	170.85	85.21	0.85	85.64	15.00%
18	12	85.64	85.64	0.43	0.00	15.00%

Sheet1 | CPM | IO ...

Figure 4.12 Example One (1)

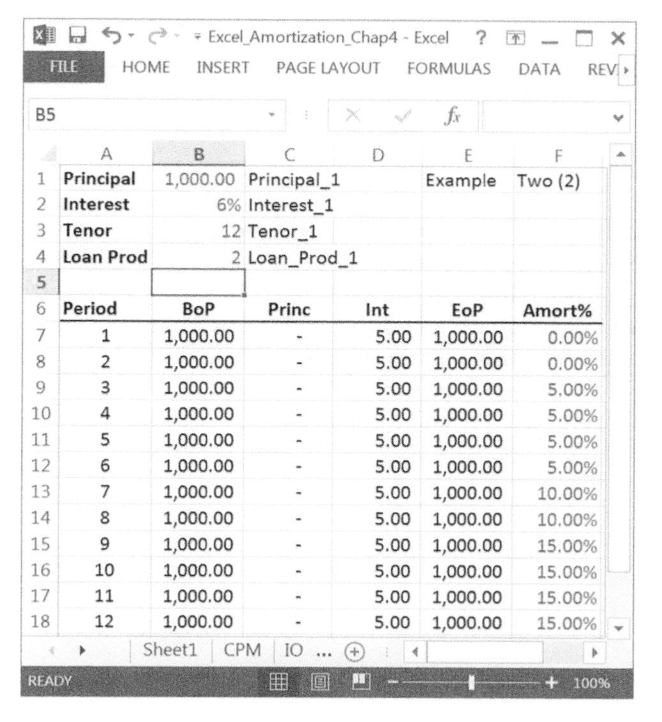

Excel_Amortization_Chap4 - Excel

B5

	A	B	C	D	E	F
1	Principal	1,000.00	Principal_1		Example	Two (2)
2	Interest	6%	Interest_1			
3	Tenor	12	Tenor_1			
4	Loan Prod	2	Loan_Prod_1			
5						
6	Period	BoP	Princ	Int	EoP	Amort%
7	1	1,000.00	-	5.00	1,000.00	0.00%
8	2	1,000.00	-	5.00	1,000.00	0.00%
9	3	1,000.00	-	5.00	1,000.00	5.00%
10	4	1,000.00	-	5.00	1,000.00	5.00%
11	5	1,000.00	-	5.00	1,000.00	5.00%
12	6	1,000.00	-	5.00	1,000.00	5.00%
13	7	1,000.00	-	5.00	1,000.00	10.00%
14	8	1,000.00	-	5.00	1,000.00	10.00%
15	9	1,000.00	-	5.00	1,000.00	15.00%
16	10	1,000.00	-	5.00	1,000.00	15.00%
17	11	1,000.00	-	5.00	1,000.00	15.00%
18	12	1,000.00	-	5.00	1,000.00	15.00%

Sheet1 | CPM | IO ...

Figure 4.13 Example Two (2)

Excel_Amortization_Chap4 - Excel

FILE HOME INSERT PAGE LAYOUT FORMULAS DATA REV

Loan_Prod_1 | f_x 3

	A	B	C	D	E	F
1	**Principal**	1,000.00	Principal_1		Example	Three (3)
2	**Interest**	6%	Interest_1			
3	**Tenor**	12	Tenor_1			
4	**Loan Prod**	3	an_Prod_1			
5						
6	**Period**	**BoP**	**Princ**	**Int**	**EoP**	**Amort%**
7	1	1,000.00	-	5.00	1,000.00	0.00%
8	2	1,000.00	-	5.00	1,000.00	0.00%
9	3	1,000.00	50.00	5.00	950.00	5.00%
10	4	950.00	50.00	4.75	900.00	5.00%
11	5	900.00	50.00	4.50	850.00	5.00%
12	6	850.00	50.00	4.25	800.00	5.00%
13	7	800.00	100.00	4.00	700.00	10.00%
14	8	700.00	100.00	3.50	600.00	10.00%
15	9	600.00	150.00	3.00	450.00	15.00%
16	10	450.00	150.00	2.25	300.00	15.00%
17	11	300.00	150.00	1.50	150.00	15.00%
18	12	150.00	150.00	0.75	-	15.00%

Sheet1 | CPM | IO ...

Select destination and press E... 100%

Figure 4.14 Example Three (3)

While amortization tables for a multiyear project are larger, they all follow the same general structure constructed and represented here. It is imperative that these be built as generic as possible because when modelling collateralized mortgage obligations (CMOs), the portfolio is constructed using multiple amortizations for different product offerings. By making each of these constant, it ensures that construction of the more complex financial products is simplified (to the greatest degree possible). An example of a CMO product is provided in Chapter 13 and it incorporates 'n' amortization tables, which are summarized in a top-level roll-up table.

5 Single-family residential

The most simple pro forma which follows the same general structure as a larger one is the single-family residential home. Take, for instance, a single-family home in Columbia, Maryland. The asking price is $430,000; closing costs are estimated at $10,000 and financing costs at 1.00% of the loan amount. The property is located in a desirable area, will rent at $2,500/month, and is expected to be vacant 1 week per year as rollover. The loan produce is a traditional 80%/20% loan with a 30-year amortization. Additional inputs will be defined later in this chapter as the pro forma construction continues.

As demonstrated earlier in this text, the structure for all commercial projects follows relatively the same construction, shown in Figure 5.1.

As such, the Inputs section will be constructed first. In order to completely understand why this is constructed first, a decomposition of the Inputs section is necessary. While a single-family home is utilized as a simple example, the structure remains roughly equivalent for all real estate income-producing properties. As this is not income producing, the structure will change slightly to accommodate the 'for sale' nature of this asset class/product type.

The Inputs section is structured into three tiers: Source of Funds, Operations, and Finance/Valuation. Although each of these areas can be expanded in larger models, they are best summarized in this format. For the first section, Tier 1, the sources and uses table is constructed and inputs provided (Figure 5.2). In real estate, a sources/uses table summarizes the transaction at time of settlement. As this is at a particular time, it can be considered similar to a balance sheet at the settlement table. The top portion of the sources and uses table represents the 'Source' of funds at settlement. Typically there are two sources: equity and debt. Equity is what actually leaves the owner/investor's wallet at settlement and debt is provided, typically, from third-party capital providers (e.g. banks). The sources represent all money being injected into the project at time of settlement. The uses are the opposite of the sources – i.e. where the money goes at the settlement table. Typical line items for 'Uses' are purchase price (seller) and soft cost fees. Of course there can be significantly more uses depending upon the size and complexity of a particular project.

Tier 1 (Sources/Uses) Tier 2 (Operations) Tier 3 (Finance/Valuation)

Tier 2 (Operations) provides the inputs for how the project will 'operate' over the time frame for the analysis. Revenue is unit numbers, revenue per unit, and revenue escalation per period. Expense is maintenance, repairs, management (if applicable), homeowner's expense (if applicable), expense escalation per period, etc.

Tier 3 (Finance/Valuation) covers Finance (loan product(s) and fees) and Valuation/Sale. Whereas Finance is similar to the amortization constructed in the earlier section, Valuation/Sale covers discount rate, tenor of loan/project, sales expenses, etc. Figuring out how each

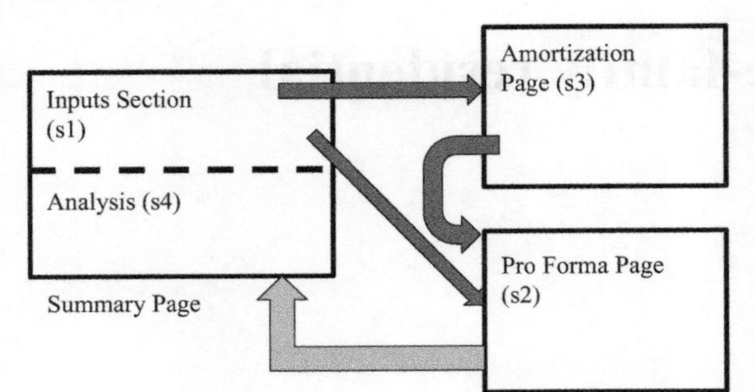

Figure 5.1

Equity / Debt	Revenue	Finance
Uses	Expenses	Valuation/ Sale

Figure 5.2

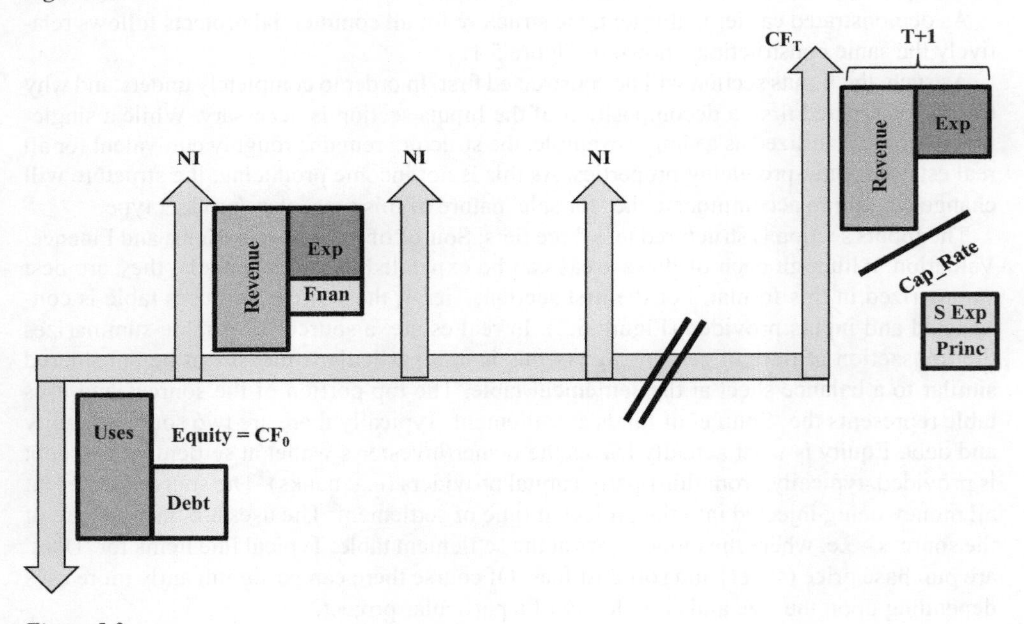

Figure 5.3

of these sections fits into creating a credible real estate pro forma model requires a mapping of input to cash flow. Remember that a real estate pro forma model is simply a tool to quantify the return and risk of a cash flow stream for a current or future real estate project.

Notice the cash flow diagram for an income-producing, in this case single-family residential, project (Figure 5.3). The first point of note is the initial cash flow out, CF_0. When valuating the cash flow stream, CF_0 is the 'sources' of the input or the delta between Uses and Debt. When valuating at the project level, CF_0 is the initial cash outflow.

The next item in the cash flow stream is the Net Income flows (i.e. cash flows) for the project during operations. Net Income is defined as Revenue less Expense less Finance

Charge. It is essentially the free cash flow for the project. As Figure 5.3 indicates, it is the delta between Revenue and Expense/Finance Charge. The Free Cash Flow, Net Income, is the cash flows utilized in the Net Present Value equation to value the project.

Finally comes the Final Cash Flow upon sale. As the diagram demonstrates, this is still governed by the input page but is slightly more involved. Remember that the gross sales price in period N is the net operating income in period $N + 1$ divided by the terminal capitalization rate. This provides gross sales value (GSV). The GSV is then reduced by sales expense and principal payback (i.e. any/all remaining principal balance to third party that capital provides). This results in the Net Sales Proceeds for the real estate project.

The first step in building a credible real estate pro forma model is therefore to construct the Inputs section as described earlier (Figure 5.4). It is recommended that columns A and B be formatted to a width of 2 and that the flexibility of potentially two investors and debt providers be constructed initially. While the models are scalable, it is generally best to build in flexibility from the beginning, avoiding potential edits later in the construction process. Also, the headers should be bolded for the benefit of the presentation.

Column D contains the inputs to Sources and Uses (Figure 5.5). The Uses are assumed to be $430,000 for purchase price, $10,000 for closing costs, and $3,400 for finance cost (1.00% of borrowed amount). Debt is a static $340,000, which is 79% of the purchase price. Equity is the remaining, the delta between total Uses and Debt provided (i.e. $103,400). Note that as these are all inputs they are to be formatted in blue.

Next (Figure 5.6) is generally formation for the roll-up line items: Sources, Equity, Debt, and Uses. A bottom border is utilized for each of these. Also, bolding roll-up items will enhance the value of the presentation. Finally, place a thick box border from A1:D13 to separate the Sources and Uses as Tier 1 for the project.

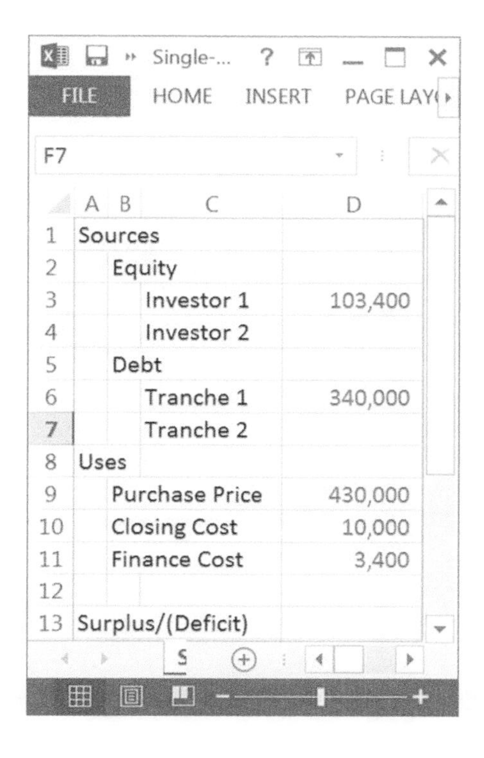

	A	B	C	D
1	Sources			
2		Equity		
3			Investor 1	103,400
4			Investor 2	
5		Debt		
6			Tranche 1	340,000
7			Tranche 2	
8	Uses			
9		Purchase Price		430,000
10		Closing Cost		10,000
11		Finance Cost		3,400
12				
13	Surplus/(Deficit)			

Figure 5.4

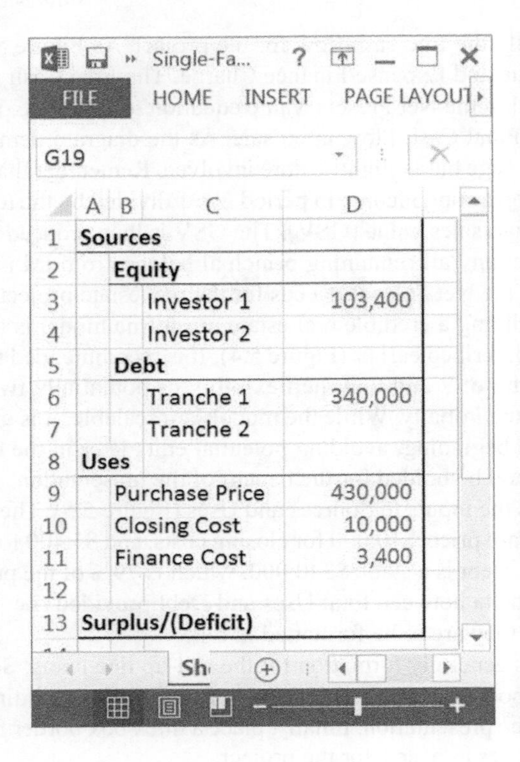

	A	B	C	D
1	Sources			
2		Equity		
3			Investor 1	103,400
4			Investor 2	
5		Debt		
6			Tranche 1	340,000
7			Tranche 2	
8	Uses			
9		Purchase Price		430,000
10		Closing Cost		10,000
11		Finance Cost		3,400
12				
13	Surplus/(Deficit)			

Figure 5.5

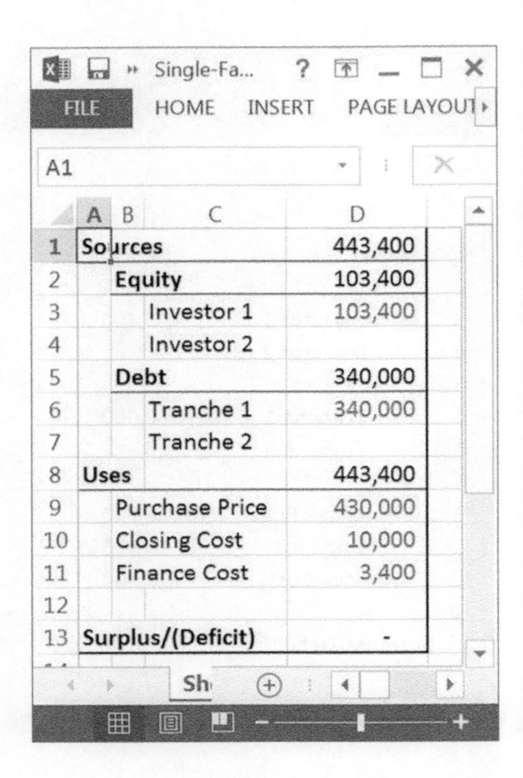

	A	B	C	D
1	Sources			443,400
2		Equity		103,400
3			Investor 1	103,400
4			Investor 2	
5		Debt		340,000
6			Tranche 1	340,000
7			Tranche 2	
8	Uses			443,400
9		Purchase Price		430,000
10		Closing Cost		10,000
11		Finance Cost		3,400
12				
13	Surplus/(Deficit)			-

Figure 5.6

Now, for the Sources and Uses section, all that remains are the roll-up formulas. Generally most analysts use the summation function, "=sum()", for addition. However, because this section has subtotals (such as Equity and Debt) underneath the 'Sources' sections, the subtotal function – "=subtotal()" – is recommended. The subtotal function negates an analyst error of double counting in that it does not 'subtotal' cells beneath that are quantified using a subtotal function.

For cell D2 the formula is "=subtotal(9,D3:D4)". The reason for the '9' as the first dependent in the function is that there is multiple functionality of the subtotal function. The '9' indicates addition. Other functionalities which require the first dependent to be adjusted are average and product, but there are more than just those two. The formula for D5 is therefore "=subtotal(9,D6:D7)" and for D1 it is "=subtotal(9,D2:D7)". Finally the Uses formula in D8 is "=subtotal(9,D9:D11)". The Surplus/(Deficit) cell, D13, is the delta between Source and Uses and demonstrates if there is sufficient capital to settle (i.e. close a transaction). If the value is negative, then additional equity is required to close the transaction. The formula for D13 is "=D1-D8".

The next section, Tier 2, is the Operations section covering revenue and expense parameters. In cell F1 (Figure 5.7) type 'Operations' and, proceeding vertically, type the headers for the revenue and expense sections. It is recommended that a blank cell be used as a separator. For revenue, three line items are utilized: Rental/mo, Rental Escl, and Vacancy. Note that Vacancy under US generally accepted accounting principles (GAAP) is a contra revenue item and therefore listed as revenue, not considered an expense. The Expense items included are Maintenance/mo, Homeowners Association Insurance (HOA), Property Tax/yr, Insurance, and Expense Escl. On larger, more complex real estate models, the list for revenue and expense will be greatly increased. Title cell I1 'Value' and then format the section using a bottom border, bold headers, and a thick box border.

Next (Figure 5.8), in column I place the inputs for each of the line items. These are assumptions based on the single-family residential stated at the beginning of this chapter. Again note that inputs should be in blue.

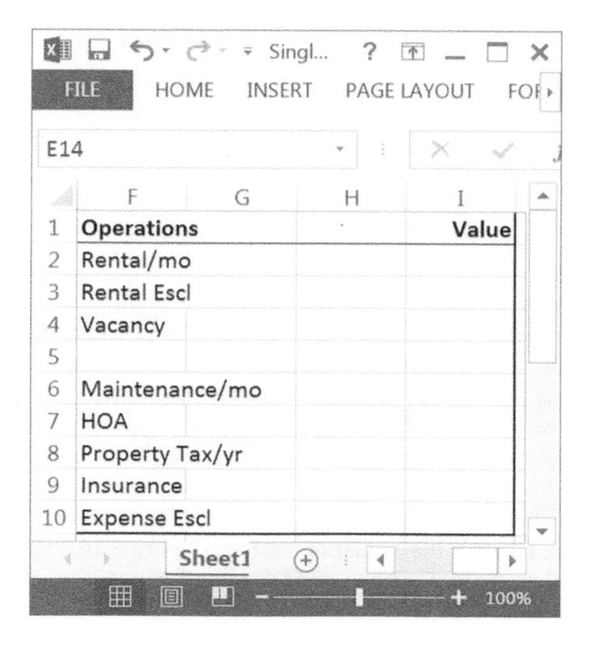

Figure 5.7

	F	G	H	I
1	Operations			Value
2	Rental/mo			2,500
3	Rental Escl			2.00%
4	Vacancy			5%
5				
6	Maintenance/mo			100
7	HOA			200
8	Property Tax/yr			4,500
9	Insurance			250
10	Expense Escl			2.00%

Figure 5.8

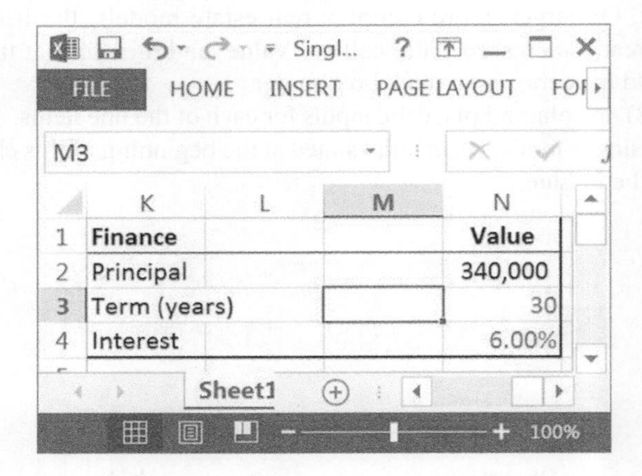

	K	L	M	N
1	Finance			Value
2	Principal			340,000
3	Term (years)			30
4	Interest			6.00%

Figure 5.9

The Tier 3 Finance and Valuation sections remain. Beginning in cell K1 is the Finance section (Figure 5.9). This will establish the loan characteristics for the project. In K1 type 'Finance' and beneath select Principal, Term (years), and Interest. Note that only a single structure is used in this basic single-family residential pro forma. Therefore structure is not being added here but will be included later in the text in larger models. In column N, place 'Value' in cell N1 as the title; the principal amount in cell N2 must be linked (i.e. the cell is "=D6"). Cells N3 and N4, Term and Interest, are 30 and 6.00%, respectively. Then format similar to the other sections to complete the Finance section.

The final section in Tier 3 is the Valuation section (Figure 5.10). Here the line items are Cap (Capitalization) Rate, Sales Expense, and Discount Rate. It is structured similarly to the Finance section in terms of formatting. The difference is that the section begins in cell K6.

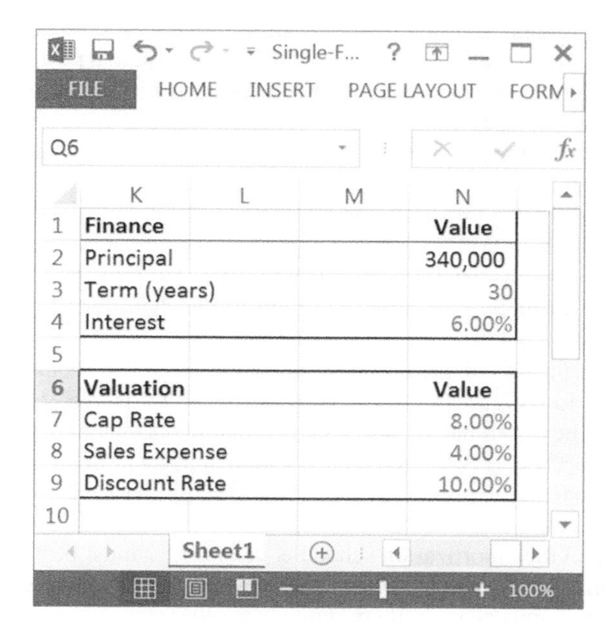

Finance	Value
Principal	340,000
Term (years)	30
Interest	6.00%

Valuation	Value
Cap Rate	8.00%
Sales Expense	4.00%
Discount Rate	10.00%

Figure 5.10

Sources	443,400
Equity	103,400
Investor 1	103,400
Investor 2	
Debt	340,000
Tranche 1	340,000
Tranche 2	
Uses	443,400
Purchase Price	430,000
Closing Cost	10,000
Finance Cost	3,400
Surplus/(Deficit)	-

Operations	Value
Rental/mo	2,500
Rental Escl	2.00%
Vacancy	5%
Maintenance/mo	100
HOA	200
Property Tax/yr	4,500
Insurance	250
Expense Escl	2.00%

Finance	Value
Principal	340,000
Term (years)	30
Interest	6.00%

Valuation	Value
Cap Rate	8.00%
Sales Expense	4.00%
Discount Rate	10.00%

Figure 5.11

The values for Cap Rate, Sales Expense, and Discount Rate are 8.00%, 4.00%, and 10.00%, respectively.

The entire Inputs section therefore appears as demonstrated in (Figure 5.11).

The second section to complete is the amortization table, as developed in the previous chapter. Add an additional sheet and label it 'Debt'. Structure it identically to what was done in the previous chapter. Principal, cell B1, is "=Inputs_Summary!N2"; Interest, cell B2, is "=Inputs_Summary!N4"; Tenor, cell B3, is "=Inputs_Summary!N3"; and Loan Prod, cell B4, is "N/A" (not applicable) in this model, that is, there is a single loan structure CPM for this model.

Because this is a 30-year loan the periods start in cell A7 as 1, but in A8 the formula is dynamic: "=A7+1". Given the simplicity of this model, only annual payments are calculated. Therefore only 30 periods are necessary, and cell A8 must be copied to cell A36. Cell B7, the Beginning of Period Balance for period 1, references the initial principal: cell B7 is "=B$1". Note that only one absolute reference was utilized in this case, prior to the row number. This is to enable the amortization table to be copied to the right in the case of the secondary, tertiary, and *n*th number of loan products.

Principal utilizes the "=PPMT" payment for principal portion for a constant payment loan. Cell C2's formula, just like Chapter 4, is "=PPMT(B$2, A7, B$3,-B$1)". Note the relative references allowing for direct copying to the right of the sheet for additional amortization tables.

Interest, cell D7, is the period interest based on the outstanding principal (i.e. BoP Balance) for the respective period (Figure 5.12). The equation in cell D7 is "=B7*B$2". End of Period Balance, cell E7, is the difference of BoP (cell B7) and the principal paydown, so cell E7 is "=B7-C7". The Amort% is N/A as there is no custom amortization product for this model's loan type. As was done in the previous chapter, the BoP for period 2 is equal to the EoP for period 1. Therefore, cell B8 is "=E7".

Just as was done in the amortization schedule, cells C7:F7 must be copied to row 8. The cells B8:F8 are copied to the final period in row 36. Although the amortization was complete at this point in the previous chapter, there remains another step. Select cells A7:F36. In the reference bar above Column A there is A7 referenced. Highlight this A7 and replace it with 'Tranche_1'. The range A7:F36 has been named 'Tranche_1'. The importance of naming this array will be highlighted in the pro forma page section following.

The third section to complete is the 'Pro_Forma' sheet. This is basically the cash-only income statement for a real estate project. Also, the calculations are all pre-tax and therefore do not include tax or depreciation.

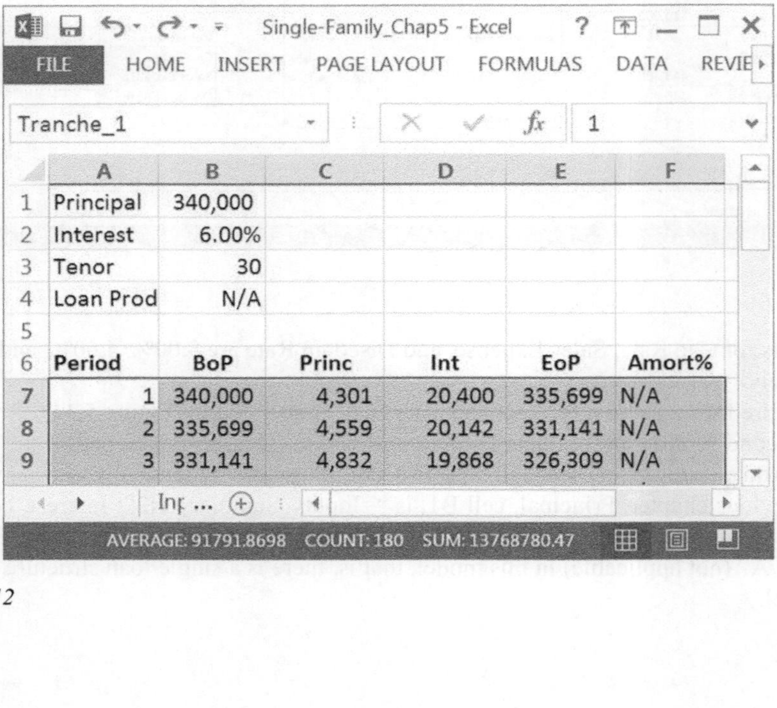

Single-Family_Chap5 - Excel

FILE HOME INSERT PAGE LAYOUT FORMULAS DATA REVIE

Tranche_1 | fx 1

	A	B	C	D	E	F
1	Principal	340,000				
2	Interest	6.00%				
3	Tenor	30				
4	Loan Prod	N/A				
5						
6	Period	BoP	Princ	Int	EoP	Amort%
7	1	340,000	4,301	20,400	335,699	N/A
8	2	335,699	4,559	20,142	331,141	N/A
9	3	331,141	4,832	19,868	326,309	N/A

Inp ...

AVERAGE: 91791.8698 COUNT: 180 SUM: 13768780.47

Figure 5.12

A new sheet must be added to the Excel workbook. This is done by selecting the circled plus sign (+) next to the Inputs_Summary tab name seen in Figure 5.11. Double click the sheet name and type 'Pro_Forma' to name the new sheet.

While general formatting will always be left to the individual analyst/developer, for the purposes of following along with this text, the formatting suggestions ought to be followed. The first adjustment to the Pro_Forma page (Figure 5.13) is therefore to change the column widths for columns A, B, D, and E to 2. Next in cell F1, type 'Period', which in this case is synonymous with 'Year'. To the right of the period beginning in cell G1 the periods will be listed (to be discussed in the next paragraph). In cells A2:C5, type the revenue headers as suggested in the example in Figure 5.13: 'Revenue', 'Rental Income', 'Vacancy', and 'Total Revenue'. In addition, format 'Vacancy' with a bottom border. Bold row 1, cell A2 (which is the Revenue summary), and row 5 (which will be total revenue for the project).

Next we complete the periods for the real estate pro forma (Figure 5.14). This is a 10-year (period) pro forma, so it is necessary to quantify net operating income for year 11 (remember that year 10 gross sales price is year 11 NOI divided by terminal capitalization rate). Cell G1 therefore inputs as "1", which is a static value. Cell H1 is "=G1+1". Note the relative reference to G1. Therefore cell H1 can be copied and dragged to cell Q1 (period 11). Column G was selected to start the timing for the cash flows specifically. Choosing an earlier column, for example column D, makes scalability more difficult because sub items may be included in revenue and/or expense. Also, when aggregating projects into a portfolio, the real estate

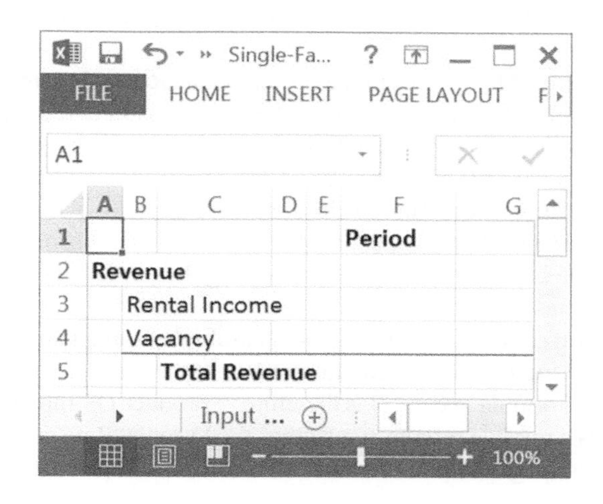

Figure 5.13

	A	B	C	D	E	F	G	H	I	J	K	L	M	N	O	P	Q
1						**Period**	**1**	**2**	**3**	**4**	**5**	**6**	**7**	**8**	**9**	**10**	**11**
2	**Revenue**																
3		Rental Income															
4		Vacancy															
5			Total Revenue														

Figure 5.14

pro formas must be built with consistency to enable ease of roll-up. Do not forget to carry the bottom border for 'Vacancy' to cell Q4 as well.

Next revenue must be calculated for each of the future periods on a nominal basis. Stating that it is calculated 'nominally' is synonymous with stating the purchasing power is not being adjusted per period to account for expected/anticipated inflation in the future.

The formula that follows in cells G3:Q3 is the Future Value formula from earlier in the text:

$$FV = PV(1+r)^t$$

where

FV: value of cells G3:Q3

PV: base rental rate as stated on the input page

r: revenue escalation rate

t: current time period

As it is imperative that each cell be consistent to the greatest extent possible when creating a real estate pro forma, the structure for each Rental Income calculation will be the same. The formula for G3 is therefore:

=Inputs_Summary!$I$2*12*(1+Inputs_Summary!$I$3)^(G1–1)

"Inputs_Summary!$I$2*12" is the annual present value Rental Income (Figure 5.15). The 'Inputs_Summary!' is the sheet location and '$I$3' is the absolute reference on the sheet. It is multiplied by 12 as the input was stated as a monthly value. The green is the escalation rate, "Inputs_Summary!$I$3" (i.e. *r*), and 'G1−1' is the escalation rate. Note that for the first period the value is zero (0). Any number to the power of zero is 1. Therefore there is no escalation in the first period. The formula, as it is structured with absolute and relative references, can therefore be copied to the right, completing the entire row. Remember to format the values using commas and zero decimals.

Vacancy, a contra revenue item, is next to be quantified. As this is functionally dependent on Rental Income, the logic is simply Rental Income in the respective period multiplied by the Vacancy Rate. The formula in cell G4 is therefore "=product(G3,Inputs_Summary!$I$4)". The Rental Income is a relative reference, whereas the Vacancy Rate is an absolute reference to the Inputs_Summary page. Total Revenue is the difference when subtracting Vacancy from

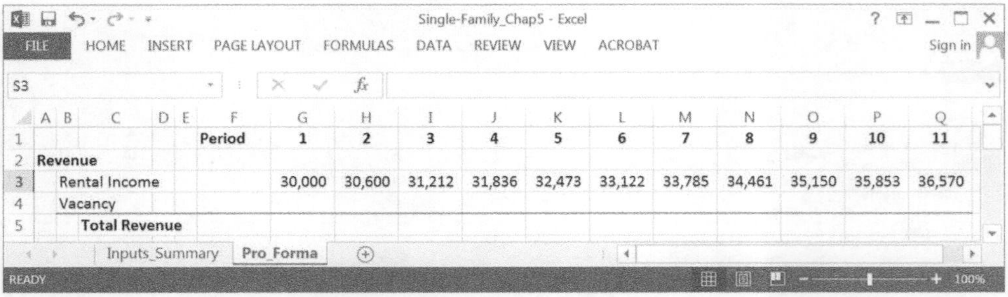

	A-E	F	G	H	I	J	K	L	M	N	O	P	Q
1		Period	1	2	3	4	5	6	7	8	9	10	11
2	Revenue												
3	Rental Income		30,000	30,600	31,212	31,836	32,473	33,122	33,785	34,461	35,150	35,853	36,570
4	Vacancy												
5	Total Revenue												

Figure 5.15

	A	B	C	D	E	F	G	H	I	J	K	L	M	N	O	P	Q
1						**Period**	**1**	**2**	**3**	**4**	**5**	**6**	**7**	**8**	**9**	**10**	**11**
2	**Revenue**																
3		Rental Income					30,000	30,600	31,212	31,836	32,473	33,122	33,785	34,461	35,150	35,853	36,570
4		Vacancy					1,500	1,530	1,561	1,592	1,624	1,656	1,689	1,723	1,757	1,793	1,828
5			**Total Revenue**				**28,500**	**29,070**	**29,651**	**30,244**	**30,849**	**31,466**	**32,096**	**32,738**	**33,392**	**34,060**	**34,741**

Figure 5.16

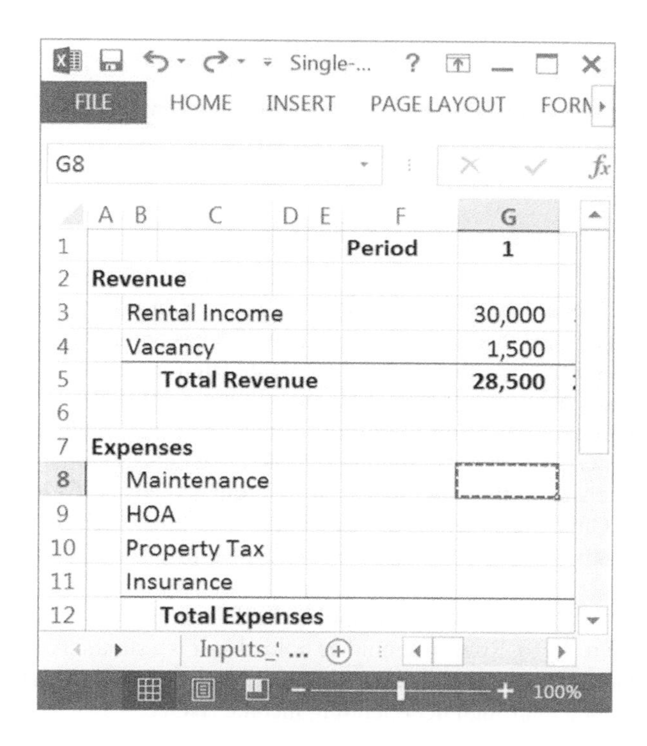

	A	B	C	D	E	F	G
1						**Period**	**1**
2	**Revenue**						
3		Rental Income					30,000
4		Vacancy					1,500
5			**Total Revenue**				**28,500**
6							
7	**Expenses**						
8		Maintenance					
9		HOA					
10		Property Tax					
11		Insurance					
12			**Total Expenses**				

Figure 5.17

Rental Income (Figure 5.16). The formula for cell G5 is "=G3-G4". Both Vacancy and Total Revenue must be formatted using commas and then copied to the right to column Q. At this point the revenue for the project has been quantified.

Expenses follow a similar structure to Revenue. In cell A7 the section is titled Expenses (Figure 5.17). In cells B8:B11 the expenses are listed in order from the input page: Maintenance, HOA, Property Tax, and Insurance. For consistency with Revenue, bold the Expense title in A7 and the row of Total Expenses, row 12. Use a top border for row 12 as well.

The expense logic per period for each line item follows the same Future Value relationship discussed for the Revenue section. The formula for Maintenance, cell G8, is "=Inputs_Summary!$I$6*12*(1+Inputs_Summary!$I$10)^(G1–1)". The reason for the '*12' at the beginning is that Maintenance on the input page has units of months but the pro forma page

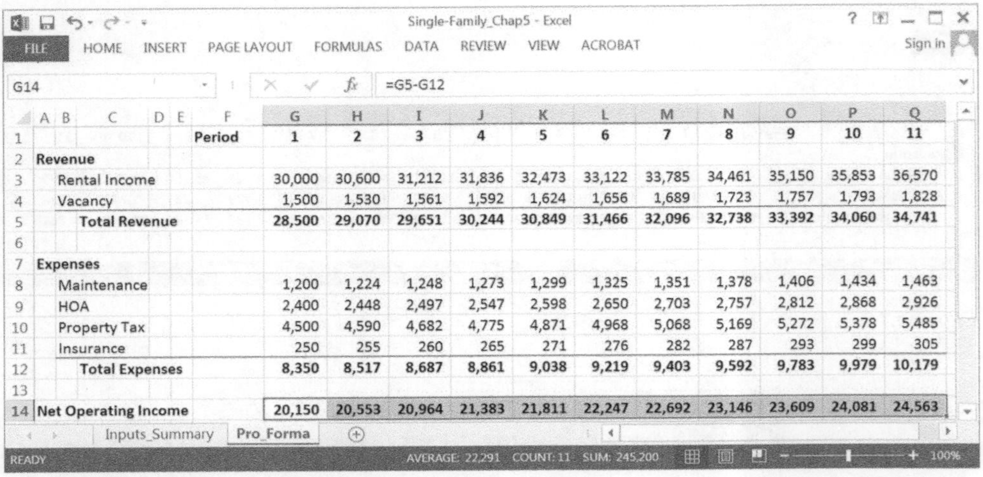

Single-Family_Chap5 - Excel

G14 =G5-G12

	Period	1	2	3	4	5	6	7	8	9	10	11
Revenue												
Rental Income		30,000	30,600	31,212	31,836	32,473	33,122	33,785	34,461	35,150	35,853	36,570
Vacancy		1,500	1,530	1,561	1,592	1,624	1,656	1,689	1,723	1,757	1,793	1,828
Total Revenue		**28,500**	**29,070**	**29,651**	**30,244**	**30,849**	**31,466**	**32,096**	**32,738**	**33,392**	**34,060**	**34,741**
Expenses												
Maintenance		1,200	1,224	1,248	1,273	1,299	1,325	1,351	1,378	1,406	1,434	1,463
HOA		2,400	2,448	2,497	2,547	2,598	2,650	2,703	2,757	2,812	2,868	2,926
Property Tax		4,500	4,590	4,682	4,775	4,871	4,968	5,068	5,169	5,272	5,378	5,485
Insurance		250	255	260	265	271	276	282	287	293	299	305
Total Expenses		**8,350**	**8,517**	**8,687**	**8,861**	**9,038**	**9,219**	**9,403**	**9,592**	**9,783**	**9,979**	**10,179**
Net Operating Income		**20,150**	**20,553**	**20,964**	**21,383**	**21,811**	**22,247**	**22,692**	**23,146**	**23,609**	**24,081**	**24,563**

Inputs_Summary | Pro_Forma

AVERAGE: 22,291 COUNT: 11 SUM: 245,200

Figure 5.18

is annual. HOA fee, cell G9, is also stated in monthly units on the input page. Therefore it has the same '*12' for the initial input: "=Inputs_Summary!$I$7*12*(1+Inputs_Summary!$I$10)^(G1–1)". Next is Property Tax which has annual units on the input page and therefore does not require the '*12' adjustment. The formula for G10, Property Tax, is therefore "=Inputs_Summary!$I$8*(1+Inputs_Summary!$I$10)^(G1–1)". Insurance, G11, is quoted in annual units and therefore follows the same methodology as Property Tax. The formula for G11 is "=Inputs_Summary!$I$9*(1+Inputs_Summary!$I$10)^(G1–1)". Finally, Total Expenses is the summation of the four expenses. Cell G12 is therefore "=sum(G8:G11)". Once the expenses are calculated they must each be formatted as commas with no decimals. At that point they can be copied to column Q. This completes the Expenses.

In cell A14 the title 'Net Operating Income' is the difference between Total Revenue and Total Expense (Figure 5.18). Row 14 should be bolded as it is a summary row. Also, cell G14 is "=G5-G12". Once this is complete, cell G14 must be copied to column Q. The pro forma page has been completed through net operating income (NOI).

After net operating income, debt charge must be addressed. In cell A16 type 'Debt'. As this is a pre-tax analysis, this represents the total payment, that is principal and interest, for the period. Select row 16 and make the entire row bold for formatting consistency. In cell G16 the first payment will be addressed.

Because debt was quantified on the 'Debt' page it must be linked to this Pro_Forma page. This is not a direct link as with other references since the amortization table is constructed vertically while the Pro_Forma page is constructed in a horizontal time series. As such, vertical data must be translated to a horizontal format. This requires the use of a Lookup function – i.e. "=VLOOKUP()". In general there are three types of Lookup functions in Excel: "=VLOOKUP()", "=HLOOKUP()", and "=INDEX()". The analyst is encouraged to be familiar with all three. However, only VLOOKUP will be discussed in this text.

As stated, VLOOKUP is utilized to translate vertical data to a horizontal format. The VLOOKUP has four arguments, but only the first three will be used at this point: "=Vlookup (lookup_value, table_array, col_Index_num)". The 'lookup_value' is the first reference point. In this case it is the period for the payment – i.e. for cell G16 it is period 1 (G1). The

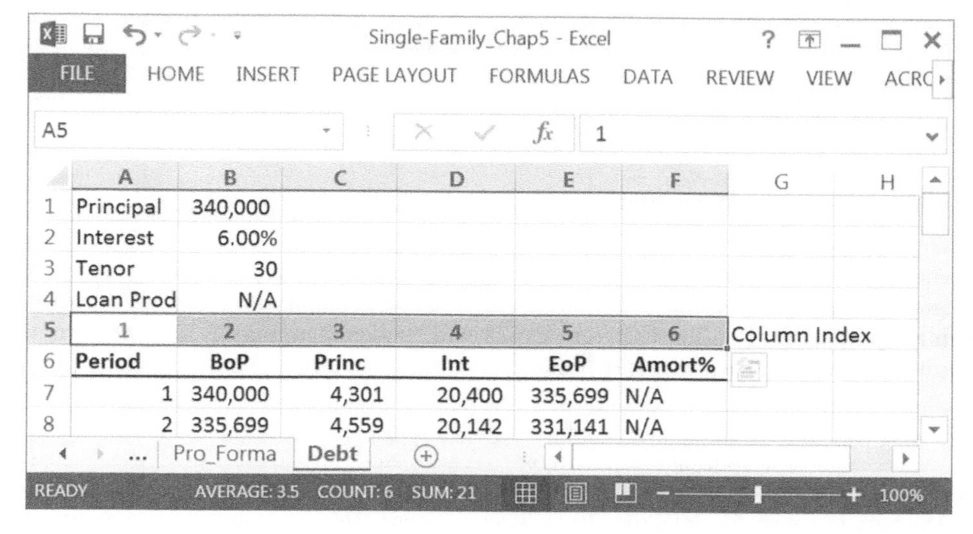

	A	B	C	D	E	F	G
1	Principal	340,000					
2	Interest	6.00%					
3	Tenor	30					
4	Loan Prod	N/A					
5	1	2	3	4	5	6	Column Index
6	Period	BoP	Princ	Int	EoP	Amort%	
7	1	340,000	4,301	20,400	335,699	N/A	
8	2	335,699	4,559	20,142	331,141	N/A	

Figure 5.19

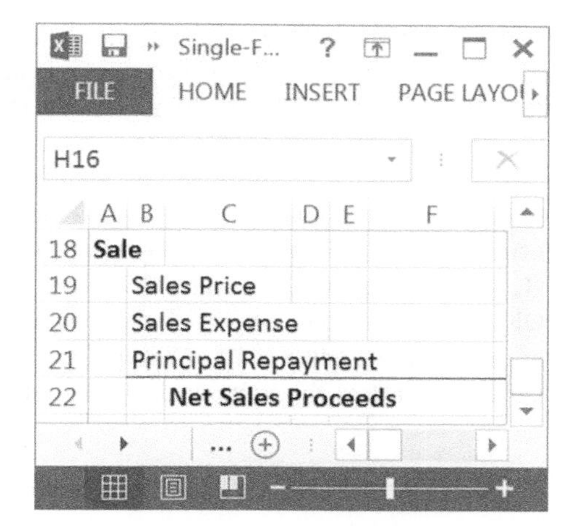

	A	B	C
18	Sale		
19		Sales Price	
20		Sales Expense	
21		Principal Repayment	
22			Net Sales Proceeds

Figure 5.20

'Table_Array' is the name of the array being referenced. In this case it is 'Tranche_1'. This is why earlier we named the Debt array 'Tranche_1', so that we could reference it now. Finally, the 'col_index_num' is the column being referenced in the array. Beginning with the first column, column A, count across to the principal and interest columns which are 3 and 4, respectively (see Figure 5.19).

Now, step 1 for the debt payment in cell G16 is to reference the Principal payment from the Debt page for the equivalent period. The formula is "=Vlookup(G1,Tranche_1,3)". The value is 4,300 but that only references the principal. For the complete payment the interest charge must also be included. The formula must then be adjusted to include interest (i.e. step 2 of a two-step process. It therefore becomes "=Vlookup(G1,Tranche_1,3)+Vlookup(G1,Tranche_1,4)". As a relative reference was utilized for the lookup_value, the formula can then be copied across to column Q. Note: in reality, it need only be copied to column P as the analysis is only 10 years.

Next is the sales summary for the project (Figure 5.20). Cell A18 is 'Sale' with row headers starting in B19 as follows: Sales Price, Sales Expense, and Principal Repayment. Finally in cell C22, 'Net Sales Proceeds' represents the net, final sales cash after expenses. Note the formatting is similar to the previous sections. Net Sales Proceeds has a top border and is bold, indicating a summary section.

Sales price is the gross sales value at time of sale. It is calculated as follows:

$$\text{Sale Price}_{10} = \frac{\sum NOI_{11}}{\text{Cap Rate}_{\text{Terminal}}}$$

As this is stated to be a 10-year pro forma, the calculation is therefore placed in cell P19 (Figure 5.21), which is the year 10 sales price. The actual Excel formula is "=Q14/Inputs_Summary!N7". 'Q14' is the NOI for year 11 and 'Inputs_Summary!N7' is the Terminal Capitalization Rate. The calculation follows:

$$\text{Sale Price}_{10} = \frac{\sum NOI_{11}}{\text{Cap Rate}_{\text{Terminal}}} = \frac{24,563}{0.08} = 307,034$$

The next line item for Sale, row 20, is Sales Expense. This is a percentage of gross sales proceeds from the Sales Price. Cell P20 is therefore "=-P19*Inputs_Summary!N8" where the 'N8' reference is the sales percentage expense from the Inputs_Summary page. Note the negative (-) at the beginning of the equation. This is to ensure the value is negative (i.e. a cash outflow). Finally, the Principal Repayment, row 21, is the principal ending balance at the end of year 10 that must be repaid through sales proceeds. Given that the Debt page, the amortization table, is constructed vertically, a "=Vlookup()" function will be utilized to translate vertical data to a horizontal format. Additional Excel functions with which analysts should be familiar but are not utilized in this chapter include "=HLOOKUP()" and "=INDEX()".

In Figure 5.21, the formula for the Principal Repayment at the end of year 10 is "=-Vlookup(P1,Tranche_1,5)". The negative, as with Sales Expense, demonstrates a cash outflow. Briefly, the VLOOKUP function contains these three arguments: P1 is the referenced period, Tranche_1 is the referenced array, and 5 is the column indexed in the referenced array, Tranche_1. Net Sales Proceeds is therefore the summation of the Sale items. Cell P22 is thus "=sum(P19:P21)".

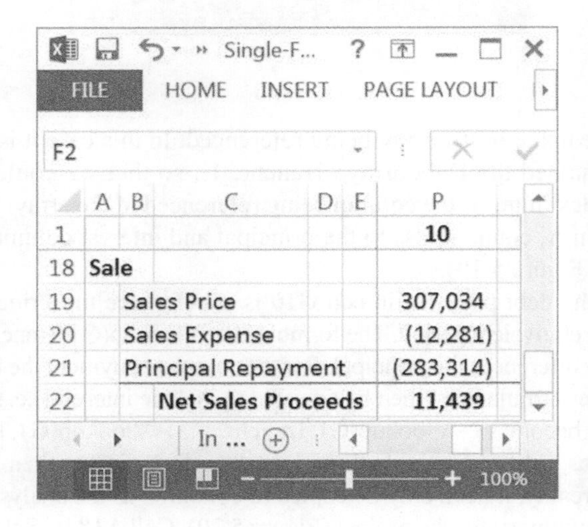

Figure 5.21

To conclude the pro forma page, Free Cash Flow (FCF) must be quantified. To best assess risk for a project, the total cash flow is bifurcated into reversion and operating (operations). The reversion cash flow represents the free cash from sale and is largely dictated by market conditions, that is inflation. Operating (operations) cash flow is the cash flow(s) created through operating (managing) the asset. Operating cash flow is therefore actively managed/created by the owner/manager.

Cell A24 is therefore titled 'FCF Reversions' (see Figure 5.22). Cell A25 is titled 'FCF Operations'. Cell C26 is the summation of both, 'Total FCF'. As Reversions only has one cash flow in the final period, period 10, most values are zero. For consistency, cell G24 will reference Net Sales Proceeds, "=G22". Operations, cell G25, is the net difference of net operating income, G14, and debt in the respective period, G16. Thus cell G25 is "=G14-G16". In the case of this pro forma, the value is negative meaning that a capital infusion will be required by the owner/manager (in other words, debt is greater than operating income). The total free cash flow is therefore the summation of reversions and operating. Cell G26 is then "=sum(G24:G25)". Cells G24:G26 are copied to column P to complete the pro forma page. Note that in this analysis, operating income is always less than debt payment.

Now we return to the Inputs_Summary sheet to complete the pro forma and value the cash flows (Figure 5.23). Select cells A15:N15 and fill the row with grey. This separates the top (inputs) from the bottom of the sheet (summary cash flows and valuation). It is strictly to format the page for superior presentation.

In cell A16 type 'Valuation'. In cell D16 type 'Period' and in E16 type '1', following with F16 as "=E16+1". Then copy cell F16 to G16:N16, creating the periods for the analysis.

The summary of the pro forma page, in this simple model, will be essentially identical to the Pro_Forma page. In cell A17 type 'Revenue' and follow with row headers in B18 and B19 as

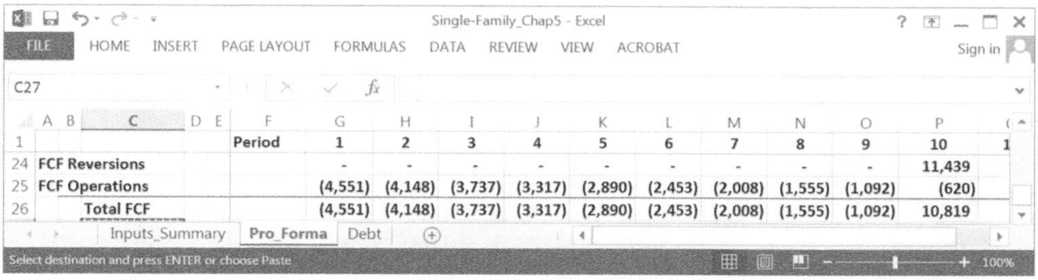

		Period	1	2	3	4	5	6	7	8	9	10
24	FCF Reversions		-	-	-	-	-	-	-	-	-	11,439
25	FCF Operations		(4,551)	(4,148)	(3,737)	(3,317)	(2,890)	(2,453)	(2,008)	(1,555)	(1,092)	(620)
26	Total FCF		(4,551)	(4,148)	(3,737)	(3,317)	(2,890)	(2,453)	(2,008)	(1,555)	(1,092)	10,819

Figure 5.22

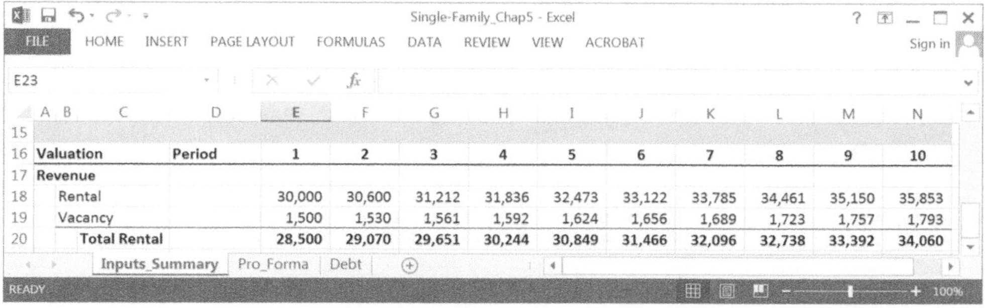

		Period	1	2	3	4	5	6	7	8	9	10
16	Valuation	Period	1	2	3	4	5	6	7	8	9	10
17	Revenue											
18	Rental		30,000	30,600	31,212	31,836	32,473	33,122	33,785	34,461	35,150	35,853
19	Vacancy		1,500	1,530	1,561	1,592	1,624	1,656	1,689	1,723	1,757	1,793
20	Total Rental		28,500	29,070	29,651	30,244	30,849	31,466	32,096	32,738	33,392	34,060

Figure 5.23

Rental and Vacancy, respectively. Then in cell C20 type 'Total Rental' which, as on the Pro_Forma page, will summarize the total Rental Income. Format as demonstrated in Figure 5.23.

Next select cell E18 and link this with cell G3 on the Pro_Forma page – i.e. "=Pro_Forma!G3". It is recommended that this be formatted using the comma format with zero decimals. Copy cell G3:N3 from the Pro_Forma page and then paste it to row 20 on the Inputs_Summary page. Note that it may be tempting to quantify Total Rental on the summary page. However, this should not be done; Total Rental should be linked. Calculations, as much as possible in a model, should be completed once and then carried through – linked – within a model. A model becomes error prone when the same calculation is completed multiple times in multiple places.

Next copy Expenses, Net Operating Income, and Debt to the Inputs_Summary page (Figure 5.24). To simplify, Net Sale will only be copied to the summary page. This negates the need to copy the expense and principal repays from the Pro_Forma page. This also proves a cleaner presentation for the analyst and manager. Finally FCF Reversions, FCF Operations, and Total FCF are linked to the Pro_Forma page. At this point the pro forma is summarized by period (year) on the Inputs_Summary page beneath the inputs. What remains is summary valuation characteristics by period, for example Cash-on-Cash return and debt service coverage ratio (DSCR), and project valuation metrics. This will be completed beneath the summary and upon the summary values. Note that in later models (larger models), these valuations will be completed on the Pro_Forma page. Valuation is completed on the summary page because this model is simplified and annual.

In cell A39 type 'DSCR' for debt service coverage ratio. (See Figure 5.25.) This metric will indicate the number of times that debt service is covered by NOI. In this example the metric will be less than 1.00, indicating that the project, as modelled, does not cover debt service. Without a greater equity infusion, higher rental income, or less expenses, the project will not go forward with third-party lending.

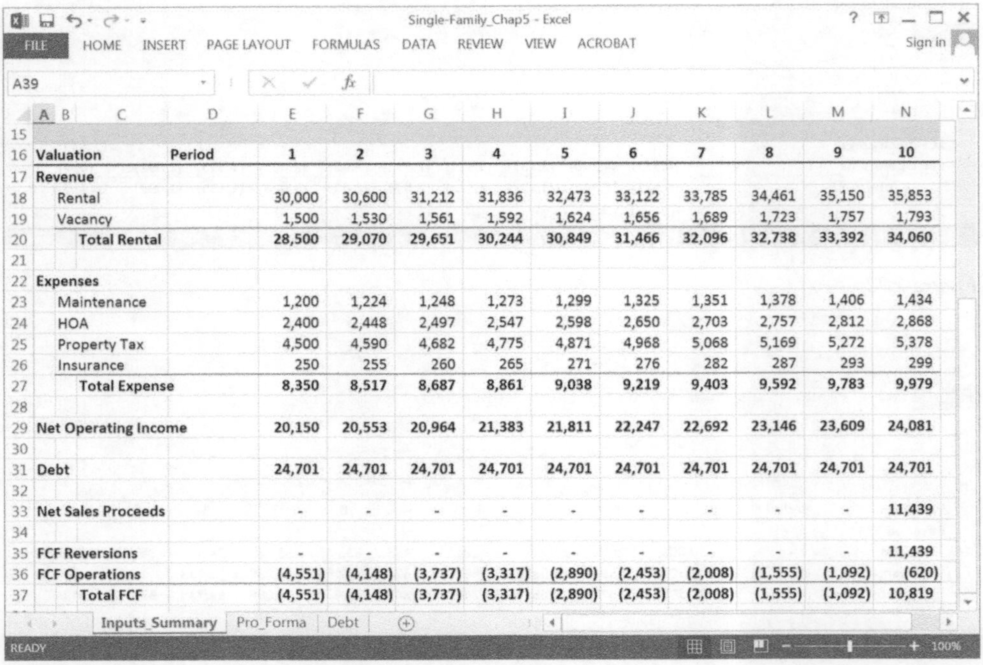

Single-Family_Chap5 - Excel

FILE HOME INSERT PAGE LAYOUT FORMULAS DATA REVIEW VIEW ACROBAT Sign in

A39

	A B C	D	E	F	G	H	I	J	K	L	M	N
15												
16	**Valuation**	**Period**	**1**	**2**	**3**	**4**	**5**	**6**	**7**	**8**	**9**	**10**
17	**Revenue**											
18	Rental		30,000	30,600	31,212	31,836	32,473	33,122	33,785	34,461	35,150	35,853
19	Vacancy		1,500	1,530	1,561	1,592	1,624	1,656	1,689	1,723	1,757	1,793
20	**Total Rental**		**28,500**	**29,070**	**29,651**	**30,244**	**30,849**	**31,466**	**32,096**	**32,738**	**33,392**	**34,060**
21												
22	**Expenses**											
23	Maintenance		1,200	1,224	1,248	1,273	1,299	1,325	1,351	1,378	1,406	1,434
24	HOA		2,400	2,448	2,497	2,547	2,598	2,650	2,703	2,757	2,812	2,868
25	Property Tax		4,500	4,590	4,682	4,775	4,871	4,968	5,068	5,169	5,272	5,378
26	Insurance		250	255	260	265	271	276	282	287	293	299
27	**Total Expense**		**8,350**	**8,517**	**8,687**	**8,861**	**9,038**	**9,219**	**9,403**	**9,592**	**9,783**	**9,979**
28												
29	**Net Operating Income**		**20,150**	**20,553**	**20,964**	**21,383**	**21,811**	**22,247**	**22,692**	**23,146**	**23,609**	**24,081**
30												
31	**Debt**		**24,701**	**24,701**	**24,701**	**24,701**	**24,701**	**24,701**	**24,701**	**24,701**	**24,701**	**24,701**
32												
33	**Net Sales Proceeds**		-	-	-	-	-	-	-	-	-	**11,439**
34												
35	**FCF Reversions**		-	-	-	-	-	-	-	-	-	**11,439**
36	**FCF Operations**		**(4,551)**	**(4,148)**	**(3,737)**	**(3,317)**	**(2,890)**	**(2,453)**	**(2,008)**	**(1,555)**	**(1,092)**	**(620)**
37	**Total FCF**		**(4,551)**	**(4,148)**	**(3,737)**	**(3,317)**	**(2,890)**	**(2,453)**	**(2,008)**	**(1,555)**	**(1,092)**	**10,819**

Inputs_Summary | Pro_Forma | Debt

READY 100%

Figure 5.24

	A B	C	D	E	F	G	H	I	J	K	L	M	N
39	DSCR			0.82	0.83	0.85	0.87	0.88	0.90	0.92	0.94	0.96	0.97
40			Minimum	0.82									
41													
42	Cash-on-Cash			(0.04)	(0.04)	(0.04)	(0.03)	(0.03)	(0.02)	(0.02)	(0.02)	(0.01)	(0.01)
43			Minimum	(0.04)									

Figure 5.25

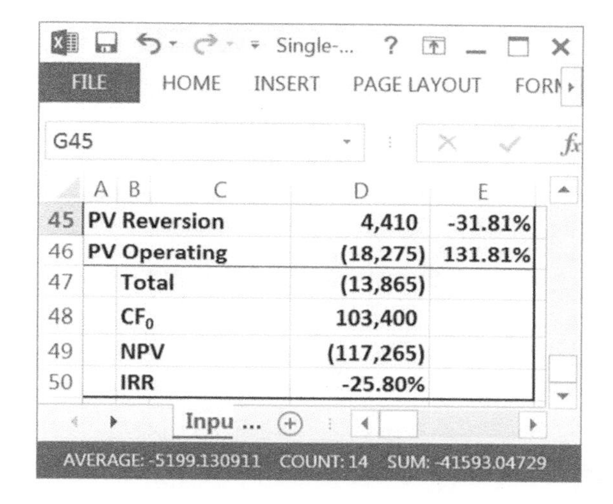

	A B	C	D	E
45	PV Reversion		4,410	-31.81%
46	PV Operating		(18,275)	131.81%
47		Total	(13,865)	
48		CF_0	103,400	
49		NPV	(117,265)	
50		IRR	-25.80%	

Figure 5.26

In cell E39 DSCR is quantified for the period as NOI divided by Debt Service (i.e. principal and interest due in the respective period). Cell E39 is quantified as "=E29/E31". Again, the value is 0.82 which indicates, as it is below 1.00, that net operating income is *not* covering total debt service. Cell E39 is then copied to cell N39.

In cell D40 type 'Minimum'. Then in cell E40 use the 'Minimum' function to quantify the lowest DSCR across the range of periods 1–10. The formula for cell E40 is "=min(E39:N39)". In this analysis, the minimum period is period 1 with a value of 0.82.

Cash-on-Cash return is the next valuation metric and is of concern for project managers and equity investors. It indicates the periodic return on equity. It is calculated as Free Cash Flow Operations divided by total equity, a project valuation metric.

In cell A42 type 'Cash-on-Cash', which is the row header for this metric. In cell E42 the formula is FCF Operations divided by total equity: "=E36/$D$2". As the metric is negative (0.04), this indicates that the project in period 1 has a negative return. The '$D$2' portion is an absolute reference because the equity cell does not change in the Sources/Uses section. Now, copy cells E42 through N42 to quantify the Cash-on-Cash return for each period. As with DSCR, type 'Minimum' in cell D43 and place the minimum function in cell E43 as "=min(E42:N42)". All values for Cash-on-Cash are negative, indicating that the project, as it currently is modelled, does not return the initial capital invested.

The final section is the valuation of the project (Figure 5.26). While this section may be located in the Inputs section in later models, it is placed at the bottom of the summary section

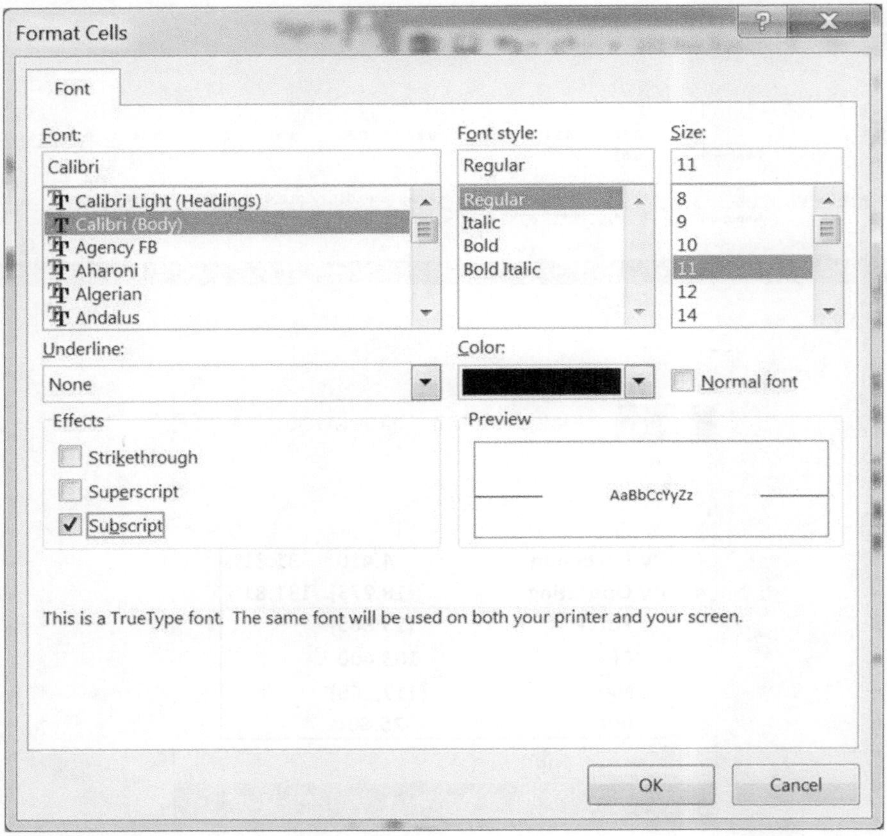

Figure 5.27

here. The purpose is to quantify project returns, Net Present Value (NPV), and Internal Rate of Return (IRR). In addition, value is bifurcated into reversion and operating to allow the analyst and manager to determine the source of value – i.e. operations (Hands-On) or reversion (Market).

In cell A45 type 'PV Reversion' (Present Value Reversion) and in cell A46 type 'PV Operating' (Present Value Operations). In cells B47–B50 type 'Total', 'CF_0', 'NPV', and 'IRR', respectively. For formatting purposes the zero (0) in CF_0 should be subscript. Therefore, select cell C48 and highlight the zero (0) only. Then in the Ribbon select 'Home' and then 'Font' (Figure 5.27). The 'Format Cells' menu will be displayed on the screen. Select 'Subscript' in the lower left and the zero will be subscript.

Next the present value for the reversion cash flows is quantified (Figure 5.26). In cell D45 type "=NPV(N9,E35:N35)". Cell N9 is the discount rate stated in the inputs and utilized for discounting. The range E35:N35 contains the cash flows attributable to reversion. As there is only one sale period, reversion values are zero for periods 1–9.

In cell D46 type "=NPV(N9,E36:N36)" which is the present value of operating cash flows. The value is (18,275) which is negative, indicating that, operationally, the cash value of future operating cash flows does not exceed the discount rate on a yield basis. Cell D47 is the summation of the two present value calculations, that is "=sum(D45:D46)". Cell D48 is the initial cash outflow for the project and because this is valued at the project level it is total equity from the Sources section – i.e. cell D48 is "=D2". Cell D49, NPV, is Present Value, D47, less the initial cash outflow (CF_0), D48. Cell D49 is therefore the total present value less initial cash flow, "=D47-D48". As stated, the net present value is negative, (117,265), and therefore returns less than the stated discount rate; in other words this is not a good project as valued at the project level only.

Prior to quantifying the Internal Rate of Return, cell C50, we will quantify the percentage of value attributable to Reversion and Operating, cells E45 and E46. Present Value Reversion is the percentage of value attributable to reversion. Cell E45 is therefore "=D45/D47". Cell E46 is therefore "=D46/D47".

Finally Internal Rate of Return for the project is quantified in cell D50. As IRR requires an initial cash flow, cell D37 must be modified to include the initial cash outflow for the Total Free Cash Flow (FCF) series (see Figure 5.28). Cell D37 therefore is "=-D2", which represents the initial equity. The negative is an indication that the equity is an outflow of capital at T = 0. Finally, cell D50 is therefore "=IRR(D37:N37)". Cell D50, IRR, returns a value of (25.80%), indicating that it has a negative return and therefore does not return the initial project equity. The valuation section is as shown.

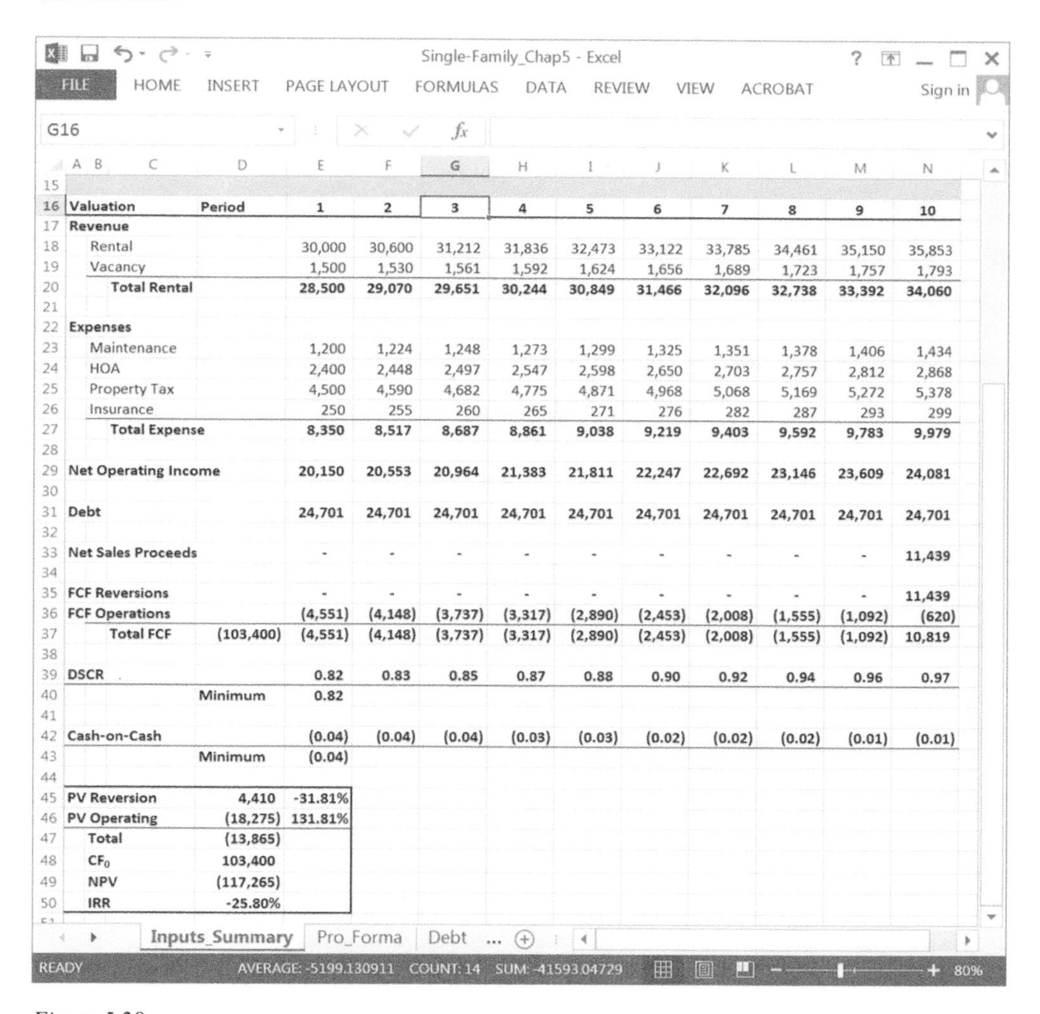

Valuation	Period	1	2	3	4	5	6	7	8	9	10
Revenue											
Rental		30,000	30,600	31,212	31,836	32,473	33,122	33,785	34,461	35,150	35,853
Vacancy		1,500	1,530	1,561	1,592	1,624	1,656	1,689	1,723	1,757	1,793
Total Rental		28,500	29,070	29,651	30,244	30,849	31,466	32,096	32,738	33,392	34,060
Expenses											
Maintenance		1,200	1,224	1,248	1,273	1,299	1,325	1,351	1,378	1,406	1,434
HOA		2,400	2,448	2,497	2,547	2,598	2,650	2,703	2,757	2,812	2,868
Property Tax		4,500	4,590	4,682	4,775	4,871	4,968	5,068	5,169	5,272	5,378
Insurance		250	255	260	265	271	276	282	287	293	299
Total Expense		8,350	8,517	8,687	8,861	9,038	9,219	9,403	9,592	9,783	9,979
Net Operating Income		20,150	20,553	20,964	21,383	21,811	22,247	22,692	23,146	23,609	24,081
Debt		24,701	24,701	24,701	24,701	24,701	24,701	24,701	24,701	24,701	24,701
Net Sales Proceeds		-	-	-	-	-	-	-	-	-	11,439
FCF Reversions		-	-	-	-	-	-	-	-	-	11,439
FCF Operations		(4,551)	(4,148)	(3,737)	(3,317)	(2,890)	(2,453)	(2,008)	(1,555)	(1,092)	(620)
Total FCF	(103,400)	(4,551)	(4,148)	(3,737)	(3,317)	(2,890)	(2,453)	(2,008)	(1,555)	(1,092)	10,819
DSCR		0.82	0.83	0.85	0.87	0.88	0.90	0.92	0.94	0.96	0.97
	Minimum	0.82									
Cash-on-Cash		(0.04)	(0.04)	(0.04)	(0.03)	(0.03)	(0.02)	(0.02)	(0.02)	(0.01)	(0.01)
	Minimum	(0.04)									
PV Reversion	4,410	-31.81%									
PV Operating	(18,275)	131.81%									
Total	(13,865)										
CF_0	103,400										
NPV	(117,265)										
IRR	-25.80%										

Figure 5.28

As an example of employing the model for a single-family residence with the stated expenses, utilize the following input adjustments:

Equity: $200,000
Debt: $243,400
Rental/mo: $3,000

This provides the following result:

NPV: (69,798)
IRR: 4.20%

This indicates that the project has a negative value for the discount rate selected and that the Internal Rate of Return is 4.20%. Adjusting the Rental Income to $3,500 per month only provides a positive NPV (i.e. 113) and an Internal Rate of Return of 10.01%. Although rental income and equity percentages were changed in this example, it was for demonstration purposes only. In a project analysis rental income is determined by market and equity is, generally, determined by a sponsor's investment limitations. These examples were completed for the purpose of demonstrating how this simplistic single-family residential model can and is utilized for an investment decision.

6 5-unit multifamily

Building upon the lessons learned in Chapter 5, for a single-family residential model, a 5-unit multifamily pro forma will be constructed here. Inputs will be added to represent additional expenses, named ranges will be utilized as more superior absolute references, and the pro forma will contain monthly periods as opposed to annual. However, the structure will remain consistent with the single-family residential model. The purpose of this chapter is twofold: (1) demonstrate scalability of the initial model and (2) demonstrate that the process of construction remains consistent even when adding complexities.

This chapter will construct the 5-unit multifamily pro forma in four sections: (1) Inputs, (2) Debt/Amortization, (3) Pro Forma, and (4) Summary/Valuation. The process is kept as linear as possible to ensure fluidity of design and build.

Inputs

The 5-unit multifamily model that will be constructed is as follows:

Unit Total: 5

Sources/uses

Purchase Price: $500,000
Equity: $100,000
Capital Improvements: $50,000
Legal (Personal): $5,000
Legal (Lender): $4,000
Finance Expense: $4,000 (approximately 1.00% of loan origination)
Escrow: $6,000
Contingency: $15,000

As this chapter and model begins to embark on more complicated models, a step-wise process will be implemented for tackling the models. Please note that while these steps are recommended, they must be modified for individual projects.

Step 1: sources and uses

The first step, obviously, is to open Excel and name the first sheet. As before, the first sheet is 'Summary_Inputs'. On very large models 'Summary' may be placed on its own sheet but for now the feedback approach – to starting with Inputs and ending with the Summary on the same page – will be used.

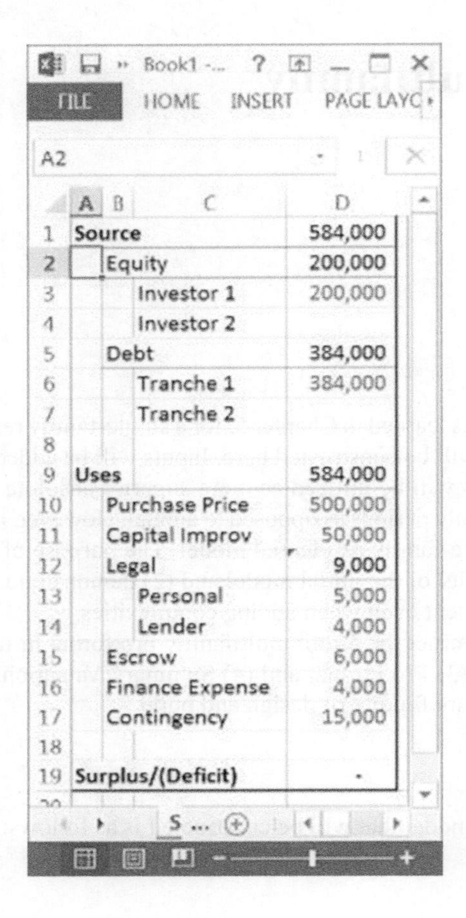

	A	B	C	D
1	Source			584,000
2		Equity		200,000
3			Investor 1	200,000
4			Investor 2	
5		Debt		384,000
6			Tranche 1	384,000
7			Tranche 2	
8				
9	Uses			584,000
10		Purchase Price		500,000
11		Capital Improv		50,000
12		Legal		9,000
13			Personal	5,000
14			Lender	4,000
15		Escrow		6,000
16		Finance Expense		4,000
17		Contingency		15,000
18				
19	Surplus/(Deficit)			-

Figure 6.1

The Sources and Uses are categorized as best estimated at the time of project projection. Beginning in cell A1, construct the Sources and Uses section (Figure 6.1). It is recommended that two categories for Equity and Debt be constructed for later scaling of the model, as was demonstrated in Chapter 5.

As stated earlier, all summations should be calculated using the "=subtotal(9,)" function. This negates double counting when summing values. Also, inputs must be coloured blue to differentiate from calculated numbers.

As discussed, Sources and Uses is a summary of the transaction at time of closing. In this case it indicates a Loan-to-Cost of 65.75% or 384,000/584,000. This is a relatively conservative loan with $200,000 in equity from Investor 1.

Step 2: operations

The revenue section for market rents will generally be provided by a market study and/or broker. The expense items will be provided by property managers with experience in similar-sized buildings/assets in the local area. Begin by constructing the Operations section with the following headings and values, as shown in Figure 6.2.

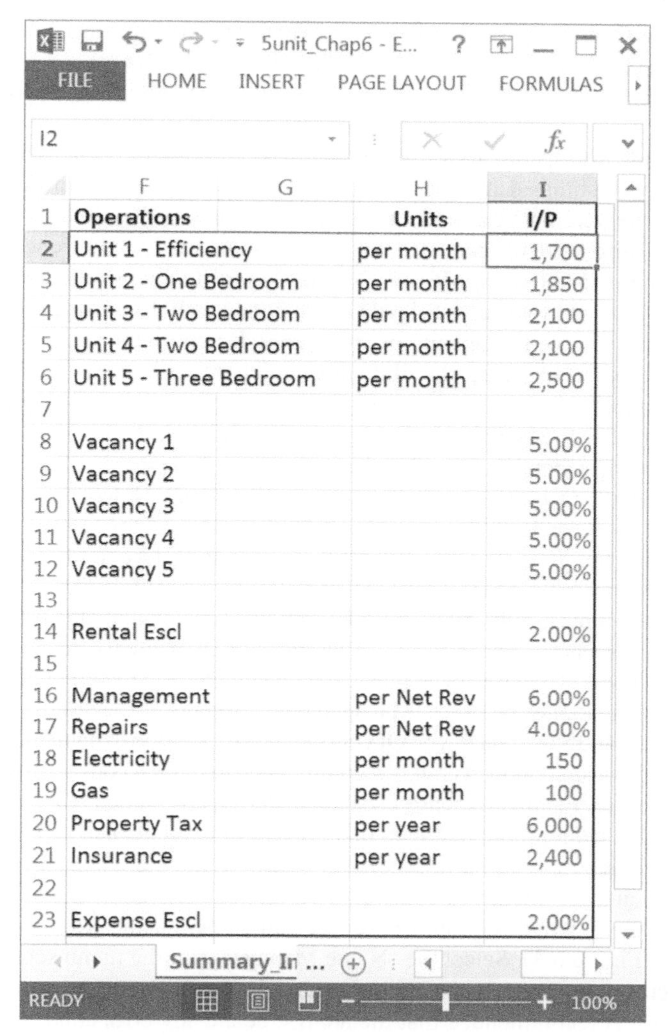
5unit_Chap6 - E...

	F	G	H	I
1	**Operations**		**Units**	**I/P**
2	Unit 1 - Efficiency		per month	1,700
3	Unit 2 - One Bedroom		per month	1,850
4	Unit 3 - Two Bedroom		per month	2,100
5	Unit 4 - Two Bedroom		per month	2,100
6	Unit 5 - Three Bedroom		per month	2,500
7				
8	Vacancy 1			5.00%
9	Vacancy 2			5.00%
10	Vacancy 3			5.00%
11	Vacancy 4			5.00%
12	Vacancy 5			5.00%
13				
14	Rental Escl			2.00%
15				
16	Management		per Net Rev	6.00%
17	Repairs		per Net Rev	4.00%
18	Electricity		per month	150
19	Gas		per month	100
20	Property Tax		per year	6,000
21	Insurance		per year	2,400
22				
23	Expense Escl			2.00%

Figure 6.2

Unit 1 – Efficiency: $1,700/mo
Unit 2 – One Bedroom: $1,850/mo
Unit 3 – Two Bedroom: $2,100/mo
Unit 4 – Two Bedroom: $2,100/mo
Unit 5 – Three Bedroom: $2,500/mo

Management: 6.00% of Net Revenue
Repairs: 4.00% of Net Revenue
Electricity (Common Area): $150/mo
Gas (Common Area): $100/mo
Property Tax: $6,000/yr
Insurance: $2,400/yr

Notice that the Operations section has been expanded from the single-family residential model. Column H now includes the title 'Units' in cell H1. This is to accommodate operating assumptions that will have different unit quotations. Also, all of the inputs ought to be coloured blue.

Notice that while the Operations section is completed in Figure 6.2, not all the inputs were provided. Vacancy is estimated at 5.00%, which is a plug utilized in many models. A more accurate plug may be one week per year, 1/52, or another value supported by the analyst. Also, revenue and expense escalation were not provided but were estimated at 2.00%. These are conservative values and based on the US Federal Reserve's inflation target of at or near 2.00%. Again, these can be adjusted by the analyst/user but should be supported with documented backup.

Note that in the example for Operations, cell I2 is selected. In the upper left portion of the diagram 'I2' is shown as the selected cell. In this model, as it is larger and more complicated, inputs will be named rather than referenced as the sheet and cell location. Each of the individual rental units (i.e. 1–5) will be named as follows: Unit1_Rent, Unit2_Rent, . . . , Unit5_Rent. To name a cell, select the cell reference in the upper left portion of the spreadsheet, for example the 'I2' in the Operations section shown in Figure 6.2. Delete the cell reference and type in the name for the cell. In the case of cell I2, the name is Unit1_Rent. The underscore between the '1' and the 'R' is there to separate the words and addresses because named cells/arrays in Excel cannot have a blank space ('Unit1 Rent' is disallowed).

Next the vacancy inputs, cells I8:I12, must be named. Utilizing the same process, name cells I8:I12 as follows: Vacancy1, Vacancy2, . . . , Vacancy5. Next name the Rental Escalation cell, cell I14, 'Rental_Escl'. The Uses cells must also be named. It is recommended that for cells I16:I21 the following names be used: 'Management', 'Repairs', 'Electricity_mo', 'Gas_mo', 'Property_Tax_yr', and 'Insurance_yr'. Of course analysts are free to name the cells using any naming convention desired; however, it is recommended to remain consistent with this text for instructional purposes. Finally, Expense Escalation, cell I23, is named 'Expense_Escl'.

Next is a word of caution. Inevitably in this process of naming each cell in the Operations section a mistake occurs – e.g. a misspelling, double-type, etc. Regardless of the reason for the error, the analyst intends to correct it. Corrections and adjustments can be made in the 'Name Manager', which is found in the Formulas tab on the Ribbon in the 'Defined Names' section (Figure 6.3). Select the 'Name Manager' and the menu screen is displayed. In this menu screen current names can be edited and deleted or new names created. At this point in the text it is recommended that the analyst delete any errors and rename manually. However, for the more advanced Excel user, names can be modified in this menu screen.

When 'Edit . . .' is selected for Electricity_mo, a new menu screen is displayed: 'Edit Name' (Figure 6.4). It is here that edits can be made, comments for rationale provided, and reference location adjusted.

As a final note to naming cells in the Operations section, naming a cell is equivalent to an absolute reference. The cell will always be defined by the name. Naming cells is absolutely necessary when building models of any real size and complexity.

Step 3: finance

Following are the parameters for the tranches used in this example.

Tranche 1 Tenor: 30 Years payable monthly
Tranche 1 Interest: 5.00% Fixed
Tranche 1 Structure: Constant Payment Mortgage (CPM)

Tranche 1 Origination Fee: 1.00% (estimated)
Tranche 2: None but building to accommodate later revisions

Typically the terms of the loan will be summarized in a non-binding Term Sheet provided by a lender. The ability to negotiate terms is related to the leverage for the project, the financial

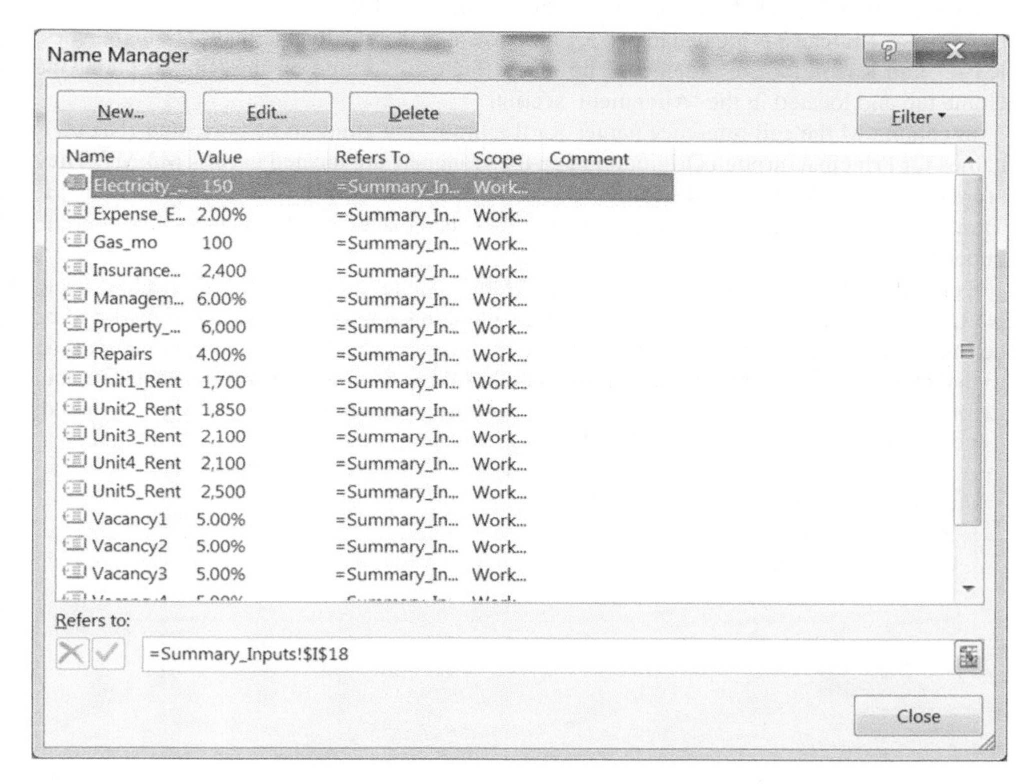

Figure 6.3

Figure 6.4

backing of the guarantor (if applicable), and the length of any relationship with the current lender. Cell K1 is the start of the Finance section (Figure 6.5).

Cell K1 is therefore 'Finance' and cell N1 is 'I/P' for 'Input'. Beginning in cell K2, as was done in Chapter 4, Tranche 1 is titled. Then beginning in cell K3 and continuing to K7, the following terms of a loan are listed in order: Principal, Term (Years), Interest Rate, Structure, and Origination Fee. The same is repeated beginning in cell K9 for Tranche 2. The Origination Fee for Tranche 2 is located in cell K14. Please note that the loan terms for both tranches, K3:K7 and K10:K14, are indented. The indent feature is found on the Ribbon under the Home tab and located in the 'Alignment' section.

In column M the cell reference names for the Inputs are going to be presented. The cells names for Principal through Origination Fee for Tranche 1 are located in cells M3:M7. They are as follows: Principal_T1, Term_T1, Interest_T1, Structure_T1, and Orig_Fee_T1. Tranche 2 has the same naming structure, only 'T1' is replaced by 'T2' as the suffix after the underscore.

Cells K6 and K13 are the structure of each of the tranches and are selected in cells N6 and N13, respectively (Figure 6.6). As was demonstrated in Chapter 4, cells N6 and N13 will be defined using Data Validation from a list of only three inputs: 1, 2, 3. The notable addition to that chapter is that the Data Validation menu has three tabs: Settings, Input Message, and Error Alert. For this model the Input Message tab will be utilized. The title will be 'Structure Legend' and the Input Message will state as follows:

1 Constant Payment Mortgage (CPM)
2 Interest Only (IO)
3 Custom Amortization.

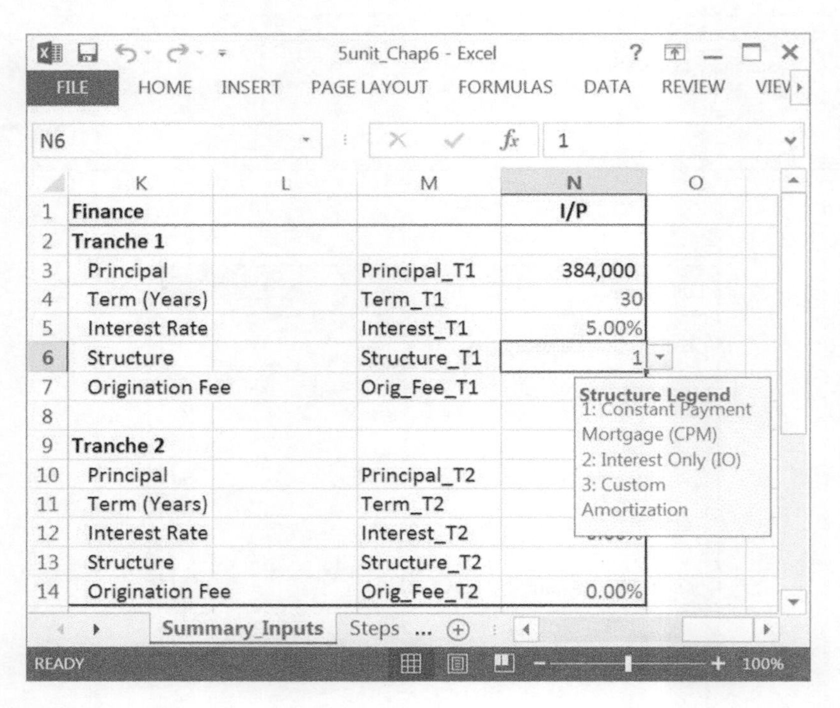

Figure 6.5

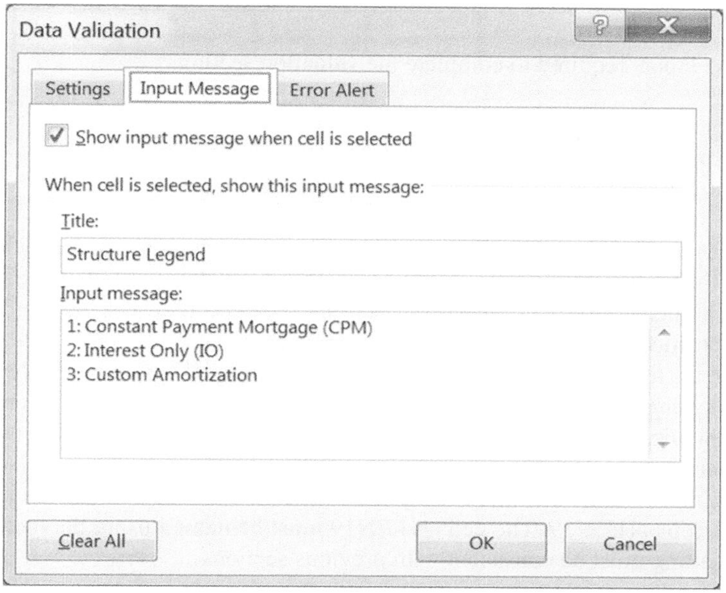

Figure 6.6

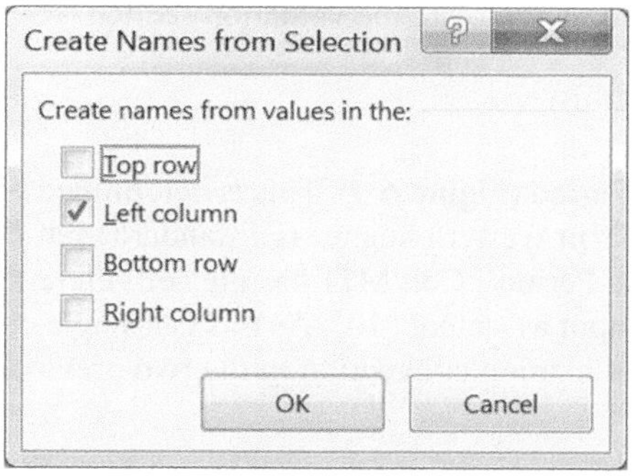

Figure 6.7

This will provide the analyst/user a clear understanding of debt structures constructed within the model. Notice in Figure 6.6 how the Inputs page appears with the new Input Message. Note that cell N13 – i.e. Structure for Tranche 2 – remains incomplete. Select cell N6 and copy, then select cell N13 and paste. The cell formatting with the Data Validation and Input Message will copy identically.

Now each of the cells in N3:N14 must be named. The process is much more simple than naming each individual cell as completed prior. Select sells M3:N14. Under the Formulas tab on the Ribbon, go to 'Defined Names' and select 'Create from Selection'. The 'Create Names from Selection' menu will appear (Figure 6.7). Ensure that 'Left column' is selected and hit 'OK'. This will then name the cells in column N the names placed in column M. Though not recommended here, if the analyst desires that the naming convention in column M be removed, it can be completed in two ways: (1) simply delete the contents or (2) colour the cells white, thus 'hiding' the contents without removal (recommended).

Step 4: valuation

Below are the inputs required to complete the valuation section.

Capitalization Rate: 6.00%
Sales Expense: 6.00%
Discount Rate: 8.00%

The valuation assumptions will be provided to the analyst or can be provided by large brokerage houses. The Capitalization Rate should be a forecast which is supportable. Sales Expense is 600 basis points and based on past transactions. The Discount Rate is the rate used for similar risk-profiled projects in the same geographic area.

In cell K16 the title of the section is placed: Valuation (Figure 6.8). In cell N16 'I/P' (input) is placed indicating the values of each criteria. In cells K17:K19 place the item references as follows: 'Capitalization Rate', 'Sales Expense', and 'Discount Rate'. Then in column M, as previously demonstrated, the intended name of the input cells must be entered: 'Cap_Rate', 'Sales_Exp', and 'Discount_Rate'. Finally the values for each, as noted in the previous list, are placed in cells N17:N19. The cells N17:N19 must be named using the methodology in Step 3. Formatting must be consistent with previous sections.

Step 5: sales

Below is the input required to complete the valuation section.

Sales Period: Year 10 (10)

The final section is Sales Period (Figure 6.9). This is determined by the investment horizon; however, a 10-year analysis in Western finance is a standard metric. Cell K21 has the section title and input name 'Sales Period'. Cell M21 has the cell reference name 'Sale_Term_Yr'. Finally, cell N21 has the input as stated, "10". As this is a single cell, the name can either be manually completed or the naming convention in the two previous sections utilized. Either is appropriate.

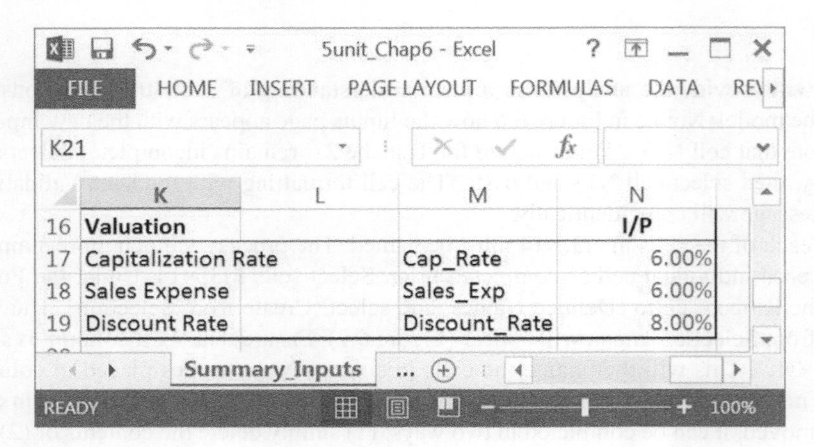

Figure 6.8

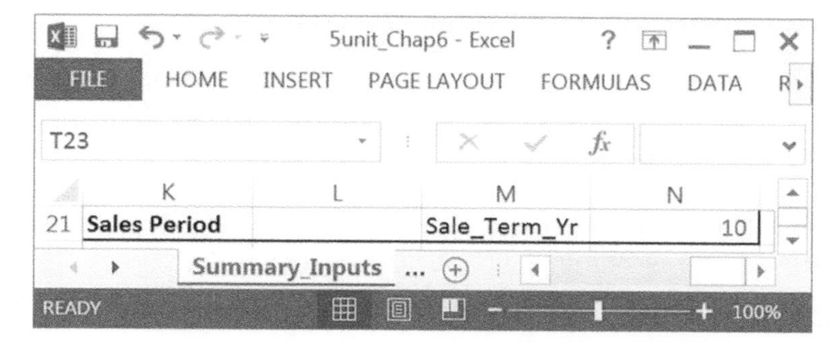

Figure 6.9

	A B C	D	E	F G	H	I	J	K L	M	N
1	**Source**	584,000		**Operations**	**Units**	**I/P**		**Finance**		**I/P**
2	Equity	200,000		Unit 1 - Efficiency	per month	1,700		**Tranche 1**		
3	Investor 1	200,000		Unit 2 - One Bedroom	per month	1,850		Principal	Principal_T1	384,000
4	Investor 2			Unit 3 - Two Bedroom	per month	2,100		Term (Years)	Term_T1	30
5	Debt	384,000		Unit 4 - Two Bedroom	per month	2,100		Interest Rate	Interest_T1	5.00%
6	Tranche 1	384,000		Unit 5 - Three Bedroom	per month	2,500		Structure	Structure_T1	1
7	Tranche 2							Origination Fee	Orig_Fee_T1	1.00%
8				Vacancy 1		5.00%				
9	**Uses**	584,000		Vacancy 2		5.00%		**Tranche 2**		
10	Purchase Price	500,000		Vacancy 3		5.00%		Principal	Principal_T2	-
11	Capital Improv	50,000		Vacancy 4		5.00%		Term (Years)	Term_T2	-
12	Legal	9,000		Vacancy 5		5.00%		Interest Rate	Interest_T2	0.00%
13	Personal	5,000						Structure	Structure_T2	
14	Lender	4,000		Rental Escl		2.00%		Origination Fee	Orig_Fee_T2	0.00%
15	Escrow	6,000								
16	Finance Expense	4,000		Management	per Net Rev	6.00%		**Valuation**		**I/P**
17	Contingency	15,000		Repairs	per Net Rev	4.00%		Capitalization Rate	Cap_Rate	6.00%
18				Electricity	per month	150		Sales Expense	Sales_Exp	6.00%
19	**Surplus/(Deficit)**	-		Gas	per month	100		Discount Rate	Discount_Rate	8.00%
20				Property Tax	per year	6,000				
21				Insurance	per year	2,400		**Sales Period**	Sale_Term_Yr	10
22										
23				Expense Escl		2.00%				

Figure 6.10

At this point the Inputs section of the pro forma has been completed (Figure 6.10).

Debt/amortization

A review/read of Chapter 4 is recommended prior to completing this section. The amortization table constructed here will follow the same structure as demonstrated in that chapter.

A new sheet must be created in the existing model. This sheet should be named 'Debt' (Figure 6.11). As with the previous amortization tables structured, cells A1:A4 are the loan characteristics: Principal, Interest, Tenor, and Structure. Cells B1:B4 are then linked to the Inputs sheet using the names defined. In order, B1:B4 are as follows: '=Principal_T1', '=Interest_T1', '=Term_T1', and '=Structure_T1'. For consistency of modelling and easy interpretation, it is recommended that cell C4 include "CPM: 1, IO: 2, Custom: 3".

Row 6 contains the amortization titles: Period, BoP, Princ, Int, Payment, EoP, and Amort%. Note that an additional header has been included – Payment. This is simply the total payment

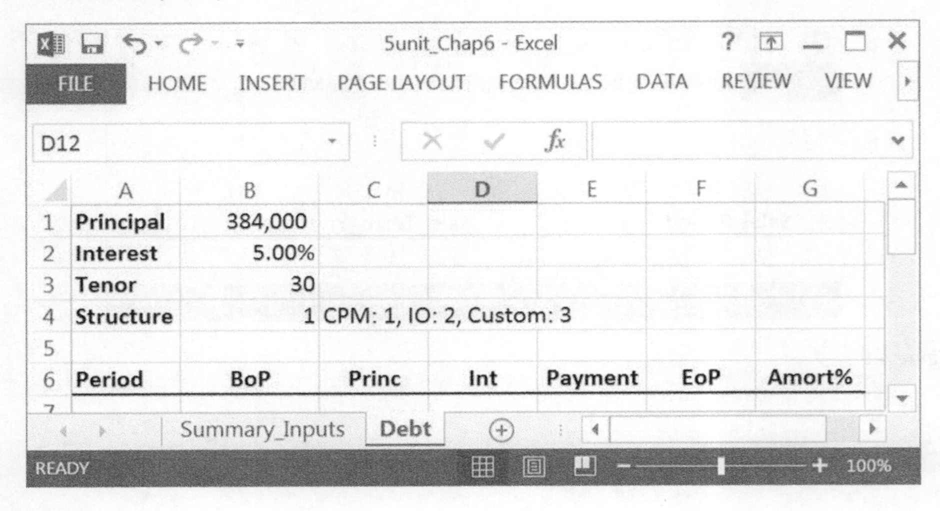

5unit_Chap6 - Excel

D12

	A	B	C	D	E	F	G
1	**Principal**	384,000					
2	**Interest**	5.00%					
3	**Tenor**	30					
4	**Structure**	1	CPM: 1, IO: 2, Custom: 3				
5							
6	**Period**	**BoP**	**Princ**	**Int**	**Payment**	**EoP**	**Amort%**

Summary_Inputs | Debt

Figure 6.11

5unit_Chap6 - Excel

A5

	A	B	C	D	E	F	G	H
1	**Principal**	384,000						
2	**Interest**	5.00%						
3	**Tenor**	30						
4	**Structure**	1	CPM: 1, IO: 2, Custom: 3					
5								
6	**Period**	**BoP**	**Princ**	**Int**	**Payment**	**EoP**	**Amort%**	
7	1	384,000	461	1,600	2,061	383,539	0.28%	
8	2	383,539	463	1,598	2,061	383,075	0.28%	
9	3	383,075	465	1,596	2,061	382,610	0.28%	

Summary_Inputs | Debt

Figure 6.12

per period and is the summation of principal and interest payments each period. It is included in this section to simplify the Pro Forma page, which is constructed later in this chapter.

The amortization table is then completed, with column A having the periods through 360 (i.e. accommodating a 30-year amortization schedule). The periods begin in cell A7 and continue through cell A366 (Figure 6.12). Cell B7, BoP for period 1, is "=B$1". It is important that column B be a relative reference while row 7 is absolute. Cell B8, BoP for period 2, has a different logic and is "=F7"; in other words, BoP for period 2 is equal to EoP for period 1. The logic stated in cell B8 will be consistent for the remainder of the amortization table.

The principal logic, cells C7:C366, is the trickiest of this section. It requires the use of the Choose Function to enable different amortization structures. Basically the structure is Choose(Structure Number Selected, CPM methodology, IO methodology, Custom methodology). The formula for cell C7 is as follows:

=choose(b$4,PPMT(B$2/12,A7,B$3*12,-B$1),0,G7*B$1)

While the custom amortization has not yet been inserted into cells G7:G366, it can still be referenced. Interest, cell D7, is the interest per period based on the Beginning of Period balance. Cell D7 is therefore "=B7*B$2/12". Cell E7 – which is also new to the structure of amortization tables in this text – is "=sum(C7:D7)". Again, this represents the total payment, principal and interest, paid per period. Cell F7, End of Period, is the Beginning of Period less any principal repaid. Cell F7 is therefore "=B7-C7". Column G is the custom amortization schedule. This is an input column and is determined by the lending institution or estimated by the analyst. For the purpose of this text it will be assumed that the loan product is equally amortizing over the tenor. Therefore cell G7 is "=1/(B$3*12)". Note that the colour of column G, again, must be blue because these are inputs.

Next, copy cells C7:G7 down one row and paste to row 8. Copy cells B8:G8 to the end of the amortization table (i.e. through row 366). Select cells H1:H366 and colour the entire column grey. It is recommended that the width for column H be adjusted to 2. The amortization table for Tranche 1 is now complete.

The next step is to create the amortization table for Tranche 2. Because Tranche 1 was carefully completed utilizing absolute and relative references, the process is straightforward and simple.

Select columns A:G and copy. Select cell I1 and paste. Tranche 1 amortization table has therefore been identically copied to the right of column H. Next adjust cells J1:J4 to reflect the reference for Tranche 2 – i.e. Principal_T2, Interest_T2, Term_T2, and Structure_T2. While the table is completed with '#Num!' (Figure 6.13), it actually works but is referencing a zero-balance Tranche 2; that is the Tranche 2 amortization table does work correctly if values are selected on the Inputs section.

Finally, the arrays for both amortization tables, which will later be referenced on the Pro Forma page, must be named. Select cells A7:F366 and replace the label 'A7' with 'Tranche_1', thus naming the first tranche array 'Tranche_1'. Next select the second Tranche array (select I7:N366). Name the array selected 'Tranche_2'.

Both amortization tables have been completed.

5unit_Chap6 - Excel

FILE HOME INSERT PAGE LAYOUT FORMULAS DATA REVIEW VIEW ACROBAT Sign in

J4 =Structure_T1

	A	B	C	D	E	F	G	H	I	J	K	L	M	N	O
1	Principal	384,000							Principal	-					
2	Interest	5.00%							Interest	0.00%					
3	Tenor	30							Tenor	0					
4	Structure	1	CPM: 1, IO: 2, Custom: 3						Structure	1	CPM: 1, IO: 2, Custom: 3				
5															
6	Period	BoP	Princ	Int	Payment	EoP	Amort%		Period	BoP	Princ	Int	Payment	EoP	Amort%
7	1	384,000	461	1,600	2,061	383,539	0.28%		1	-	#NUM!	-	#NUM!	#NUM!	#DIV/0!
8	2	383,539	463	1,598	2,061	383,075	0.28%		2	#NUM!	#NUM!	#NUM!	#NUM!	#NUM!	#DIV/0!

Summary_Inputs Debt

READY 100%

Figure 6.13

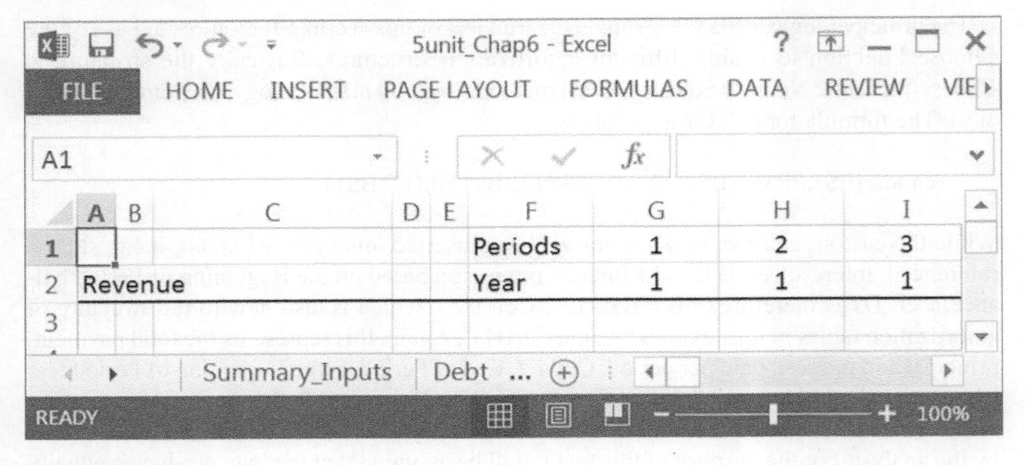

Figure 6.14

Pro forma

The Pro Forma section is the guts and glory of the financial model. This Pro Forma will be slightly more complicated than in Chapter 5 as the units will be months rather than years. However, the structure remains the same. It is best to consider the Pro Forma as being constructed in sections. The logical sections are Schedule, Total Net Revenue, Total Expense, Net Operating Income, Debt, Net Sales Proceeds, Free Cash Flow, and finally Valuation. The flow of this text will follow this structure.

A final sheet must be added to the pro forma. This sheet is appropriately titled 'Pro_Forma' (Figure 6.14). Columns A, B, D, and E should be formatted with a column width of 2. Cells F1 and F2 should be labeled 'Periods' and 'Year', respectively. Cell A2 is labeled 'Revenue'. Rows 1 and 2 are therefore the schedule for the pro forma. It is critical that all schedules begin in the same column for consistency. In this modeller's experience, column G is a good one to select and will be used for the remainder of this text.

Schedule

Cell G1 has the value of "1" as it is the beginning period. Cell H1 is therefore "=G1+1". Cell H1 is then copied to cell EH1, which is period 132. Note that a 10-year pro forma analysis – i.e. 120 periods – requires an 11-year (132-period) pro forma to be developed. This results because the end of period 10 gross sales proceeds, by definition, is equal to year 11 NOI divided by the terminal capitalization rate.

Next the year must be calculated from the period. The ROUNDUP function is utilized to provide the accurate year per period. Cell G2 is "=roundup(G1/12,0)". This divides the period by 12 and rounds up to the next integer. The zero (0) is the number of decimals after the integer value (yes, this author understands that by definition if there were decimals after an integer the value is no longer an integer). Format row 2 with a thick bottom border beginning in cell A2, and copy G2 to every cell in row 2 through column EH. The structure with the schedule for the Pro Forma page has been completed.

Total net revenue

In cell B3 the unit labels are placed (Figure 6.15). Rather than rekeying the Unit Names, cell B3 is best linked to the Inputs page. Therefore cell B3 is "=Summary_Inputs!F2". Cell B2 is then copied to cells B4:B7, labeling each of the unit revenues on the Pro Forma page. The individual unit revenues per period are then begun in column G.

The logic for unit rent is similar to Chapter 5 and follows the Future Value logic. Cell G3 is "=Unit1_Rent*(1+Rental_Escl)^(G$2–1)". Rather than retyping the logic for Units 2–5, cell G3 should be copied and pasted through G7. Then adjust the unit number (i.e. "Unit1_Rent") for the appropriate number. For example, cell G7 should be "=Unit5_Rent*(1+Rental_Escl)^(G$2–1)". Finally, cell C8 should be labeled 'Total Rent'. Cell G8 is therefore the summation of the five unit period rents: "=sum(G3:G7)". After providing formatting and the top border to row 8, copy G3:G8 to EH3:EH. Total Rent is completed.

As a Vacancy Rate was established by unit in the Inputs section, Vacancy will be modelled by unit here on the Pro Forma page. In cell B10, the Vacancy titles per unit are linked to the Input page. Cell B10 is therefore "=Summary_Inputs!F8". This is then copied to cells B11:B14. As with Unit Revenue, cell C15 is titled 'Total Vacancy'. The Vacancy values are calculated beginning in cell G10. Cell G10 is "=Vacancy1*G3". This formula is then copied

5unit_Chap6 - Excel

FILE HOME INSERT PAGE LAYOUT FORMULAS DATA REVIEW VIEW ACROB

A1 fx

	A B C D E	F	G	H	I	J
1		Periods	1	2	3	4
2	Revenue	Year	1	1	1	1
3	Unit 1 - Efficiency		1,700	1,700	1,700	1,700
4	Unit 2 - One Bedroom		1,850	1,850	1,850	1,850
5	Unit 3 - Two Bedroom		2,100	2,100	2,100	2,100
6	Unit 4 - Two Bedroom		2,100	2,100	2,100	2,100
7	Unit 5 - Three Bedroom		2,500	2,500	2,500	2,500
8	Total Rent		10,250	10,250	10,250	10,250
9						
10	Vacancy 1		85	85	85	85
11	Vacancy 2		93	93	93	93
12	Vacancy 3		105	105	105	105
13	Vacancy 4		105	105	105	105
14	Vacancy 5		125	125	125	125
15	Total Vacancy		513	513	513	513
16						
17	Total Revenue		9,738	9,738	9,738	9,738

... Debt Pro_Forma

READY 100%

Figure 6.15

to each cell in that column through cell G14. Remember to adjust the Vacancy reference for the appropriate unit; for example, cell G14 is "=Vacancy5*G7". Cell G15 is the summation of the vacancy, "=sum(G10:G14)". Copy cells, with the appropriate formatting, to the corresponding rows in column EH.

Total Revenue, total rent less vacancy, is therefore calculated in row 17. Cell A17 is 'Total Revenue'. Cell G17 is therefore "=G8-G15". Copy cell G17 to column EH. It is recommended that the Total Revenue row be bolded as it is a summary row.

Total expense

The next section is Expenses, where each of the expenses for the project are quantified (Figure 6.16). The methodology and process are identical to Chapter 5 and, as such, will be discussed briefly. There will be a few adjustments given that several of the expense items have different compounding methodologies, presented here.

Cell A19 is Expenses. The expenses are then listed, one column indented from C19. These expenses are linked to the Inputs section to disallow rewriting of titles. Cells B20 and B25 are therefore "=Summary_Inputs!F16" and "=Summary_Inputs!F21", respectively. As Management and Repairs are a percentage of net revenue, the logic is the same for each – i.e. a percentage of Total Revenue. Management and Repairs, cells G20 and G21, are therefore "=management*G17" and "=repairs*G18", respectively. Electricity and Gas, both commodities, are modelled as 'per month' and thus are escalated on a monthly basis. Cells G22 and G23 are "=Electricity_mo*(1+Rental_Escl/12)^(G1–1)" and "=Gas_mo*(1+Rental_Escl/12)^(G1–1)", respectively. Finally Property Tax and Insurance, both of which are escalated annually, must be quantified per period. Cells G24 and G25 are "=(Property_Tax_Yr/12)*(1+Rental_Escl)^(G2–1)" and "=(Insurance_Yr/12)*(1+Rental_Escl)^(G2–1)", respectively. Note the values for Property Tax and Insurance are divided by 12 because they are paid monthly. It could be considered that the values are accrued and

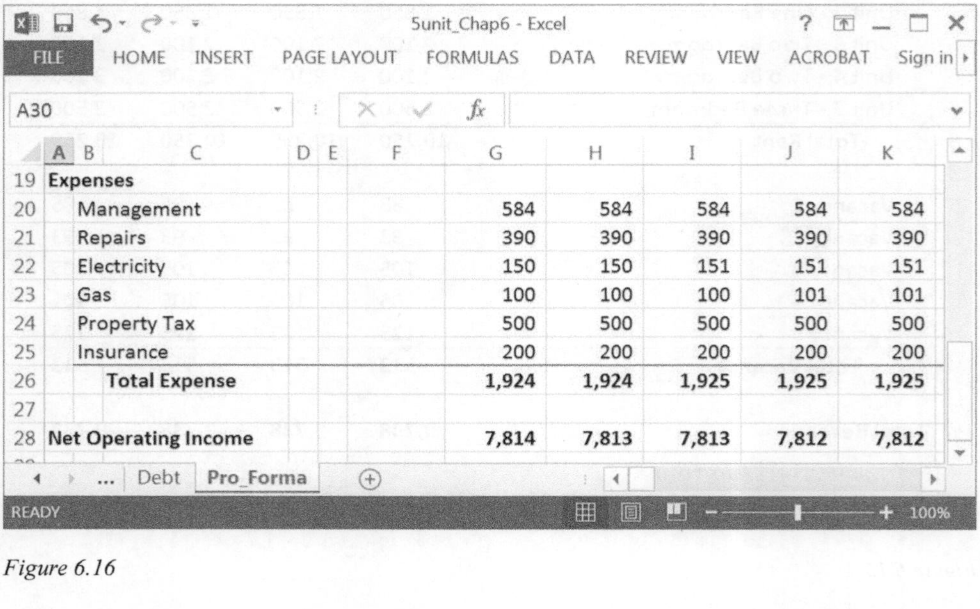

5unit_Chap6 - Excel

FILE HOME INSERT PAGE LAYOUT FORMULAS DATA REVIEW VIEW ACROBAT Sign in

A30

	A B C	D E F	G	H	I	J	K
19	**Expenses**						
20	Management		584	584	584	584	584
21	Repairs		390	390	390	390	390
22	Electricity		150	150	151	151	151
23	Gas		100	100	100	101	101
24	Property Tax		500	500	500	500	500
25	Insurance		200	200	200	200	200
26	**Total Expense**		**1,924**	**1,924**	**1,925**	**1,925**	**1,925**
27							
28	**Net Operating Income**		**7,814**	**7,813**	**7,813**	**7,812**	**7,812**

... Debt Pro_Forma

READY 100%

Figure 6.16

then paid at the end of each respective year. However, for modelling purposes, the expenses are modelled monthly with annual escalation. Finally, total expense, cell G26, is the summation of cells G20:G25: "=sum(G20:G25)". Cells G20:G26 are then copied to the corresponding cells in column EH.

Net operating income

Net Operating Income is labeled in cell A28. Net Operating Income is quantified as the difference in Revenue and Expense in cell G28, i.e. "=G17-G26". Cell G28 is then copied to the corresponding cells in column EH, completing Net Operating Income.

Debt

Debt references the amortization table on the 'Debt' page. The referencing is similar to Chapter 5 but for the addition of the payment column in the amortization table and debt existing assumption.

The IFERROR function is introduced in this model because the capital structure for a model may not include debt and a credible and scalable model must be able to accommodate an all-equity-financed capital structure.

Cell A30 is labeled 'Debt' and cells B31 and B32 are Tranche 1 and Tranche 2. Cell C33 is then total debt. Cell G31, first period for Tranche 1 debt, is the total payment in the first period. The VLOOKUP function, as completed in the previous chapter, is utilized to reference the vertically constructed amortization table to the horizontally structured pro forma page. Cell G31 is therefore "=vlookup(g1,tranche_1,5)". Note that the column reference five (5) is the column for total payment, that is Principal plus Interest. Cell G32 follows the same methodology but references 'Tranche_2': "=vlookup(g1,tranche_2,5)".

Note that as the model is currently constructed, #NUM is present in cell G31. This is because the principal, interest, and tenor for Tranche 2 are all zero. To accommodate a structure without debt, cells G31 and G32 must be adjusted. So cell G31 is modified as follows: "=IFERROR(VLOOKUP(G1,Tranche_1,5),0)". Cell G32 follows the same logic and is therefore "=IFERROR(VLOOKUP(G1,Tranche_2,5),0)". Note that cell G32 is now zero rather than #NUM. Cell G33 is then the summation of the two tranches: "=SUM(G31:G32)". Then copy cells G31:G33 to the corresponding cells in column EH to complete the Debt section.

Net sales proceeds

Cell A35 is the label for 'Net Sales Proceeds' (Figure 6.17). Cells B36:B39 are labeled 'Sales Price', 'Sales Expense', 'Principal Repay – Tranche 1', and 'Principal Repay – Tranche 2', respectively. Cell C40 is then labeled 'Net Sales Proceeds'.

Sales Price is quantified as Net Operating Income at t + 1 divided by terminal capitalization rate:

$$\text{Sales Price} = \frac{\text{NOI}_{t+1}}{\text{Capitalization Rate}}$$

To determine the Sales Price, the first step is determining the appropriate sales period. As such, logic comparing the current period to the sales period must be created using an If/Then

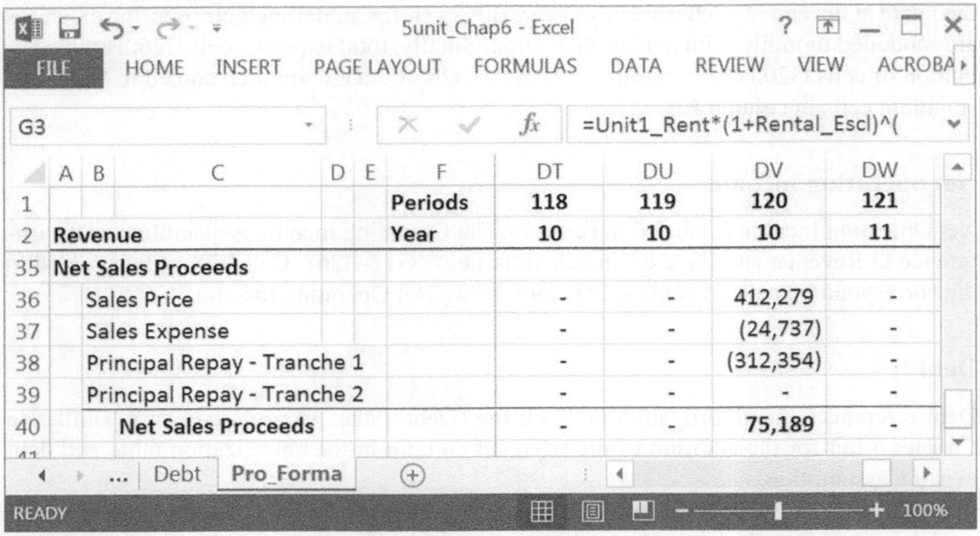

5unit_Chap6 - Excel

FILE HOME INSERT PAGE LAYOUT FORMULAS DATA REVIEW VIEW ACROBA

G3 =Unit1_Rent*(1+Rental_Escl)^(

	A	B	C	D	E	F	DT	DU	DV	DW
1						**Periods**	**118**	**119**	**120**	**121**
2	**Revenue**					**Year**	**10**	**10**	**10**	**11**
35	**Net Sales Proceeds**									
36		Sales Price					-	-	412,279	-
37		Sales Expense					-	-	(24,737)	-
38		Principal Repay - Tranche 1					-	-	(312,354)	-
39		Principal Repay - Tranche 2					-	-	-	-
40		**Net Sales Proceeds**					-	-	**75,189**	-

... Debt Pro_Forma

READY 100%

Figure 6.17

function. This logic will compare the current period divided by 12 (i.e. annualize the current period) to the sales term (Sale_Term_Yr). Upon a positive, True, result, the Sales Price will be quantified by adding the next 12 periods of Net Operating Income and dividing that summation by the Capitalization Rate. Cell G36 is therefore "=IF(G1/12=Sale_Term_Yr,SUM(H28:S28)/Cap_Rate,0)". This ensures that a value will be placed for Sales Price only in the period of sale. As an example, copy cell G36 to column EH. Note in column DV, period 120 (year 10), the value is 412,279 (Figure 6.17).

Sales Expense is the expense attributable to sale (Figure 6.18). As logic for sales period was embedded in the Sales Price row, none is needed here. Therefore cell G37 is simply "=-G36*Sales_Exp". Note the negative (-) prior to 'G36'. This is because the Sales Expense is a net outflow.

Principal Repay, rows 38 and 39, are the end of period balance, that is, the amount that must be repaid to the lender at time of sale. The logic is equivalent to the debt tranche, rows 31 and 32, except for the column reference in the required VLOOKUP function: i.e. column 6 (EoP) is referenced rather than total payment. Cell G38 is therefore "=-IFERROR(VLOOKUP(G$1,Tranche_1,6),0)*IF(G36=0,0,1)". Note the 'IF' statement at the end of the function. This utilizes the logic in Sales Price, row 36, by negating the Principal Repay if there is no sale. Cell G39 is equivalent to this but for the reference to 'Tranche_1'. The negative (-) prior to the formula, as with Sales Expense, is to delineate Principal Repay as an outflow. Net Sales Proceeds, row 40, is the summation of Sales Expense through Principal Repay – Tranche 2. As the outflows are all negative, only a summation is needed. The formula for cell G40 is therefore "=SUM(G36:G39)". Finally, copy cells G36:G40 through the corresponding cells in column EH to complete the section. Note that only column DV, period 120, has values. This is due to the logic embedded in the Net Sales Proceeds section.

The not-so-final section of the Pro Forma page is Free Cash Flow (Figure 6.18). It is "not-so-final" because the valuation portion remains. However, functionally, this will complete the pro forma cash flow.

5unit_Chap6 - Excel

FILE HOME INSERT PAGE LAYOUT FORMULAS DATA REVIEW VIEW ACROBAT Sign in

A47

	A–E	F	G	H	I	J	K	L	M	N	O	P	Q	R	S	T
1		**Periods**	**1**	**2**	**3**	**4**	**5**	**6**	**7**	**8**	**9**	**10**	**11**	**12**	**13**	**14**
2	**Revenue**	**Year**	**1**	**1**	**1**	**1**	**1**	**1**	**1**	**1**	**1**	**1**	**1**	**1**	**2**	**2**
3	Unit 1 - Efficiency		1,700	1,700	1,700	1,700	1,700	1,700	1,700	1,700	1,700	1,700	1,700	1,700	1,734	
4	Unit 2 - One Bedroom		1,850	1,850	1,850	1,850	1,850	1,850	1,850	1,850	1,850	1,850	1,850	1,850	1,887	
5	Unit 3 - Two Bedroom		2,100	2,100	2,100	2,100	2,100	2,100	2,100	2,100	2,100	2,100	2,100	2,100	2,142	
6	Unit 4 - Two Bedroom		2,100	2,100	2,100	2,100	2,100	2,100	2,100	2,100	2,100	2,100	2,100	2,100	2,142	
7	Unit 5 - Three Bedroom		2,500	2,500	2,500	2,500	2,500	2,500	2,500	2,500	2,500	2,500	2,500	2,500	2,550	
8	**Total Rent**		**10,250**	**10,250**	**10,250**	**10,250**	**10,250**	**10,250**	**10,250**	**10,250**	**10,250**	**10,250**	**10,250**	**10,250**	**10,455**	**10**
9																
10	Vacancy 1		85	85	85	85	85	85	85	85	85	85	85	85	87	
11	Vacancy 2		93	93	93	93	93	93	93	93	93	93	93	93	94	
12	Vacancy 3		105	105	105	105	105	105	105	105	105	105	105	105	107	
13	Vacancy 4		105	105	105	105	105	105	105	105	105	105	105	105	107	
14	Vacancy 5		125	125	125	125	125	125	125	125	125	125	125	125	128	
15	**Total Vacancy**		**513**	**513**	**513**	**513**	**513**	**513**	**513**	**513**	**513**	**513**	**513**	**513**	**523**	
16																
17	**Total Revenue**		**9,738**	**9,738**	**9,738**	**9,738**	**9,738**	**9,738**	**9,738**	**9,738**	**9,738**	**9,738**	**9,738**	**9,738**	**9,932**	**9**
18																
19	**Expenses**															
20	Management		584	584	584	584	584	584	584	584	584	584	584	584	596	
21	Repairs		390	390	390	390	390	390	390	390	390	390	390	390	397	
22	Electricity		150	150	151	151	151	151	152	152	152	152	153	153	153	
23	Gas		100	100	100	101	101	101	101	101	101	102	102	102	102	
24	Property Tax		500	500	500	500	500	500	500	500	500	500	500	500	510	
25	Insurance		200	200	200	200	200	200	200	200	200	200	200	200	204	
26	**Total Expense**		**1,924**	**1,924**	**1,925**	**1,925**	**1,925**	**1,926**	**1,926**	**1,927**	**1,927**	**1,928**	**1,928**	**1,928**	**1,962**	**1**
27																
28	**Net Operating Income**		**7,814**	**7,813**	**7,813**	**7,812**	**7,812**	**7,812**	**7,811**	**7,811**	**7,810**	**7,810**	**7,810**	**7,809**	**7,970**	**7**
29																
30	**Debt**															
31	Tranche 1		2,061	2,061	2,061	2,061	2,061	2,061	2,061	2,061	2,061	2,061	2,061	2,061	2,061	
32	Tranche 2		-	-	-	-	-	-	-	-	-	-	-	-	-	
33	**Total Debt**		**2,061**	**2,061**	**2,061**	**2,061**	**2,061**	**2,061**	**2,061**	**2,061**	**2,061**	**2,061**	**2,061**	**2,061**	**2,061**	**2**
34																
35	**Net Sales Proceeds**															
36	Sales Price		-	-	-	-	-	-	-	-	-	-	-	-	-	
37	Sales Expense		-	-	-	-	-	-	-	-	-	-	-	-	-	
38	Principal Repay - Tranche 1		-	-	-	-	-	-	-	-	-	-	-	-	-	
39	Principal Repay - Tranche 2		-	-	-	-	-	-	-	-	-	-	-	-	-	
40	**Net Sales Proceeds**		-	-	-	-	-	-	-	-	-	-	-	-	-	
41																
42	**Operating FCF**		**5,752**	**5,752**	**5,752**	**5,751**	**5,751**	**5,750**	**5,750**	**5,749**	**5,749**	**5,749**	**5,748**	**5,748**	**5,909**	**5**
43	**Reversion FCF**		-	-	-	-	-	-	-	-	-	-	-	-	-	
44	**Total FCF**		**5,752**	**5,752**	**5,752**	**5,751**	**5,751**	**5,750**	**5,750**	**5,749**	**5,749**	**5,749**	**5,748**	**5,748**	**5,909**	**5**
45																

... Debt Pro_Forma

READY 70%

Figure 6.18

Free Cash Flow is bifurcated into components: Operations and Reversion. Operating Cash Flow represents cash flows created from active management of the asset. Reversion is the net sales proceeds and is determined by market factors at time of sale. Reversion is not actively managed but determined by the market.

Cells A42 and A43 are therefore 'Operating FCF' and 'Reversion FCF'. Cell C44 is the summation of both or 'Total FCF'. Operating FCF, cell G42, is Net Operating Income less Total Debt Service or "=G28-G33". Reversion FCF is the Net Sales Proceeds or "=G40". Total FCF is the summation of both or "=SUM(G42:G43)". Cells G42:G44 are then copied to column EH.

Summary/valuation

Returning to the Summary_Inputs page, the Valuation section, the final section, is completed (Figure 6.19). As this is a functionally separate section, cells A25:N25 are coloured grey, delineating the Valuation section as separate.

Cell A26 is titled 'Valuation:' and D27 'Year'. Cell E27 is "1" and F27 is "=E27+1". Copy cell F27 to all the cells in that row through N27, thus setting up the section summary header.

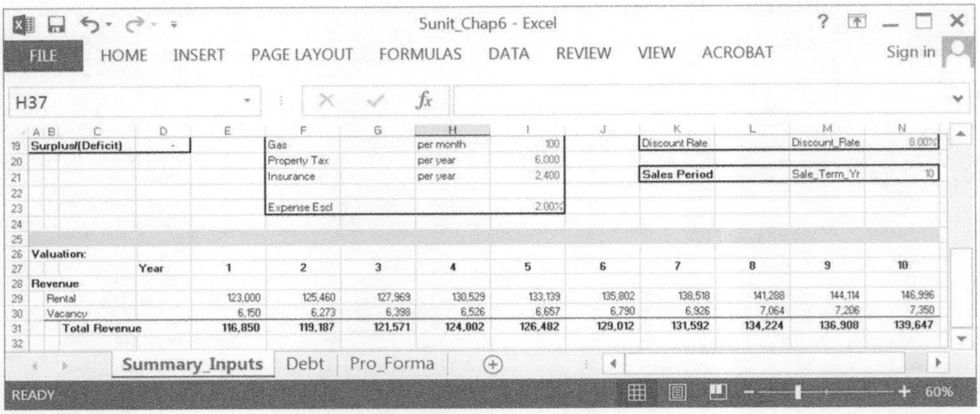

	A	B	C	D	E	F	G	H	I	J	K	L	M	N
19	**Surplus/(Deficit)**			-		Gas		per month	100		Discount Rate		Discount_Rate	8.00%
20						Property Tax		per year	6,000					
21						Insurance		per year	2,400		**Sales Period**		Sale_Term_Yr	10
22														
23						Expense Escl			2.00%					
24														
25														
26	**Valuation:**													
27				**Year**	**1**	**2**	**3**	**4**	**5**	**6**	**7**	**8**	**9**	**10**
28	**Revenue**													
29		Rental			123,000	125,460	127,969	130,529	133,139	135,802	138,518	141,288	144,114	146,996
30		Vacancy			6,150	6,273	6,398	6,526	6,657	6,790	6,926	7,064	7,206	7,350
31			**Total Revenue**		**116,850**	**119,187**	**121,571**	**124,002**	**126,482**	**129,012**	**131,592**	**134,224**	**136,908**	**139,647**
32														

Figure 6.19

For formatting consistency it is recommended that columns E–N be formatted with an equivalent column width. In the current example, the width is 12.

The majority of this section is summarizing the cash flows annually from the Pro Forma page, which has quantified cash flows on a periodic (monthly) basis. Cell A28 is 'Revenue' and B29 and B30 are 'Rental' and 'Vacancy', respectively. Cell C31 is therefore 'Total Revenue'. The SUMIF function is utilized in this section. The total rental income for year 1, cell E29, is the sum of total rents on the Pro_Forma page for the first year. Therefore the formula for cell E29 is "=SUMIF(Pro_Forma!2:2, E$27,Pro_Forma!8:8)". The SUMIF formula has three arguments: 'Range', 'Criteria', and '[Sum Range]'. The first argument, 'Pro_Forma!2:2', references the periodic year on the Pro_Forma page. The second argument, 'Criteria', is the year being referenced and is located on the summary page. The third argument, '[Sum Range]', is the values to be added for each instance where the criteria equals the range. Vacancy and Total Revenue are calculated using the same methodology but referencing the relative rows on the Pro_Forma page. Cells G30 and G31 are "=SUMIF(Pro_Forma!2:2,E$27,Pro_Forma!15:15)" and "=SUMIF(Pro_Forma!2:2,E$27,Pro_Forma!17:17)", respectively. Note that it is not recommended to calculate Total Revenue on the summary page (e.g. cell G31 "=G29-G30"). This is due to consistency when modelling. It is best to quantify numbers in one single location and reference throughout a model rather than calculate the same value several times and in several places. Cells E29:E31 are then copied to N29:N31, providing the Revenue annual summary for each of the 10 years.

The next section summary is expenses (Figure 6.20). The methodology will be similar to revenue but for the use of absolute and relative referencing for more simplified referencing. Cells B34:B39 are the title headers for the expenses. While these can be rekeyed or referenced from the Input section above, it is recommended that the headers be referenced from the Pro_Forma page for consistency. Therefore, cell E34 is "=Pro_Forma!B20". Cell E34 is then copied down to cells E35:E39. The row headers are displayed and linked to the Pro_Forma page. Cell C40 is then 'Total Expenses'.

Management, cell E34, is "=SUMIF(Pro_Forma!$2:$2,e$27,Pro_Forma!20:20)". Note the absolute reference for the first argument, 'Pro_Forma!$2:$2', and the relative reference for the third argument, 'Pro_Forma!20:20'. As the expense items are contiguous rows on both the Pro_Forma and Summary_Inputs pages, cell E34 is copied down through E40. Cells

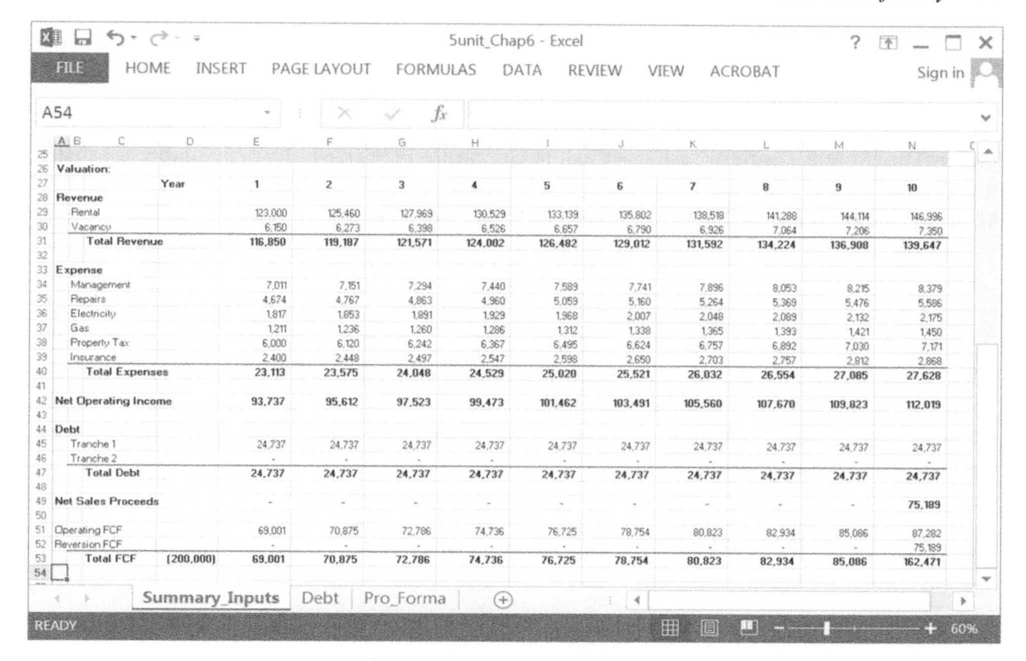

	Year	1	2	3	4	5	6	7	8	9	10
Valuation:											
Revenue											
Rental		123,000	125,460	127,969	130,529	133,139	135,802	138,518	141,288	144,114	146,996
Vacancy		6,150	6,273	6,398	6,526	6,657	6,790	6,926	7,064	7,206	7,350
Total Revenue		**116,850**	**119,187**	**121,571**	**124,002**	**126,482**	**129,012**	**131,592**	**134,224**	**136,908**	**139,647**
Expense											
Management		7,011	7,151	7,294	7,440	7,589	7,741	7,896	8,053	8,215	8,379
Repairs		4,674	4,767	4,863	4,960	5,059	5,160	5,264	5,369	5,476	5,586
Electricity		1,817	1,853	1,891	1,929	1,968	2,007	2,048	2,089	2,132	2,175
Gas		1,211	1,236	1,260	1,286	1,312	1,338	1,365	1,393	1,421	1,450
Property Tax		6,000	6,120	6,242	6,367	6,495	6,624	6,757	6,892	7,030	7,171
Insurance		2,400	2,448	2,497	2,547	2,598	2,650	2,703	2,757	2,812	2,868
Total Expenses		**23,113**	**23,575**	**24,048**	**24,529**	**25,020**	**25,521**	**26,032**	**26,554**	**27,085**	**27,628**
Net Operating Income		**93,737**	**95,612**	**97,523**	**99,473**	**101,462**	**103,491**	**105,560**	**107,670**	**109,823**	**112,019**
Debt											
Tranche 1		24,737	24,737	24,737	24,737	24,737	24,737	24,737	24,737	24,737	24,737
Tranche 2		-	-	-	-	-	-	-	-	-	-
Total Debt		**24,737**	**24,737**	**24,737**	**24,737**	**24,737**	**24,737**	**24,737**	**24,737**	**24,737**	**24,737**
Net Sales Proceeds		-	-	-	-	-	-	-	-	-	**75,189**
Operating FCF		69,001	70,875	72,786	74,736	76,725	78,754	80,823	82,934	85,086	87,282
Reversion FCF		-	-	-	-	-	-	-	-	-	75,189
Total FCF	**(200,000)**	**69,001**	**70,875**	**72,786**	**74,736**	**76,725**	**78,754**	**80,823**	**82,934**	**85,086**	**162,471**

Figure 6.20

E34:E40 are then copied to the corresponding cells in column N, completing the Expense section of the summary page.

Net Operating Income and Debt are summarized utilizing the same methodology. Cell A42 is 'Net Operating Income' and, using the SUMIF function, uses row 28 for its third argument. Cell A44 is 'Debt', and cells B45 and B46 are 'Tranche 1' and 'Tranche 2', respectively. Cell C47 is 'Total Debt'. The formulas for these sections are not stated here in the text to reduce redundancy; however, they follow the same logical progression.

Cell A49 is 'Net Sales Proceeds', which summarizes only the net of the sales transaction. Cell E49 is therefore "=SUMIF(Pro_Forma!2:2,E$27,Pro_Forma!40:40)". Cells A51 and A52 are then 'Operating FCF' and 'Reversion FCF', respectively. Cell C53 is 'Total FCF'. Following the same logic as the expenses, cell E51 is "=SUMIF(Pro_Forma!$2:$2,E$27,Pro_Forma!42:42)". Cell E51 is copied to E53 and then E51:E53 are copied to column N, thus completing the summary cash flows on an annual basis. The final adjustment occurs in cell D53, which is "=-D2". Cell D53 in the Total FCF line represents the initial equity, which is an outflow at T = 0. It is therefore accounted for in the Total FCF line.

The pro forma has now been summarized on the Summary_Inputs page. What remains is to value the real estate asset to determine the attractiveness of the project to a potential investor. The valuation methodologies demonstrated are debt service coverage ratio (DSCR), Cash-on-Cash, Bifurcation of cash flows (Operating versus Reversion), Net Present Value (NPV), and Internal Rate of Return (IRR).

Debt service coverage ratio will be quantified first in row 55 (see Figure 6.21). DSCR is a metric most often used by lenders to determine the 'coverage' of total payment by the project's net operating income. Although the required value is lender specific and risk specific, typically values of 1.25–1.50X are acceptable.

Cell A55 is titled 'DSCR'. Cell E55 is "=E42/E47" and has a value in this example of 3.79. The value is high enough that, in reality, the project should cause the analyst to pause. Cell E55 is copied through N55. Cell D56 is 'Minimum' and cell E56 is "=MIN(E55:N55)".

Cash-on-Cash is a similar metric, only it is focused on the equity investor rather than the lender. The logic is similar. Cash-on-Cash is Operating FCF (cash flows available for distribution) divided by equity invested. Cell A58 is 'Cash-on-Cash' and cell E58 is "=E51/$D$2". As prior, '$D$2' is the initial equity invested in the project. Cell E58 is then copied to column N. Following the same structure as DSCR, cell D59 is 'Minimum' and cell E59 is "=MIN(E58:N58)".

The bifurcation of operating and reversion, net present value, and internal rate of return are the coda for the valuation section (Figure 6.22). Cell A61 is 'PV Operating' and A62 is 'PV Reversion'. Cells B63:B66 are then Total, CF_0, NPV, and IRR, respectively. The zero (0) in CF_0 is subscript in Excel. This is done by selecting just the zero (0) in the formula bar and selecting the Home tab on the Ribbon, 'Font', and then subscript. Though it is not necessary, it adds to the formatting.

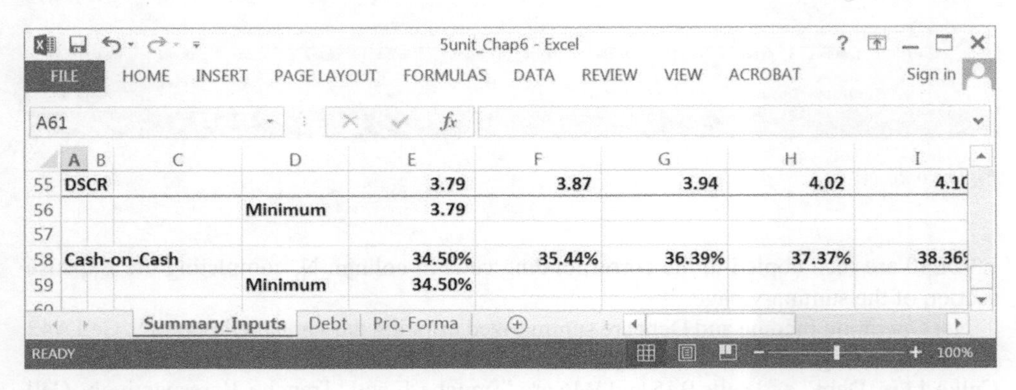

	A B	C	D	E	F	G	H	I
55	DSCR			3.79	3.87	3.94	4.02	4.10
56			Minimum	3.79				
57								
58	Cash-on-Cash			34.50%	35.44%	36.39%	37.37%	38.36
59			Minimum	34.50%				

Figure 6.21

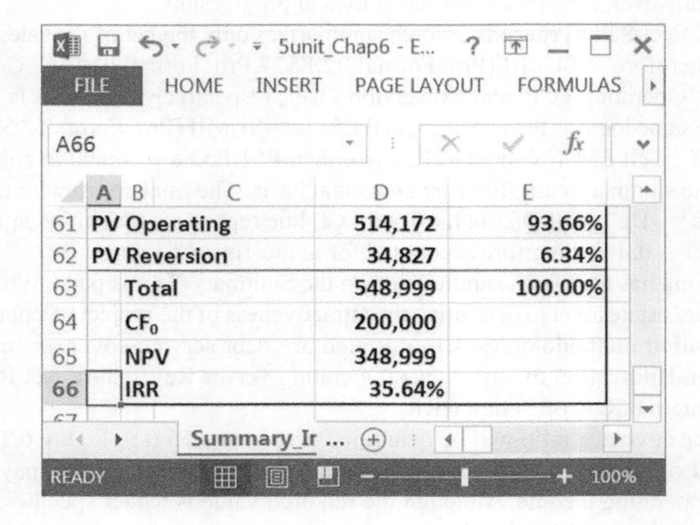

	A B C	D	E
61	PV Operating	514,172	93.66%
62	PV Reversion	34,827	6.34%
63	Total	548,999	100.00%
64	CF_0	200,000	
65	NPV	348,999	
66	IRR	35.64%	

Figure 6.22

Cell D61, PV Operating, is the present value of the operating cash flows. It is "=NPV(discount_Rate,E51:N51)". The 'NPV' (net present value) function was used in Excel despite quantifying a present value. The NPV formula in Excel is incorrectly labeled; it is actually a present value equation. The discount rate was referenced in the Inputs section and is set by the analyst, presumably, for an equivalent risk-based project.

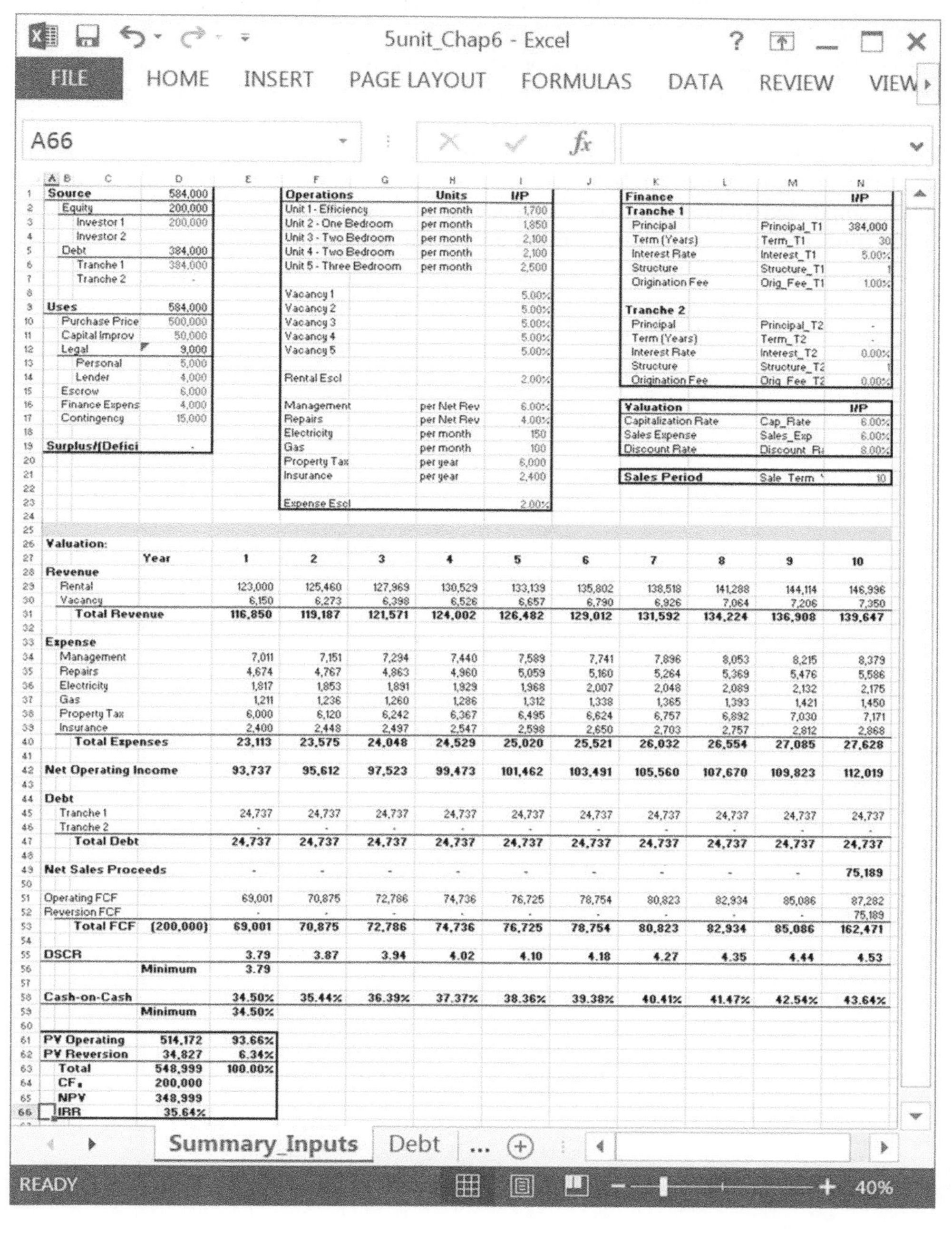

Source	584,000
Equity	200,000
Investor 1	200,000
Investor 2	
Debt	384,000
Tranche 1	384,000
Tranche 2	-
Uses	584,000
Purchase Price	500,000
Capital Improv	50,000
Legal	9,000
Personal	5,000
Lender	4,000
Escrow	6,000
Finance Expens	4,000
Contingency	15,000
Surplus/(Defici	-

Operations	Units	I/P
Unit 1 - Efficiency	per month	1,700
Unit 2 - One Bedroom	per month	1,850
Unit 3 - Two Bedroom	per month	2,100
Unit 4 - Two Bedroom	per month	2,100
Unit 5 - Three Bedroom	per month	2,500
Vacancy 1		5.00%
Vacancy 2		5.00%
Vacancy 3		5.00%
Vacancy 4		5.00%
Vacancy 5		5.00%
Rental Escl		2.00%
Management	per Net Rev	6.00%
Repairs	per Net Rev	4.00%
Electricity	per month	150
Gas	per month	100
Property Tax	per year	6,000
Insurance	per year	2,400
Expense Escl		2.00%

Finance		I/P
Tranche 1		
Principal	Principal_T1	384,000
Term (Years)	Term_T1	30
Interest Rate	Interest_T1	5.00%
Structure	Structure_T1	1
Origination Fee	Orig_Fee_T1	1.00%
Tranche 2		
Principal	Principal_T2	-
Term (Years)	Term_T2	-
Interest Rate	Interest_T2	0.00%
Structure	Structure_T2	1
Origination Fee	Orig_Fee_T2	0.00%

Valuation		I/P
Capitalization Rate	Cap_Rate	6.00%
Sales Expense	Sales_Exp	6.00%
Discount Rate	Discount_R	8.00%

Sales Period	Sale_Term	10

Valuation:

Year		1	2	3	4	5	6	7	8	9	10
Revenue											
Rental		123,000	125,460	127,969	130,529	133,139	135,802	138,518	141,288	144,114	146,996
Vacancy		6,150	6,273	6,398	6,526	6,657	6,790	6,926	7,064	7,206	7,350
Total Revenue		**116,850**	**119,187**	**121,571**	**124,002**	**126,482**	**129,012**	**131,592**	**134,224**	**136,908**	**139,647**
Expense											
Management		7,011	7,151	7,294	7,440	7,589	7,741	7,896	8,053	8,215	8,379
Repairs		4,674	4,767	4,863	4,960	5,059	5,160	5,264	5,369	5,476	5,586
Electricity		1,817	1,853	1,891	1,929	1,968	2,007	2,048	2,089	2,132	2,175
Gas		1,211	1,236	1,260	1,286	1,312	1,338	1,365	1,393	1,421	1,450
Property Tax		6,000	6,120	6,242	6,367	6,495	6,624	6,757	6,892	7,030	7,171
Insurance		2,400	2,448	2,497	2,547	2,598	2,650	2,703	2,757	2,812	2,868
Total Expenses		**23,113**	**23,575**	**24,048**	**24,529**	**25,020**	**25,521**	**26,032**	**26,554**	**27,085**	**27,628**
Net Operating Income		**93,737**	**95,612**	**97,523**	**99,473**	**101,462**	**103,491**	**105,560**	**107,670**	**109,823**	**112,019**
Debt											
Tranche 1		24,737	24,737	24,737	24,737	24,737	24,737	24,737	24,737	24,737	24,737
Tranche 2		-	-	-	-	-	-	-	-	-	-
Total Debt		**24,737**	**24,737**	**24,737**	**24,737**	**24,737**	**24,737**	**24,737**	**24,737**	**24,737**	**24,737**
Net Sales Proceeds		-	-	-	-	-	-	-	-	-	**75,189**
Operating FCF		69,001	70,875	72,786	74,736	76,725	78,754	80,823	82,934	85,086	87,282
Reversion FCF		-	-	-	-	-	-	-	-	-	75,189
Total FCF	**(200,000)**	**69,001**	**70,875**	**72,786**	**74,736**	**76,725**	**78,754**	**80,823**	**82,934**	**85,086**	**162,471**
DSCR		**3.79**	**3.87**	**3.94**	**4.02**	**4.10**	**4.18**	**4.27**	**4.35**	**4.44**	**4.53**
	Minimum	**3.79**									
Cash-on-Cash		**34.50%**	**35.44%**	**36.39%**	**37.37%**	**38.36%**	**39.38%**	**40.41%**	**41.47%**	**42.54%**	**43.64%**
	Minimum	**34.50%**									

PV Operating	**514,172**	**93.66%**
PV Reversion	**34,827**	**6.34%**
Total	**548,999**	**100.00%**
CF_0	**200,000**	
NPV	**348,999**	
IRR	**35.64%**	

Figure 6.23

Cell D62, PV Reversion, is the present value of revision cash flow which, essentially, is the final net sales proceeds value. The formula is "=NPV(discount_rate,E52:N52)". Cell D63, Total, is the summation of the two present values, that is "=SUM(D61:D62)". Cells E61:E62 are the percentage of value attributable to Operating and Reversion. Cell E61 is therefore "=D61/D63" and cell E62 is "=D62/D63". Cell E63 is the summation of the two cells atop it and must always be 100.00% or there is an issue with a calculation.

Cell D64 is the initial cash flow and is therefore the initial equity, "=D2". Row 65, Net Present Value, by definition is the total present value less the initial equity. Cell D65 is thus "=D63-D64". The final valuation value, Internal Rate of Return, is cell D66. The formula is "=IRR(D53:N53)". The project, as modelled, has an IRR of 35.64%, which is abnormally high but not insane. A complete review of the assumptions should be done prior to undertaking this project given the high IRR.

The completed Summary_Inputs page appears in Figure 6.23.

The model can now be utilized to make investment decisions.

7 "N"-unit multifamily

As demonstrated in the previous chapter, the general structure for a pro forma is three sheets: Summary Inputs, Amortization (Debt), and Pro Forma. The limiting factor of the previous model is that, as structured, it is not easily scalable. For instance, the previous chapter modelled a 5-unit multifamily. However, if the project was to be expanded to "N" units, the Inputs section would require significant adjustments to formatting to accommodate "N" units. As such, the pro forma is separated into a functional unit – i.e. Rent Roll – to accommodate "N" units and increased complexity when modelling unit cash flows (e.g. rent, vacancy, tenant allowances, collection allowances, etc.) (see Figure 7.1). The Inputs section, specifically the top half of the Operations section, is modified to be a summary of unit characteristics rather than inputs. So a separate sheet is added to accommodate the Rent Roll for "N" units. This sheet is aptly titled "Unit_Characteristics" but could also be titled "Rent_Roll" depending upon analyst preference.

Once the Unit_Characteristics page has been constructed and integrated within the pro forma, the valuation section will be moved from the Summary_Inputs page to the Pro_Forma page and modified to discount/value on a monthly basis rather than an annual basis, as was done on the summary page. Advanced valuation functions, XNPV and XIRR, will be utilized in models going forward.

Returning to the Summary_Inputs page, an adjustment must be made to the Inputs to accommodate the new "Unit_Characteristics" sheet. Select cells K21:N21, place the cursor on the bottom horizontal bar, and click the left mouse button. While keeping the left mouse button depressed, pull down the cells one row. This will place 'Sales Period' in cells K22:N22. It is important to 'drag' the range down one row rather than inputting a row or inserting cells. Either of these processes will harm the formatting and possibly the calculations in the existing pro forma.

In cell K21 place 'Comm Operating Date' to indicate the first month of operations. For consistency, place COD, the name of cell N21, in cell M21. Cell N21 then becomes the input cell for the first date of operations. In this case type "1 January 2013" and colour blue because it is an input. Name cell N21 'COD'.

Returning to the Unit_Characteristics page, in cells F1:F3 place the schedule headers: Periods, Months, and Year, respectively (Figure 7.2). In cell G1 place the value Question: "1", and cell H1 is "=G1+1". As cell G1 is an input, colour it blue. Cell G2 is first month of operation. This input was newly created on the Inputs sheet as COD. The formula for G2 is "=EOMONTH(COD,0)". The formula is 'EO' for end of month and is a conservative estimate, recognizing that rents are typically collected at the beginning of the month. The zero (0), the second argument, indicates that there is no adjustment to the date COD. Cell H2 is then "=EOMONTH(G2,1)". The second argument has a one (1) because it adds a month to

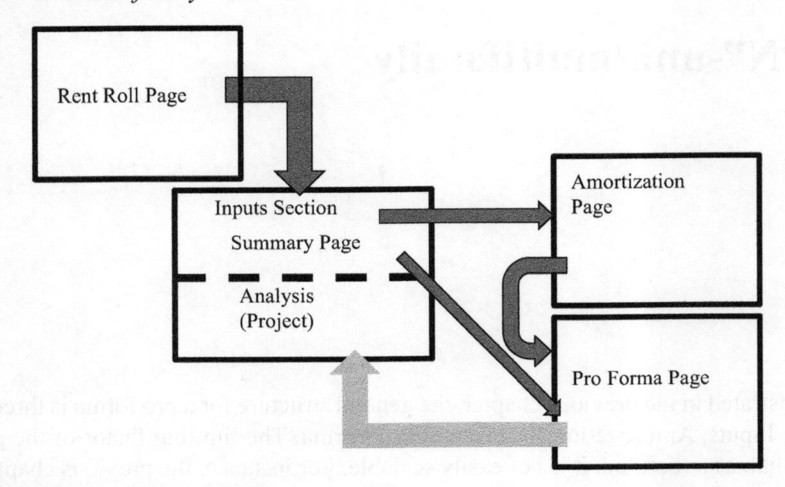

Figure 7.1 Rent Roll Added

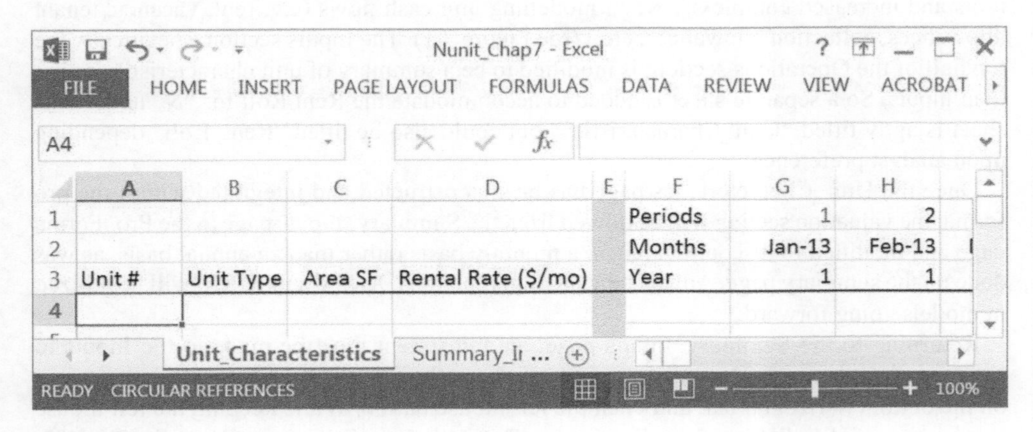

Figure 7.2

the previous date. Row 3, Year, uses the ROUNDUP function as discussed in the previous chapter. Cell G3 is therefore "=ROUNDUP(G1/12,0)". Cell G3 is then copied to cell H3. Cells H1:H3 are copied to EH1:EH3 (period 132).

Cells A3:D3 are column headers and are listed as Unit #, Unit Type, Area SF, and Rental Rate ($/mo), respectively. Depending upon the project and the analyst, these suggested column headings can be modified according to project and requirements. Column E is coloured grey to differentiate inputs and calculations.

For this demonstration the "N" unit is actually going to be thirty-nine (39) but the process will be consistent with any "N"-unit asset (Figure 7.3). As such, cells A4:A42 will be unit numbers 1–39. Cell A4 is "1" and cell A5 is "=A4+1", which is then copied down to each cell through row 42.

Unit #	Unit Type	Area SF	Rental Rate ($/mo)
1	Efficiency	800	1200
2	Efficiency	825	1250
3	Efficiency	850	1275
4	Efficiency	875	1300
5	Efficiency	800	1200
6	Efficiency	825	1250
7	Efficiency	850	1275
8	Efficiency	875	1300
9	Efficiency	800	1200
10	Efficiency	825	1250
11	Efficiency	850	1275
12	Efficiency	875	1300
13	One Bed	950	1500
14	One Bed	975	1550
15	One Bed	1000	1600
16	One Bed	950	1500
17	One Bed	975	1550
18	One Bed	1000	1600
19	One Bed	950	1500
20	One Bed	975	1550
21	One Bed	1000	1600
22	One Bed	950	1500
23	One Bed	975	1550
24	One Bed	1000	1600
25	One Bed + Den	1200	1750
26	One Bed + Den	1250	1800
27	One Bed + Den	1300	1850
28	One Bed + Den	1200	1750
29	One Bed + Den	1250	1800
30	One Bed + Den	1300	1850
31	Two Bed	1400	2200
32	Two Bed	1450	2300
33	Two Bed	1500	2400
34	Two Bed	1400	2200
35	Two Bed	1450	2300
36	Two Bed	1500	2400
37	Two Bed	1400	2200
38	Two Bed	1450	2300
39	Two Bed	1500	2400

Figure 7.3

There will be four Unit Types with quantities to follow: Efficiency (12), One Bedroom (12), One Bedroom + Den (6), and Two Bedrooms (9). The square footage (SF) will be unit specific but will range from 950 to 1500. The inputs for the Rent Roll by Unit are input into the table in B4:D42.

The Area SF and Rental Rate both should be coloured blue because they are inputs. The unit number and square footage will be provided by the designer/architect firm. The Rental Rate will be provided in a market study or supported by market comparables.

This section can be greatly increased to include individual lease characteristics for a building. Start/Stop dates for a lease, probability of renewal, tenant concessions, and so forth can be added. Note, however, if updating the Rent Roll to include increased characteristics, consistency of schedule among sheets may be sacrificed or it may need to be modified on additional pages, for example the Pro_Forma page.

Further, while not modelled in this pro forma, lease-up considerations must be considered at the beginning of any project. How this is modelled is individual to different investors and companies; however, it is a very important consideration for the first 2 years of any project. After 2 years the project is generally considered stabilized.

In column F starting in row 4, link the Unit numbers from the corresponding row in column A (Figure 7.4). This is just formatting for the Rent Roll page.

Beginning in cell G4, the monthly rents for units are calculated. The methodology is consistent and follows the future value formula. The calculation for cell G4 is therefore "=$D4*(1+Rent_Escl)^(G$3–1)". Note that the average square foot, column C, was not utilized in the calculations. Some projects have rental rates quoted as price per square foot. In this case column C, Area SF, is necessary in the calculation.

As cell G4 utilized relative and absolute references appropriately, it can be copied all the way down to the 39th unit (row 42) and across to the 132nd period (year 11). The Rent Roll by unit has been completed.

Row 43 separates logical sections and is therefore coloured grey (see Figure 7.5). A summary table for the unit characteristics is completed in A44:D48 (Figure 7.6). The column headers for row 44 are Unit #s, Unit Type, Avg SF, and Avg Monthly Rent. As a start, beginning in cell B45, list the Unit Types, being *very* careful to not misspell or create any error in the classification. The Unit Types in our model are Efficiency, One Bed, One Bed + Den, and Two Bed. It is imperative that these not have any misspellings or mistakes, as they are the functional references for the rest of the table. These are also to be coloured blue as inputs to the section.

	A	B	C	D	E	F	G	H	I	J
1						Periods	1	2	3	4
2						Months	Jan-13	Feb-13	Mar-13	Apr-13
3	Unit #	Unit Type	Area SF	Rental Rate ($/mo)		Year	1	1	1	1
4	1	Efficiency	800	1200		1	1,200	1,200	1,200	1,200
5	2	Efficiency	825	1250		2	1,250	1,250	1,250	1,250
6	3	Efficiency	850	1275		3	1,275	1,275	1,275	1,275
7	4	Efficiency	875	1300		4	1,300	1,300	1,300	1,300

Figure 7.4

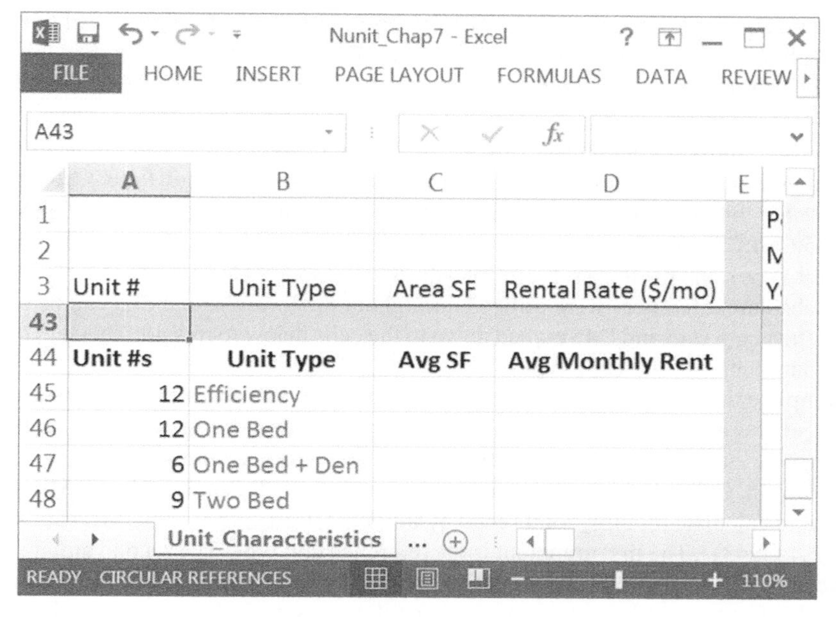

Unit #	Unit Type	Area SF	Rental Rate ($/mo)
Unit #s	**Unit Type**	**Avg SF**	**Avg Monthly Rent**
12	Efficiency		
12	One Bed		
6	One Bed + Den		
9	Two Bed		

Figure 7.5

Unit #	Unit Type	Area SF	Rental Rate ($/mo)
Unit #s	**Unit Type**	**Avg SF**	**Avg Monthly Rent**
12	Efficiency	837.50	1,256.25
12	One Bed	975.00	1,550.00
6	One Bed + Den	1,250.00	1,800.00
9	Two Bed	1,450.00	2,300.00

Figure 7.6

To quantify the unit numbers, the COUNTIF function is utilized. Cell A45 is "=COUNTIF($B$4:$B$43,B45)". There are two arguments to the COUNTIF function. The first argument is the target row/column. The second is the reference. In this function, cells in the array B4:B43 are counted if they agree with (are equal to) B45 – i.e. Efficiency. While the unit numbers end with row 42, the COUNTIF function references to row 43. This is to enable scalability and ensure the formulas adjust to increases/decreases in the Unit counts. Given that absolute and relative references were utilized in cell A45, the cell is copied to cells A46:A48.

Columns C and D beginning in row 44 are average summaries. Therefore, the AVERAGEIF function is utilized to calculate the average for a specific Unit Type. Cell C45 is the average value for square feet for Efficiency units. The formula is "=AVERAGEIF($B$4:$B$43,$B45,C$4:C$45)". The first argument is the reference column, the second is the reference (Unit Type), and the third is the array with square footages to be averaged. Because relative and absolute references were utilized throughout the formula, cell C45 can be copied to D45 and then both C45 and D45 copied down to the cells below to row 48. This will complete the summary table for the Unit Types.

To complete the Unit_Characteristics page, a summary by month by unit is completed to the right of the summary table. Cells F45:F48 are the Unit Types. Cell F45 is therefore "=B45" (see Figure 7.7). Cell F45 is then copied down to the following cells in column F through row 48. Cell G45 is the sum of monthly rentals for Efficiencies. The SUMIF function is utilized to calculate monthly total rental by unit type. Cell G45 is "=SUMIF($B$4:$B$43,$B45,G$4:G$43)". The first argument is the reference unit type. The second argument is the referenced unit, and the third argument the individual unit rents, which are summarized. Because this formula used absolute and relative references, it can be copied to cells G46:G48. Then cells G45:G48 are copied to column EH. At this point the Unit_Characteristics page is complete. What remains is the incorporation and modification to the existing model to reference the Unit_Characteristics.

Returning to the Pro_Forma page, adjustments to accommodate the "N" units and the additional construction are required (Figure 7.8). To start the schedule heading, rows 1 and 2 must be adjusted to incorporate the months as on the Unit_Characteristics page. Note also that the Pro_Forma page is a starting point for the Periods and Years; this must be made consistent with the flow of the model – i.e. the Unit_Characteristics page must be the starting point for the schedule.

On the Pro_Forma page, select cell A2 and insert a row. Label cell F2 'Months' as consistent with the Unit_Characteristics page. Now select cell G1 and link this to the

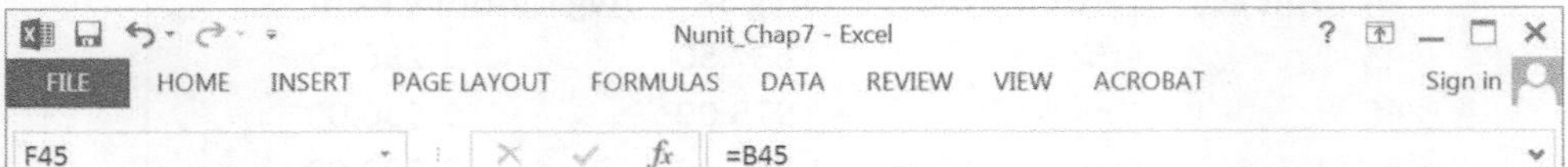

Nunit_Chap7 - Excel

FILE HOME INSERT PAGE LAYOUT FORMULAS DATA REVIEW VIEW ACROBAT Sign in

F45 fx =B45

	A	B	C	D	E	F	G	H	
1						Periods	1	2	
2						Months	Jan-13	Feb-13	M
3	Unit #	Unit Type	Area SF	Rental Rate ($/mo)		Year	1	1	
45	12	Efficiency	837.50	1,256.25		Efficiency	15,075	15,075	15
46	12	One Bed	975.00	1,550.00		One Bed	18,600	18,600	18
47	6	One Bed + Den	1,250.00	1,800.00		One Bed + Den	10,800	10,800	10
48	9	Two Bed	1,450.00	2,300.00		Two Bed	20,700	20,700	20

Unit_Characteristics | Summary_Inputs | Debt ...

READY CIRCULAR REFERENCES 110%

Figure 7.7

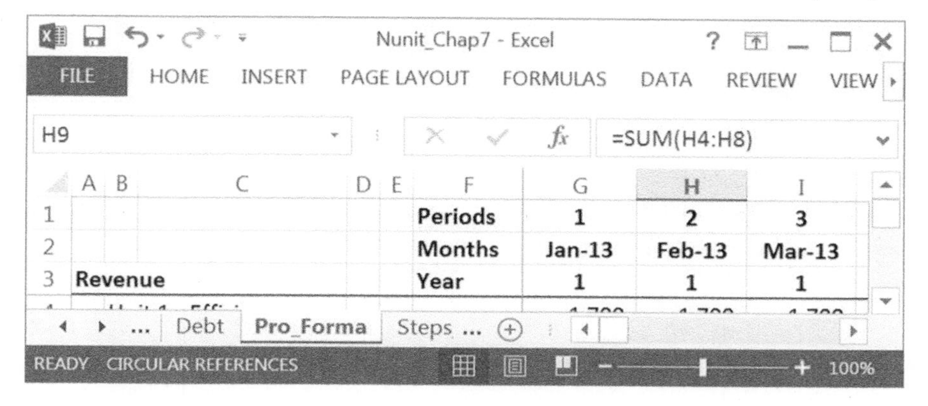

Figure 7.8

Nunit_Chap7 - Excel

FILE HOME INSERT PAGE LAYOUT FORMULAS DATA REVIEW VIE

A1

	A B	C	D E	F	G	H	I
1				**Periods**	**1**	**2**	**3**
2				**Months**	**Jan-13**	**Feb-13**	**Mar-13**
3	**Revenue**			**Year**	**1**	**1**	**1**
4		Efficiency			15,075	15,075	15,075
5		One Bed			18,600	18,600	18,600
6		One Bed + Den			10,800	10,800	10,800
7		Two Bed			20,700	20,700	20,700
8		**Total Rent**			**65,175**	**65,175**	**65,175**
9							
10		Vacancy 1			754	754	754
11		Vacancy 2			930	930	930
12		Vacancy 3			540	540	540
13		Vacancy 4			1,035	1,035	1,035
14		**Total Vacancy**			**3,259**	**3,259**	**3,259**

... Summary_Inputs | Debt ...

READY CIRCULAR REFERENCES 100%

Figure 7.9

Unit_Characteristics page (cell G1 becomes "=Unit_Characteristics!G1"). Then copy cell G1 on the Pro_Forma page to cell G3. Select cells G1:G3 and copy to column EH, thus linking the schedule to the Unit_Characteristics page.

The next section for adjustment is the Total Rent section on the pro_forma sheet (Figure 7.9). The Units have been adjusted from five types to four types in the model. As the unit

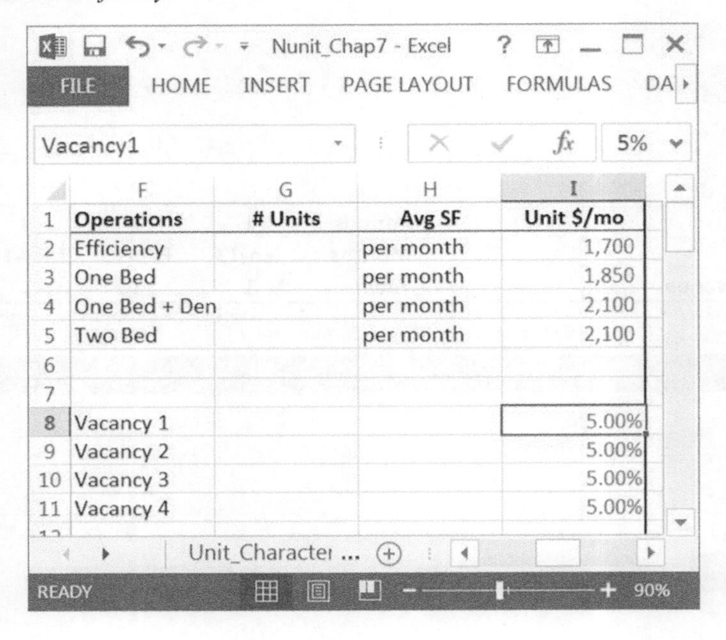

Figure 7.10

types are begun on the Unit_Characteristics page, cells B4:B7 must be linked. Therefore cell B4 is "=Unit_Characteristics!B45". Copy cell B4 to cell B7.

The only adjustments remaining on the Pro_Forma page are the actual rents for the units. Select cell G4 and link this to the Unit_Characteristics page. The formula for G4 is "=Unit_Characteristics!G45". Copy cell G4 to cells G5:G7 and then copy cells G4:G7 to column EH. This completes the modifications necessary to the Pro_Forma page for the N-Unit adjustment.

Returning to the 'Summary_Inputs' page, the Operations section in the Inputs must be adjusted as a summary rather than an input (Figure 7.10). Select cell F2 and link this to the Unit_Characteristics page as "=Unit_Characteristics!B45". Copy cell F2 and paste to F3:F5. Select cells F6:I6 and delete the contents (Unit 5 – Three Bedroom). Select cells F12 through I12 and delete the contents (Vacancy5). Next adjust the contents of cells G1 through I1 as follows: # Units, Avg SF, and Unit $/mo.

Select cell G2 and link to the Unit_Characteristics page, making this "=Unit_Characteristics!A45". Link cell H2 to the Unit_Characteristics page as well: "=Unit_Characteristics!C45". As Avg SF and Unit $/mo are contiguous on the Unit_Characteristics page, cell H2 can be copied to I2 (i.e. Unit $/mo). Then cells H2:I2 are copied down to the three rows below.

In cell G6 summarize the total units for the project, i.e. cell G6 is "=sum(G2:G5)". Cell H6 is the weighted average of the Units. Therefore the SUMPRODUCT function is utilized to quantify the weighted sum of square feet and then divide by the total number of units. Cell H6 is thus "=SUMPRODUCT($G$2:$G$5,H2:H5)/$G$6". Because absolute and relative references were used, cell H6 can be copied to I6 directly.

A further adjustment is required to base the model in reality. The expenses have not been adjusted to reflect the 39 units rather than the 5 units. As a placeholder, adjust cells I18:I21 on the Inputs section as follows:

Electricity/mo: $2,000
Gas/mo: $2,000
Property Tax/yr: $80,000
Insurance/yr: $30,000

Uses, Sources, and Debt must also be adjusted to accommodate the larger model. As placeholders, the following are recommended (see Figure 7.11):

Purchase Price: $4,000,000
Capital Improvement: $500,000
Legal (Personal): $25,000
Legal (Lender): $20,000
Escrow: $60,000
Finance Expense: $40,000
Contingency: $150,000

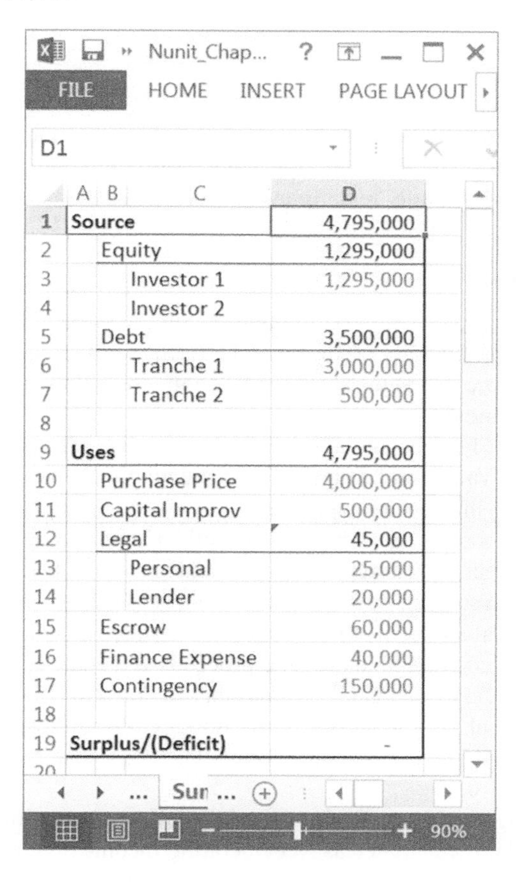

	A B C	D
1	Source	4,795,000
2	Equity	1,295,000
3	Investor 1	1,295,000
4	Investor 2	
5	Debt	3,500,000
6	Tranche 1	3,000,000
7	Tranche 2	500,000
8		
9	Uses	4,795,000
10	Purchase Price	4,000,000
11	Capital Improv	500,000
12	Legal	45,000
13	Personal	25,000
14	Lender	20,000
15	Escrow	60,000
16	Finance Expense	40,000
17	Contingency	150,000
18		
19	Surplus/(Deficit)	-

Figure 7.11

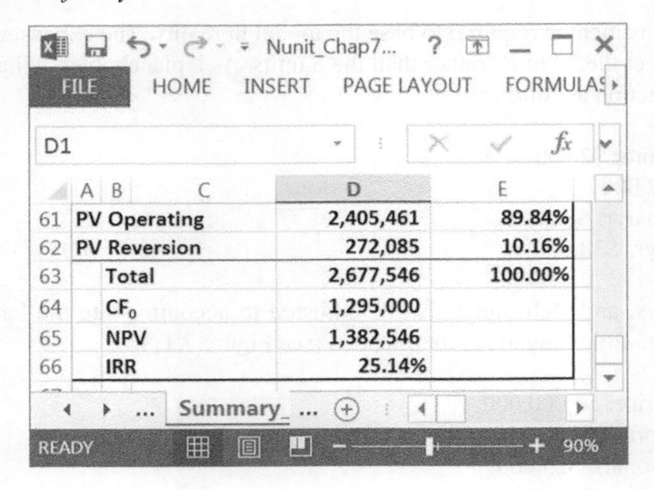

Figure 7.12

Debt Tranche 1: $3,000,000
Debt Tranche 2: $500,000

Investor 1: $1,295,000

The project returns an IRR of 25.14% with 90% value resulting from Operations (Figure 7.12). Overall, at this point, this is a respectable return for a real estate project. However, this information remains woefully incomplete to make an investment decision.

Adjusted valuation

As noted prior, the valuation section on the Inputs_Summary page used annual valuations. Annual valuations do not accurately reflect a real estate project as cash flows are received on a monthly basis. Therefore, the valuation must be completed on the Pro_Forma page, which provides cash flows by month.

In cell A45 on the Pro_Forma page, just below the grey line separating the pro forma from this new section, type "Valuation:" (Figure 7.13). In cell G45 link to the months in row 2: i.e. "=G2". Cell G45 can then be copied to the rest of row 45 through column EH. This series of dates only covers the operations of the project. Cell F45 must include the date of settlement – i.e. CF_0 – for the project. It is assumed that settlement occurs the month prior to the first month of cash flow. Therefore cell F45 is "=EOMONTH(G45,-1)".

Cell A46, beneath 'Valuation:' is then 'Total FCF', which is linked directly to the Total FCF in row 43. This includes the final net sales proceeds as it is the combination of Operating and Reversion FCF. Cell G46 is "=G43", which can be copied to H43:EH43. As was the case with the date for financial settlement (CF_0), the initial cash outflow, equity, must be represented in cell F46. Cell F46 is "=-Summary_Inputs!D2". The negative indicates that the cash flow is an outflow at T = 0.

Debt service coverage ratio and Cash-on-Cash returns are quantified next (Figure 7.14). Row 48 is 'DSCR' and it is calculated as Net Operating Income divided by Operational Cash Flow beginning in cell G48. Cell G48 is "=G27/G32". Cell G48 is then copied to the rest of

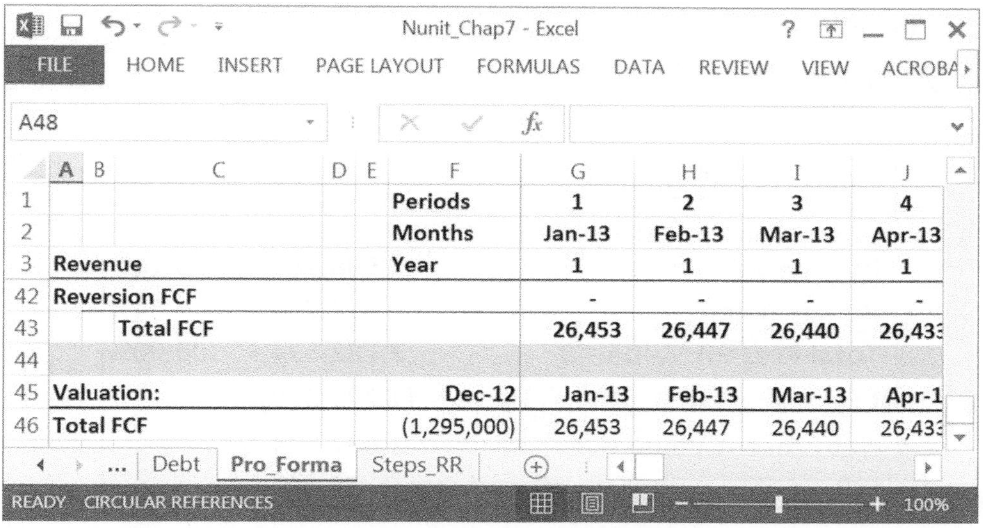

Nunit_Chap7 - Excel

FILE HOME INSERT PAGE LAYOUT FORMULAS DATA REVIEW VIEW ACROBA

A48 | fx

	A	B	C	D	E	F	G	H	I	J
1						Periods	1	2	3	4
2						Months	Jan-13	Feb-13	Mar-13	Apr-13
3	Revenue					Year	1	1	1	1
42	Reversion FCF						-	-	-	-
43			Total FCF				26,453	26,447	26,440	26,43:
44										
45	Valuation:					Dec-12	Jan-13	Feb-13	Mar-13	Apr-1
46	Total FCF					(1,295,000)	26,453	26,447	26,440	26,43:

... Debt Pro_Forma Steps_RR

READY CIRCULAR REFERENCES 100%

Figure 7.13

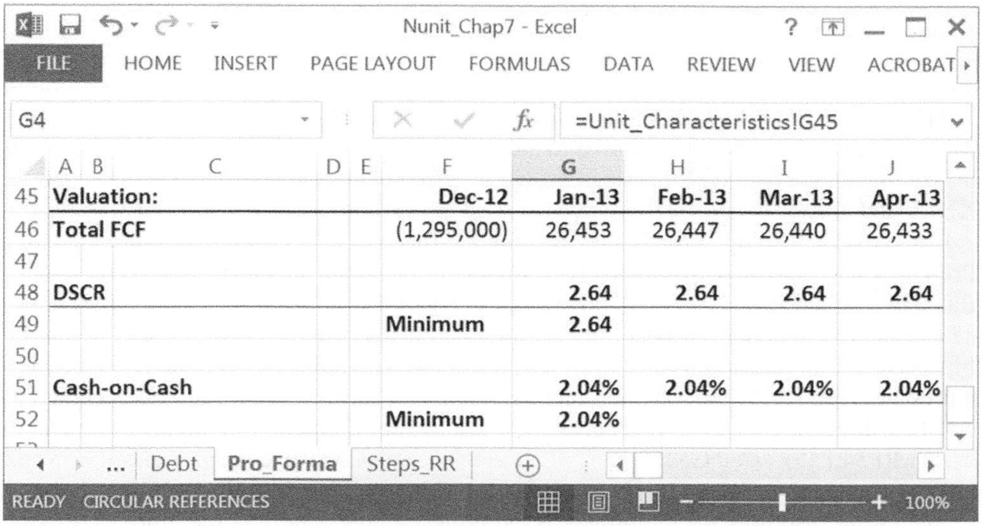

Nunit_Chap7 - Excel

FILE HOME INSERT PAGE LAYOUT FORMULAS DATA REVIEW VIEW ACROBAT

G4 | fx | =Unit_Characteristics!G45

	A	B	C	D	E	F	G	H	I	J
45	Valuation:					Dec-12	Jan-13	Feb-13	Mar-13	Apr-13
46	Total FCF					(1,295,000)	26,453	26,447	26,440	26,433
47										
48	DSCR						2.64	2.64	2.64	2.64
49						Minimum	2.64			
50										
51	Cash-on-Cash						2.04%	2.04%	2.04%	2.04%
52						Minimum	2.04%			

... Debt Pro_Forma Steps_RR

READY CIRCULAR REFERENCES 100%

Figure 7.14

the row through column EH. Cells F49 and G49 are 'Minimum' and "=min(G48:DV48)", respectively.

Cash-on-Cash is quantified similarly. Cell A51 is 'Cash-on-Cash' and cell G51 is "=G41/Summary_Inputs!$D$2". Cell G41 represents Operating Cash Flow and 'Summary_Inputs!$D$2' the total equity invested at the project level. Cell G41 also represents monthly Cash-on-Cash. Cell G51 is copied down the row to column DV, which is the 120th period. Cell G52 is "=MIN(G51:DV51)".

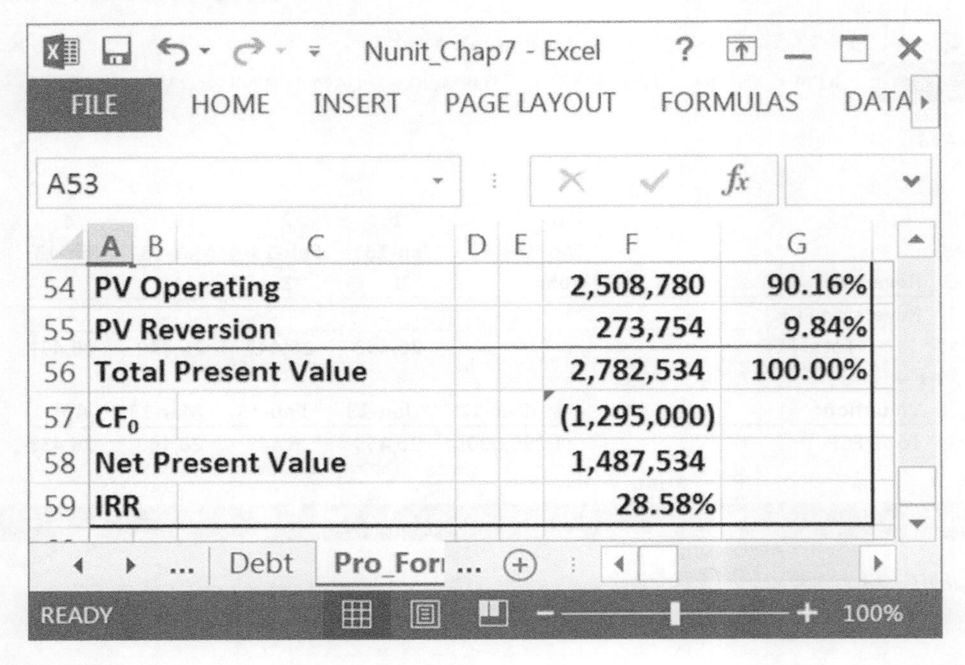

	A B C	D E	F	G
54	PV Operating		2,508,780	90.16%
55	PV Reversion		273,754	9.84%
56	Total Present Value		2,782,534	100.00%
57	CF_0		(1,295,000)	
58	Net Present Value		1,487,534	
59	IRR		28.58%	

Figure 7.15

The final section is the valuation utilizing Present Value, Net Present Value, and Internal Rate of Return. This section is similar to the earlier completion in Chapters 1–3, only it will utilize the advanced Excel functions providing discounting on a daily basis, for example XNPV and XIRR.

Cells A54:A59 are therefore PV Operating, PV Reversion, Total Present Value, CF_0, Net Present Value, and IRR, respectively. PV Operating in cell F54 is "=XNPV(Discount_Rate,G41:DV41,$G$2:$DV$2)". (See Figure 7.15.) There are three arguments for the advanced NPV function: discount rate, values, and dates. The discount rate was set on the Inputs page. The values and dates were selected through period 120 (i.e. column DV), as that represents year 10. As absolute and relative references were utilized, cell F54 is copied to cell F55 directly. Cell F56, Total Present Value, is the summation of cells F54 and F55: "=SUM(F54,F55)". Cell G54 is then "=F54/$F$56". Cell G54 can be copied to cell G55 and G56.

The initial cash flow for the project is the project equity. Cell F57 therefore is "=F46". Note that it does not link back to the summary page but references the equity outflow on the Pro_Forma page. This will be noted in further detail in Chapter 8, which discusses the construction of the Draw page in a Real Estate Model. Net Present Value, cell F58, is the summation of Present Value and CF_0, which in this case is negative. Cell F58 is "=SUM(F57:F58)".

Finally, Internal Rate of Return, cell F59, is "=XIRR(F46:DV46:F45:DV45)". This provides the Internal Rate of Return as of 31 December 2012 (the date stated in cell F45).

Returning to the Summary_Inputs page, delete rows 55:66, everything beneath the annual Total FCF summary. Then, returning to the Inputs section, select row 24 (i.e. the row just above the grey bar) and select two additional rows beneath. Add three rows just above the grey bar to the Inputs section.

Nunit_Chap7 - Excel

FILE HOME INSERT PAGE LAYOUT FORMULAS DATA REVIEW VIEW ACROBAT

F24

Row	A–B	C	D
1	**Source**		4,795,000
2	Equity		1,295,000
3		Investor 1	1,295,000
4		Investor 2	
5	Debt		3,500,000
6		Tranche 1	3,000,000
7		Tranche 2	500,000
8			
9	**Uses**		4,795,000
10	Purchase Price		4,000,000
11	Capital Improv		500,000
12	Legal		45,000
13		Personal	25,000
14		Lender	20,000
15	Escrow		60,000
16	Finance Expense		40,000
17	Contingency		150,000
18			
19	**Surplus/(Deficit)**		-
20			
21	**PV Ops**	**2,508,780**	**90.16%**
22	**PV Rev**	**273,754**	**9.84%**
23		**2,782,534**	**100.00%**
24	**CFO**	**(1,295,000)**	
25	**NPV**	**1,487,534**	
26	**IRR**	**28.58%**	

Row	F	G	H	I
1	**Operations**	**# Units**	**Avg SF**	**Unit $/mo**
2	Efficiency	12	837.50	1,256.25
3	One Bed	12	975.00	1,550.00
4	One Bed + Den	6	1,250.00	1,800.00
5	Two Bed	9	1,450.00	2,300.00
6		39	1,084.62	1,671.15
8	Vacancy 1			5.00%
9	Vacancy 2			5.00%
10	Vacancy 3			5.00%
11	Vacancy 4			5.00%
14	Rental Escl			2.00%
16	Management		per Net Rev	6.00%
17	Repairs		per Net Rev	4.00%
18	Electricity		per month	2,000
19	Gas		per month	2,000
20	Property Tax		per year	80,000
21	Insurance		per year	30,000
23	Expense Escl			2.00%
25	**DSCR Minimum**			**2.64**
26	**Cash-on-Cash (monthly) Minimum**			**2.04%**

Row	K	L	M	N
1	**Finance**			**I/P**
2	**Tranche 1**			
3	Principal		Principal_T1	3,000,000
4	Term (Years)		Term_T1	30
5	Interest Rate		Interest_T1	5.00%
6	Structure		Structure_T1	1
7	Origination Fee		Orig_Fee_T1	1.00%
9	**Tranche 2**			
10	Principal		Principal_T2	500,000
11	Term (Years)		Term_T2	-
12	Interest Rate		Interest_T2	0.00%
13	Structure		Structure_T2	1
14	Origination Fee		Orig_Fee_T2	0.00%
16	**Valuation**			**I/P**
17	Capitalization Rate		Cap_Rate	6.00%
18	Sales Expense		Sales_Exp	6.00%
19	Discount Rate		Discount_Rate	8.00%
21	**Comm Operating Date**		COD	1-Jan-13
22	**Sales Period**		Sale_Term_Yr	10

Summary_Inputs | Debt | Pro_Forma | Steps_RR

READY CIRCULAR REFERENCES

Figure 7.16

The purpose of adding these three rows is to add the valuation summary to the Inputs section. This will make analyzing the effects of input changes more simple because the result is located in the same section.

In cells A21–D26 recreate the Valuation summary at the bottom of the Pro_Forma page (Figure 7.16). This is accomplished by linking each of the cells. Beneath the Operations section in column F, the debt service coverage ratio and Cash-on-Cash (monthly) minimums are linked. The model has now been correctly modified to accommodate an N-Unit structure.

8 Development/construction page

The previous chapter added the Rent Roll (Unit_Characteristics) page to the pro forma, but the model was still a turnkey project with 'Develop' being a single line item in the Uses section of the Sources/Uses table on the Inputs page. This chapter expands the model further by adding a Construction/Draw page (Figure 8.1), which will accommodate a general contractor's schedule for cost and scheduling.

The Construction/Draw page is constructed in five sections. The first is the Header/Schedule section. The second is the individual line item cost percentages by month for the project beginning with Land and ending with Financing costs. The third is the actual costs by month. The fourth is an Inflow/Outflow cash summary of the project during construction. This Inflow/Outflow assumes that equity is injected first into the project and debt is drawn only after all equity has been injected. The final section, labeled Z-Score, is a dynamic normal distribution to provide the user/analyst a simple way to allocate costs if a general contractor provides total line item costs without a schedule.

Header/schedule

The inputs page must be adjusted for the date of construction start. On the Summary_Inputs page in the Inputs section, cells K21:N22 must be shifted down one row. Do not shift the entire row and/or add another row but simply move the section loosely referenced as 'Schedule' down one row by selecting it and dragging it down a row. In the space created, the Purchase/Dev Start input date is placed. The naming convention is 'Dev_Start'.

In cell K21 place 'Purchase/Dev Start' (Figure 8.2). In cell M21 place the naming convention for cell N21, 'Dev_Start'. Finally in cell N21 place the date 1 January 2012. As this is an input, the cell should be coloured blue.

While the Summary_Inputs page has not seen its last adjustment for this Construction/Draw addition, it is complete for the moment. Now a new sheet must be added to the workbook, labeled 'Development' (Figure 8.3). Place the sheet to the left of Unit_Characteristics (i.e. the front of the workbook), because this essentially is the starting point of the analysis.

On the Development page, cells A1:A3 are the schedule summary for Construction/Development and are Development Start, Construction Months, and Opening Date, respectively. Cell B1 is therefore "=Dev_Start". Cell B2, Construction Months, is a calculation of the months between Commercial Operating Date and Development Start. The formula for cell B2 is "=YEARFRAC(Dev_Start,COD)*12". Given that Excel uses numeric values for dates – e.g. 0 = 1/0/1900 and 30,000 = 2/18/1982 – the number of days between Commercial Operating Date and Development Start is calculated. The YEARFRAC function provides

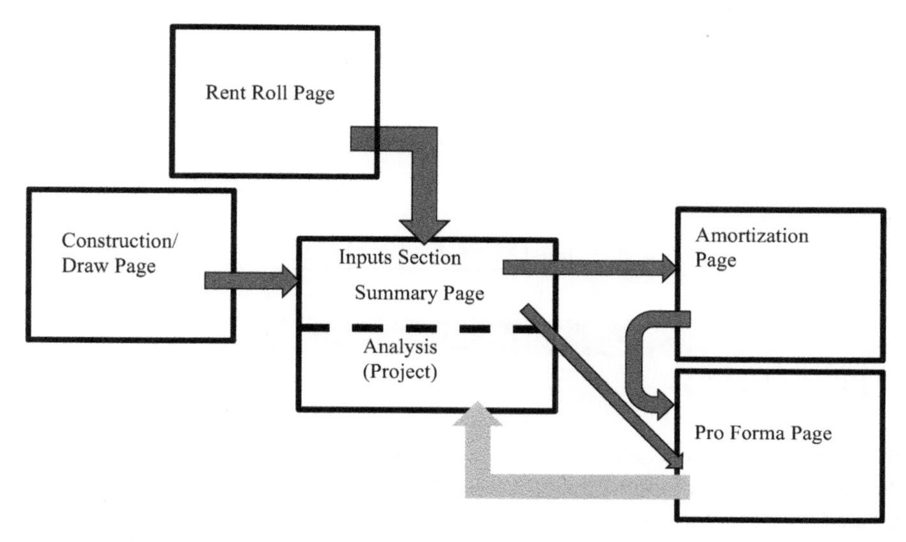

Figure 8.1 Construction Diagram

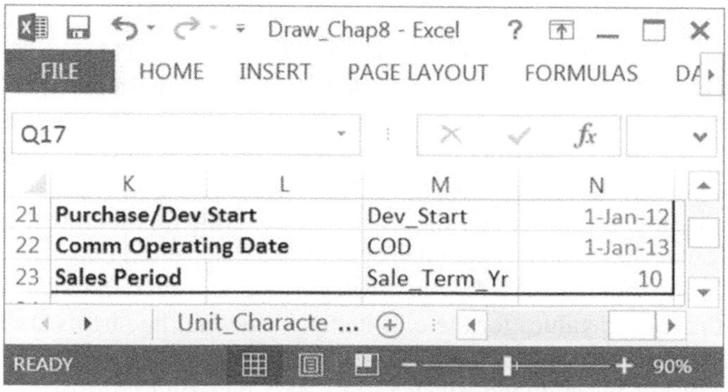

Draw_Chap8 - Excel

FILE HOME INSERT PAGE LAYOUT FORMULAS

Q17

	K	L	M	N
21	**Purchase/Dev Start**		Dev_Start	1-Jan-12
22	**Comm Operating Date**		COD	1-Jan-13
23	**Sales Period**		Sale_Term_Yr	10

Unit_Characte ...

READY 90%

Figure 8.2

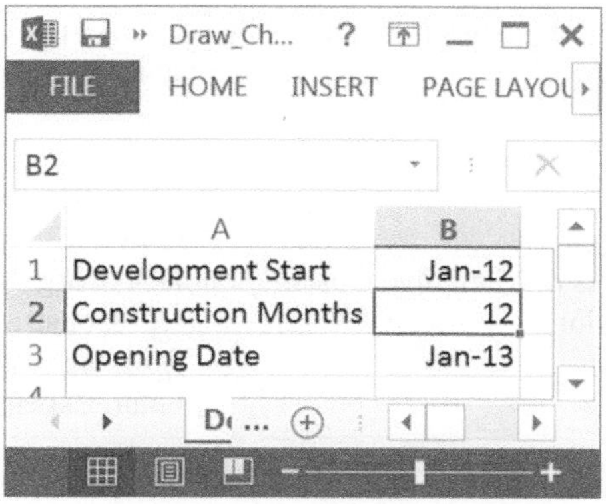

Draw_Ch...

FILE HOME INSERT PAGE LAYOU

B2

	A	B
1	Development Start	Jan-12
2	Construction Months	12
3	Opening Date	Jan-13

Figure 8.3

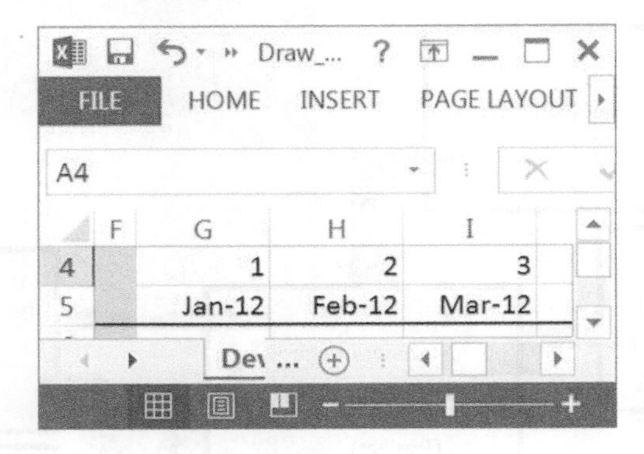

Figure 8.4

the fraction of a year between two dates. This value is then multiplied by 12 to provide a whole number value for months. Finally cell B3 is "=COD", which is the commercial operating date (i.e. lease start-up and delivery).

Colour column F all grey and place a one (1) in cell G4 (Figure 8.4). Cell G4 will be the beginning of the schedule. As prior, cell H4 is "=G4+1". This model will accommodate up to a 36-month construction schedule but could be later adjusted to accommodate longer schedules by expanding the range. As such, copy cell H4 to column AP. In row 5 the dates for each cost will be placed. Cell G5 is "=EOMONTH(Dev_Start,G4–1)". The second argument permits each cell in row 5 for the schedule to have the consistent logic. The minus 1 is to ensure that the first month is the Development Start month from the Inputs page.

The major sections for a Construction/Draw page are Land & Related, General & Administrative, Development Charges & Municipal Costs, Hard Cost Construction, Building Operations, and Financing Expense. All the costs, in general, fall within these six general categories. Of course these can be expanded or contracted depending upon a general contractor's cost item listing, but these are the basic categories and what will be utilized in this text. These are also the roll-up values for the cost items; individual line items are found beneath. As this text is not a text on construction but rather a model of construction, a detailed discussion for each will not be included. Please see a text covering construction for a greater explanation. In this text these are considered cost items provided by a general contractor and will be treated as such.

In cell A5 place 'Cost Item' to indicate the items that will be listed below for the construction costs. Cells A7:A45 are the listed line items by section (Figure 8.5). Column D has the respective cell names for roll-up time costs. These map to the list of the major sections as follows:

Land & Related	Land_Related
General & Administrative	G_A
Development Charges & Municipal Costs	Dev_Chrg_Mun_Costs
Hard Cost Construction	HC_Construction
Building Operations	Building_Operations
Financing Expense	Fnan_Expense

Column E is then the individual item cost (i.e. total cost) and the subtotal for the category. As an example, Land & Related and General & Administrative will be discussed in detail. The remaining are shown but will be left to the reader of the text to complete.

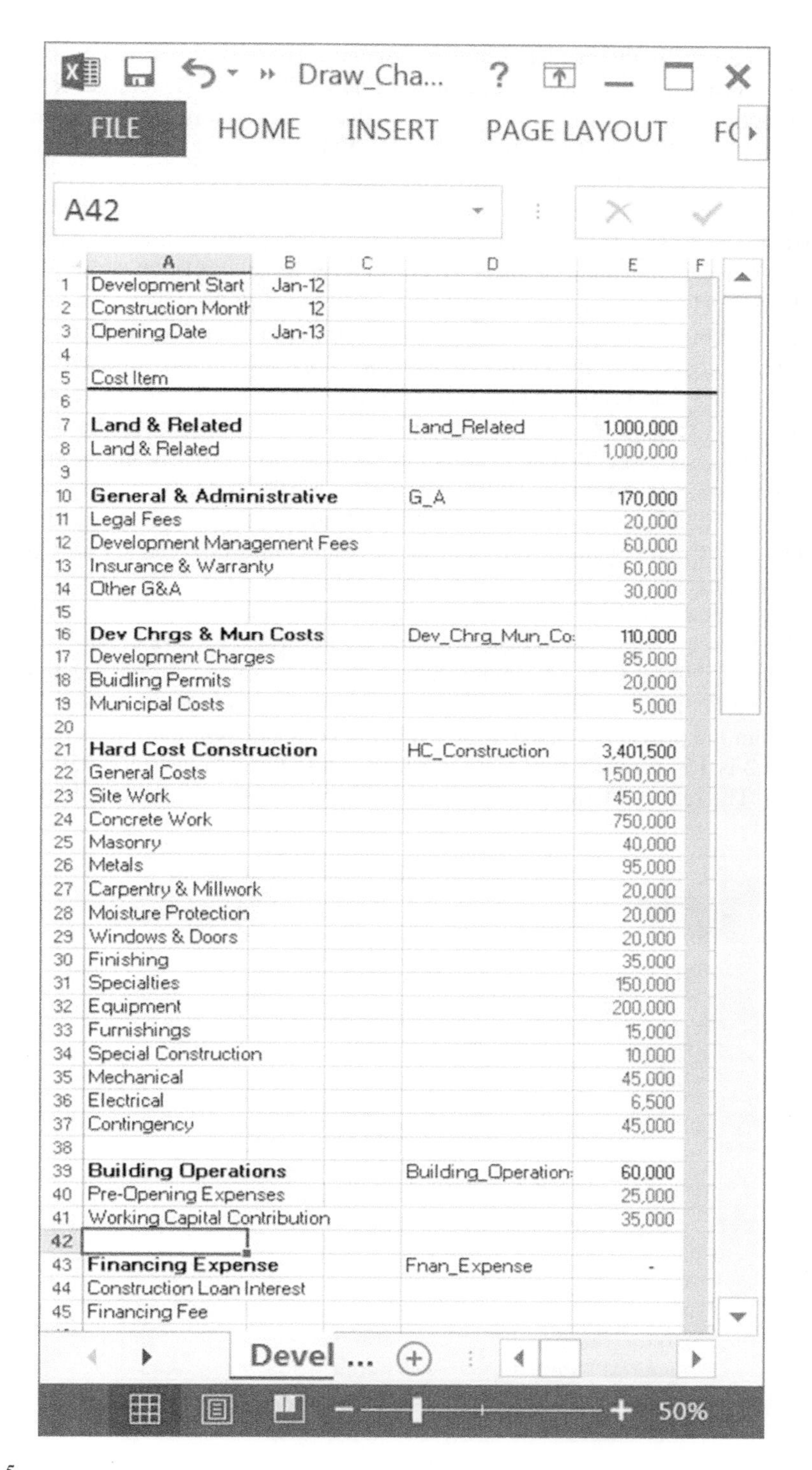

	A	B	C	D	E
1	Development Start	Jan-12			
2	Construction Montl	12			
3	Opening Date	Jan-13			
4					
5	Cost Item				
6					
7	**Land & Related**			Land_Related	1,000,000
8	Land & Related				1,000,000
9					
10	**General & Administrative**			G_A	170,000
11	Legal Fees				20,000
12	Development Management Fees				60,000
13	Insurance & Warranty				60,000
14	Other G&A				30,000
15					
16	**Dev Chrgs & Mun Costs**			Dev_Chrg_Mun_Co:	110,000
17	Development Charges				85,000
18	Buidling Permits				20,000
19	Municipal Costs				5,000
20					
21	**Hard Cost Construction**			HC_Construction	3,401,500
22	General Costs				1,500,000
23	Site Work				450,000
24	Concrete Work				750,000
25	Masonry				40,000
26	Metals				95,000
27	Carpentry & Millwork				20,000
28	Moisture Protection				20,000
29	Windows & Doors				20,000
30	Finishing				35,000
31	Specialties				150,000
32	Equipment				200,000
33	Furnishings				15,000
34	Special Construction				10,000
35	Mechanical				45,000
36	Electrical				6,500
37	Contingency				45,000
38					
39	**Building Operations**			Building_Operation:	60,000
40	Pre-Opening Expenses				25,000
41	Working Capital Contribution				35,000
42					
43	**Financing Expense**			Fnan_Expense	-
44	Construction Loan Interest				
45	Financing Fee				

Figure 8.5

As stated, cell A7 is the header 'Land & Related'. There is only one item beneath this category, so it is placed in cell A8 as equivalent, 'Land & Related'. General & Administrative is the next section and begins with the roll-up in cell A10 as 'General & Administrative'. The four line items making up G&A (listed in cells A11:A14) are Legal Fees, Development Management Fees, Insurance & Warranty, and Other G&A. In cell D7, corresponding to the section header 'Land & Related', the cell name for E7 is placed: "Land_Related". Likewise, in cell D10, the cell name for E10 is placed: "G_A". Next the inputs for the total cost of each line item are placed in the appropriate cells. For Land & Related, $1,000,000 is placed in cell E8. In cells E11:E14 the total cost is listed for Legal Fees ($20,000), Development Management Fees ($60,000), Insurance & Warranty ($60,000), and Other G&A ($30,000). Note that each of these costs will be coloured blue as they are inputs. Finally, cell E7 is "=SUBTOTAL(E8)" and cell E10 is "=SUBTOTAL(E11:E14)". This must be completed for each of the items through row 43 (i.e. Financing Expense). Also, even though Financing Expense has no costs associated at this point in the modelling process, the subtotal function should still be completed for the roll-up in cell E43. Finally, select cells D7:E45; go to the Formulas tab on the Ribbon, then from the Defined Names menu, choose 'Create from Selection'. When the 'Create Names from Selection' menu comes up, select 'Left Column' and 'OK'. This will name the cells for the roll-ups in column E.

Financing Expense, cell E45, is the sum of the financing fee and principal for the respective tranche from the Inputs page. The equation for cell E45 is "=Principal_T1*Orig_Fee_T1+Principal_T2*Orig_Fee_T2". Note that on the Inputs page Tranche 2 has principal applied, $500,000, but there is not a Term, Interest Rate, or Origination Fee. This was an oversight from the previous chapter which is being corrected. On the Inputs page the Term for Tranche 2 is 15 years, Interest Rate is 8.00%, and Origination Fee is 150 basis points (Figure 8.6). The reduced term and increased interest rate reflect that this is junior debt and

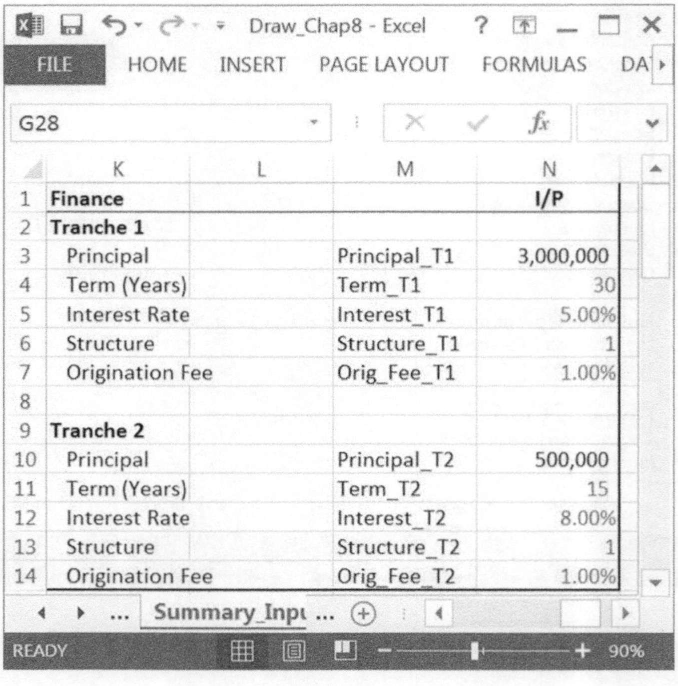

Draw_Chap8 - Excel

FILE HOME INSERT PAGE LAYOUT FORMULAS DA

G28

	K	L	M	N
1	**Finance**			**I/P**
2	**Tranche 1**			
3	Principal		Principal_T1	3,000,000
4	Term (Years)		Term_T1	30
5	Interest Rate		Interest_T1	5.00%
6	Structure		Structure_T1	1
7	Origination Fee		Orig_Fee_T1	1.00%
8				
9	**Tranche 2**			
10	Principal		Principal_T2	500,000
11	Term (Years)		Term_T2	15
12	Interest Rate		Interest_T2	8.00%
13	Structure		Structure_T2	1
14	Origination Fee		Orig_Fee_T2	1.00%

Summary_Inpu ...

READY 90%

Figure 8.6

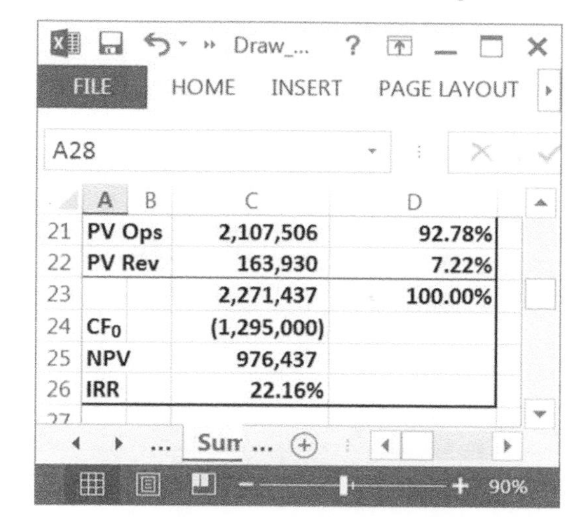

Figure 8.7

are intended to compensate the provider of Tranche 2 for the increased risk. The project IRR therefore reduced to 22.16% at the project level (Figure 8.7).

For now the Construction Loan Interest line item will be left blank (see Figure 8.5). This will be calculated below and input into the model on the Development sheet in cell E44. Finally note that cell E45, Financing Fee, should not be coloured blue; this is a calculation and not a direct input.

The next section to complete is the Draw percentages, i.e. the percentage of cost drawn per page. This directly relates to the construction interest expense and to the pricing by a lender of the construction loan. While it is *highly* recommended that the draw schedule be provided by the general contractor, there are three methodologies which will be demonstrated in this text: (1) Level, (2) Normal, and (3) Custom. 'Level' payments are equal payments through the construction process. In the case of the current model, this is payment of 1/12 each month for the 12 months of construction for a total draw of 100.0% for the line item. The second method is 'Normal'. Note that a cumulative Normal Distribution is the standard S-curve. The final methodology is 'Custom', which is neither, but has determined, specific months of draw. For example, Land & Related is 100.0% drawn in the first month. This represents the takedown of land, that is the purchase, at the start of construction.

Because the construction takedown percentages are based on the total months of construction, it is necessary to name cell B2 'Construction_Months' (Figure 8.8). Upon naming cell B2, development of the generic draw schedules for the first two methodologies are possible.

Colour row 47 grey to indicate a new section. Title this section in cell A48 'Schedule Percentage Draw'. Row 49 will be the percentages for a Level cost structure. Cell A49 is therefore 'Levelized Cost'. Cell G49 is "=IF(Construction_Months>=G4,1/Construction_Months,0)". This section, an input, will be coloured blue. Note that it will be utilized later for individual item draws. Because they will be referenced, the target cells will also be coloured blue.

Cell A51 is labeled 'Z-Score' to indicate this methodology of Draw. Cell G51 is the beginning of the methodology for the z-score. Row 51 beginning in column G is the z-score, corresponding to the normal curve with three standard deviations equaling the number of construction months in the draw. This will provide a construction schedule which has cost normally distributed over the construction period. Cell G51 is therefore "=IF(G4>Construction_Months-1,0,-3+G4*

Draw_Chap8 - Excel

K5 =EOMONTH(Dev_Start,K4-1)

	A	B	C	D	E	F	G	H	I	J
1	Development Start	Jan-12								
2	Construction Months	12								
3	Opening Date	Jan-13								
4							1	2	3	4
5	Cost Item						Jan-12	Feb-12	Mar-12	Apr-12
48	Schedule Percentage Draw									
49	Levelized Cost						8.33%	8.33%	8.33%	8.33%
50										
51	Z-Score						(2.50)	(2.00)	(1.50)	(1.00)
52	S-Curve						0.62%	2.28%	6.68%	15.87%
53	Draw				100.00%		0.62%	1.65%	4.41%	9.18%

Development | Unit_Characteristics | Summary_Inputs

Figure 8.8

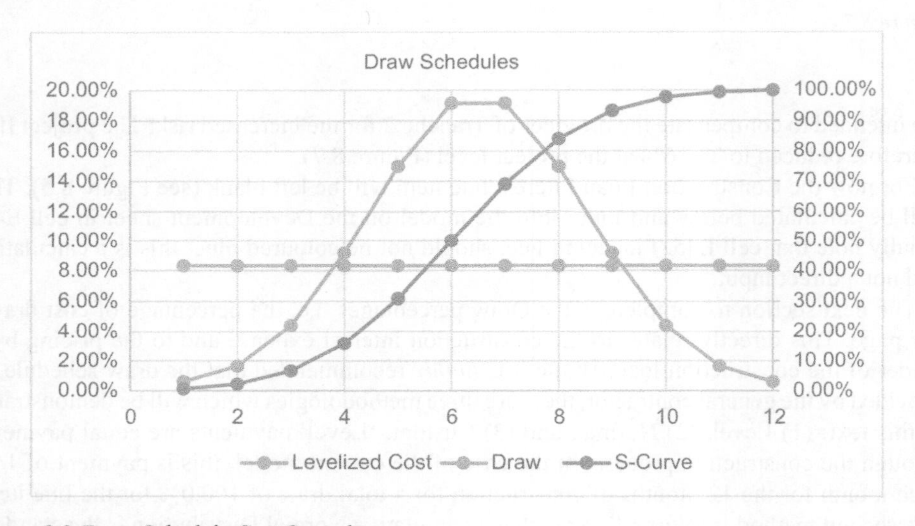

Figure 8.9 Draw Schedule Cost Spread

(6/Construction_Months))". G51 is then copied across the row to column AP, the 36th period. Row 51 provides the standard normal curve values for a normal curve by period. Row 52 is then the standard S-Curve, or rather, the cumulative normal curve values. Cell A52 is 'S-Curve' and cell G52 is "=IF(G4>=Construction_Months,1,NORMSDIST(G51))". The '=NORMSDIST()' function provides the cumulative percentage from the left of a standard normal curve value. Finally, row 53 is the actual Draw value by month. Cell A53 is 'Draw' and cell G53 is "=G52-F52" or the periodic draw for the month. Cells G52:G53 must be copied to column AP, the 36th period. Cells G53:AP53 are coloured blue; although they are not direct inputs, they are considered inputs for this model. Finally, cell E53 is the summation of G53:AP53. This is an analyst check to ensure a 100% draw for the line item.

An example of the draws are demonstrated graphically in Figure 8.9. The S-Curve is the cumulative 'Draw' and is represented on the secondary y-axis. Both the levelized Cost and Draw are represented on the primary y-axis. Levelized Cost represents row 49, the S-Curve

	A	B	C	D	E	F	G	H	I	J
1	Development Start	Jan-12								
2	Construction Months	12								
3	Opening Date	Jan-13								
4							1	2	3	
5	Cost Item						Jan-12	Feb-12	Mar-12	Apr
7	**Land & Related**			Land_Related	1,000,000					
8	Land & Related				1,000,000		100.00%			
9										
10	**General & Administrative**			G_A	170,000					
11	Legal Fees				20,000		8.33%	8.33%	8.33%	8.3
12	Development Management Fees				60,000		8.33%	8.33%	8.33%	8.3
13	Insurance & Warranty				60,000		8.33%	8.33%	8.33%	8.3
14	Other G&A				30,000		8.33%	8.33%	8.33%	8.3
15										
16	**Dev Chrgs & Mun Costs**			Dev_Chrg_Mun_Costs	110,000					
17	Development Charges				85,000		8.33%	8.33%	8.33%	8.3
18	Buidling Permits				20,000		8.33%	8.33%	8.33%	8.3
19	Municipal Costs				5,000		8.33%	8.33%	8.33%	8.3

Figure 8.10

row 52, and Draw row 53. These are provided to the analyst for ease of calculating the draw schedule; however, it is *highly* recommended that the draw schedule be provided by the general contractor or an engineer.

Now, returning to row 6, the draw schedule for each line item will be provided for the model. For row 8, Land & Related, 100% of the draw is in the first month. Cell G8 is therefore 100.0% and ought to be coloured blue. (See Figure 8.10.) General & Administrative and Development Charges & Municipal Costs, rows 11:14 and 17:19 respectively, are modelled as a levelized draw. Cells G11:G14 and G17:G19 are therefore "=G$49" and are copied to column AP.

Hard Cost Construction, rows 22:37, are modelled as a generic S-curve (Figure 8.11). Cells G22:G37 are therefore "=G$53" and then are copied to column AP. Building Operations, rows 40:41, are modelled as levelized payments – "=G$49" – and copied to column AP. Finally, Financing Fee, row 45, is modelled as Land & Related and is simply 100.00% in cell G45.

Construction Loan Interest is left without a draw schedule. This will be completed in the next section, but the total cost will be summarized in cell E44.

The next section, also titled 'Cost Item', allocates the cost to the period by multiplying the draw percentage in the period by the total cost above in column E. Row 47 through row 91 should be selected and rows added. The section will be completely grey. Then select rows 48:90 and unfill the rows. Cell A48 is "=A5", which is copied down to row 90. Note that some rows – e.g. row 49 and row 52 – have a zero where there is no value above. For formatting, these cells should be deleted: A49, A52, A58, A63, A81, A85, A89, and A90.

Cell G51, the first period for Land & Related, is the product of the line item total cost and percentage draw for the period: i.e. "=G8:$E8". Cell G51 is then copied to each individual line item through to the end, row 88. The categories are then subtotaled. Cell G50 is "=SUBTOTAL(9,G51)" and cell G53 is "=SUBTOTAL(9,G54:G57)". The subtotals are quantified for rows 50, 53, 59, 64, 82, and 86. Column G is then copied to column AP and represents the total cost by line item by period.

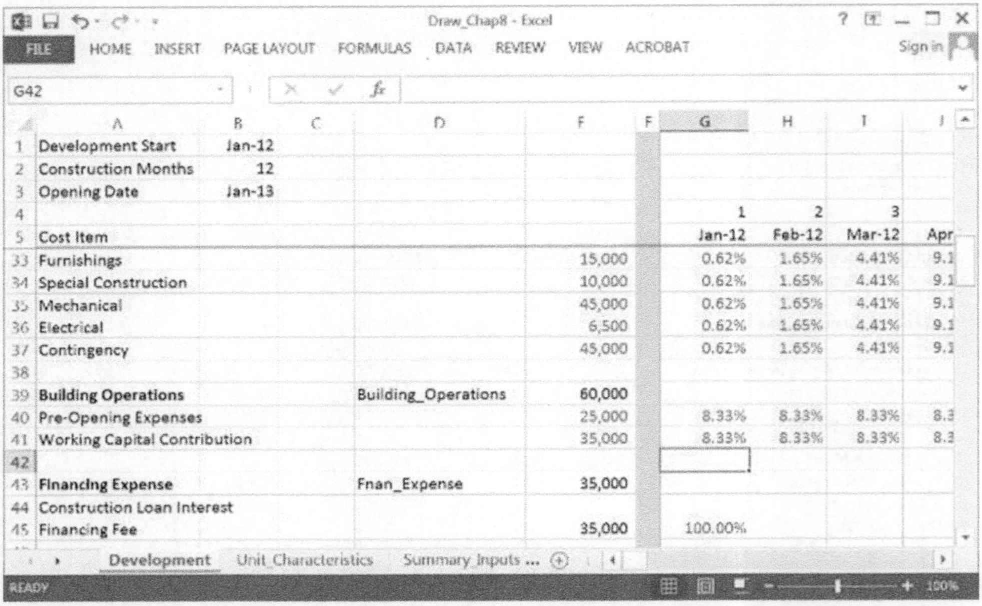

Draw_Chap8 - Excel

G42

	A	B	C	D	E	F	G	H	I	J
1	Development Start	Jan-12								
2	Construction Months	12								
3	Opening Date	Jan-13								
4							1	2	3	
5	Cost Item						Jan-12	Feb-12	Mar-12	Apr
33	Furnishings				15,000		0.62%	1.65%	4.41%	9.1
34	Special Construction				10,000		0.62%	1.65%	4.41%	9.1
35	Mechanical				45,000		0.62%	1.65%	4.41%	9.1
36	Electrical				6,500		0.62%	1.65%	4.41%	9.1
37	Contingency				45,000		0.62%	1.65%	4.41%	9.1
38										
39	**Building Operations**			Building_Operations	60,000					
40	Pre-Opening Expenses				25,000		8.33%	8.33%	8.33%	8.3
41	Working Capital Contribution				35,000		8.33%	8.33%	8.33%	8.3
42										
43	**Financing Expense**			Fnan_Expense	35,000					
44	Construction Loan Interest									
45	Financing Fee				35,000		100.00%			

Development | Unit_Characteristics | Summary_Inputs ...

Figure 8.11

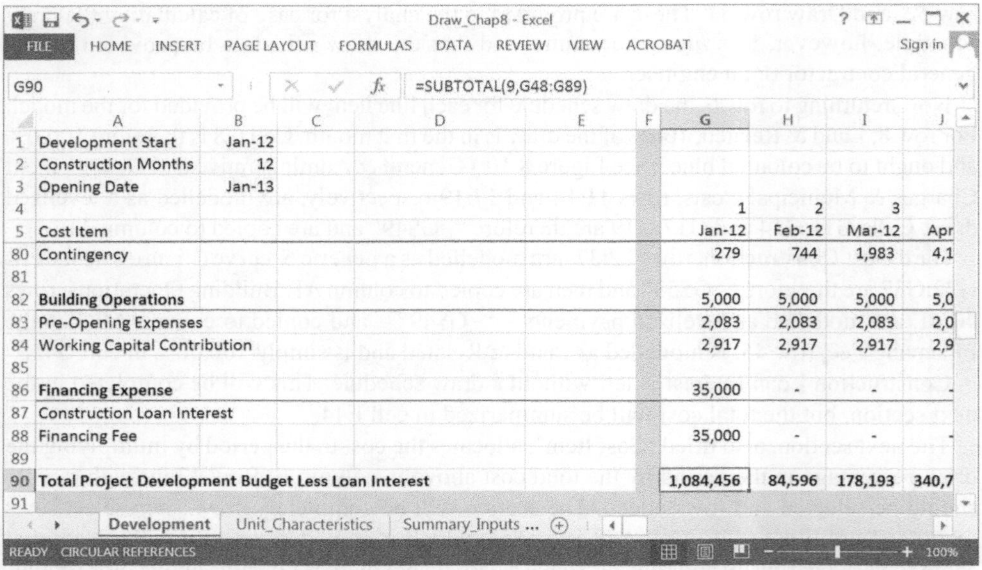

Draw_Chap8 - Excel

G90 =SUBTOTAL(9,G48:G89)

	A	B	C	D	E	F	G	H	I	J
1	Development Start	Jan-12								
2	Construction Months	12								
3	Opening Date	Jan-13								
4							1	2	3	
5	Cost Item						Jan-12	Feb-12	Mar-12	Apr
80	Contingency						279	744	1,983	4,1
81										
82	**Building Operations**						5,000	5,000	5,000	5,0
83	Pre-Opening Expenses						2,083	2,083	2,083	2,0
84	Working Capital Contribution						2,917	2,917	2,917	2,9
85										
86	**Financing Expense**						35,000	-	-	-
87	Construction Loan Interest									
88	Financing Fee						35,000	-	-	-
89										
90	**Total Project Development Budget Less Loan Interest**						1,084,456	84,596	178,193	340,7
91										

Development | Unit_Characteristics | Summary_Inputs ...

READY CIRCULAR REFERENCES

Figure 8.12

The total construction project budget must be summarized by period. Construction Loan Interest must be quantified, and this requires total draw cost by period. Cell A90 is therefore 'Total Project Development Budget Less Loan Interest' (Figure 8.12). Cell G90 is the summation of the costs for the project, "=SUBTOTAL(9,G48:G89)". Cell G90 is then copied to H90:AP90. The section is now complete. Note that, again, Construction Loan Interest will be calculated in the next section and is therefore blank in this section.

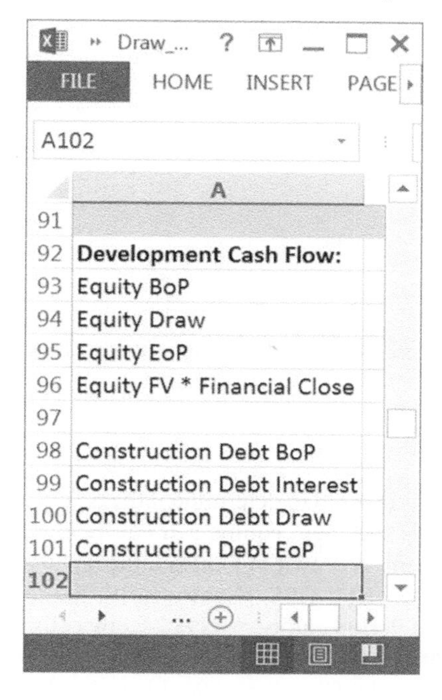

Figure 8.13

Draw_Chap8 - Excel

FILE HOME INSERT PAGE LAYOUT FORMULAS DATA REVIEW VIEW ACROBAT

A93 | *fx* Equity BoP

	A	B	C	D	E	F	G	H	I	J
92	**Development Cash Flow:**									
93	Equity BoP						1,295,000	210,544	125,949	-
94	Equity Draw						1,084,456	84,596	125,949	-
95	Equity EoP						210,544	125,949	-	-
96	Equity FV * Financial Close									

Development | Unit_Characteristics | Summary_Inputs | D ...

READY CIRCULAR REFERENCES 100%

Figure 8.14

The final section is the Inflows/Outflows section (Figure 8.13). This demonstrates how the capital flows through the project during construction. For modelling purposes, all equity is first into the project and remains in the project until final sale. After equity has been utilized in the project, debt is utilized for all shortfalls. The construction loan balance is "taken out" 100% by a Permanent Borrowing Facility at the commercial operating date (COD).

Starting in row 92, add 10 rows for this Inflow/Outflow section. Select rows 92:101 and unfill these rows. Cell A92 is 'Development Cash Flow:'. The first section is equity outflow and the second section is debt outflow. 'Equity FV * Financial Close' quantifies the equity at commercial operating date, which is the assumed T = 0 for valuation purposes. The construction section calculates the debt draw and the corresponding construction interest.

Cell G93, 'Equity BoP', is "=Summary_Inputs!D2", which is the total project equity for the project (Figure 8.14). Cell G94, 'Equity Draw', is the equity to be drawn for the project.

	A	B	C	D	E	F	G	H	I	J	K	L
92	**Development Cash Flow:**											
93	Equity BoP						1,295,000	210,544	125,949	-	-	-
94	Equity Draw						1,084,456	84,596	125,949	-	-	-
95	Equity EoP						210,544	125,949	-	-	-	-
96	Equity FV * Financial Close											
97												
98	Construction Debt BoP						-	-	-	52,245	393,217	933,013
99	Construction Debt Interest				75,999		-	-	-	218	1,638	3,888
100	Construction Debt Draw						-	-	52,245	340,754	538,158	679,593
101	Construction Debt EoP						-	-	52,245	393,217	933,013	1,616,494

Figure 8.15

An IF/THEN logic function is utilized to determine if the equity available exceeds the project draw in the period. If there is sufficient equity, just equity is drawn, if the equity is insufficient, then the remaining equity is drawn in total and the delta is drawn from debt. Cell G94 is "=IF(G93>G90,G90,G93)". Cell G95, 'Equity EoP', is the remaining equity: "=G93-G94". Cell H93 is the End of Period equity, "=G95". Copy cells G94:G95 to H94:H95. Then copy cells H93:H95 to AP93:AP95.

Leaving 'Equity FV * Financial Close', row 96, blank for the moment, Construction Debt Inflow/Outflow is quantified (Figure 8.15). Cell G98, 'Construction Debt BoP', is "=F101", or zero (0). Row 99, 'Construction Debt Interest', is the interest payable per period. As there will be no interest after construction has ended and the project has been completed, an IF/THEN statement is utilized to ensure interest calculation ends at the conclusion of construction. Cell G99 is "=G98*Interest_T1/12*IF(G4>Construction_Months,0,1)". The interest for Tranche 1 is assumed to be the same interest rate for construction. Row 100, 'Construction Debt Draw', is the total project requirement less equity provided. Cell G100 is therefore "=G90-G94". The Construction Debt EoP, cell G101, is then the summation of the Construction payments – i.e. cell G101 is "=SUM(G98:G100)". Cells G98:G101 are then copied to column AP. The total construction interest is then placed in cell E99, which is "=SUM(G99:AP99)".

As cell E99 is the total Construction Debt Interest, it is then linked to cell E44: cell E44 is "=E99". Now that the Development page has been complete, the costs must be added to the Uses section on the Inputs_Summary page (Figure 8.16). Cells B10:B17 must be adjusted in the Sources/Uses section on the Inputs page. This section is then linked to the Development page totals. Replace the Uses headers in column B by linking the roll-up titles on the Development section. Cell B10:B15 are linked to the Development page as "=Development!A7", "=Development!A10", "=Development!A16", "=Development!A21", "=Development!A39", and "=Development!A43", respectively. Cells D10:D15 are then linked to the subtotal values for each section: "=Land_Related", "=G_A", "=Dev_Chrg_Mun_Costs", "=HC_Construction", "=Building_Operations", and "=Fnan_Expense", respectively. The Surplus/(Deficit) then demonstrates a $57,599 shortfall in the project. Note that the Uses section should not be formatted as blue because these are no longer inputs but rather a summary from the Development page (see Figure 8.16).

Note the $57,499 shortfall; in other words, there is insufficient capital to close. Also note that equity affects the amount of debt required, which contributes to the amount of Financing Expense, which adds to Uses, which affects the amount of equity required: the logic is circular. Therefore, select the File tab from the Ribbon, then choose 'Options'. The menu 'Excel

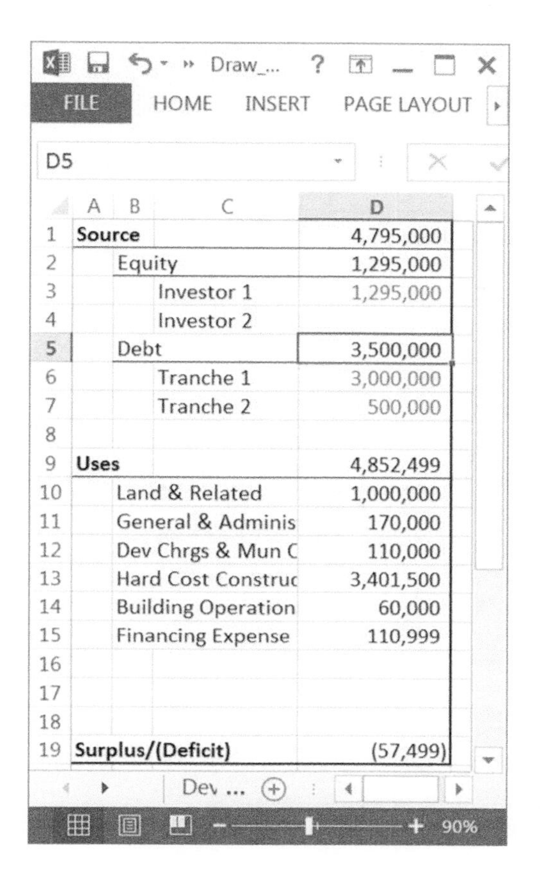

	A B C	D
1	**Source**	4,795,000
2	Equity	1,295,000
3	Investor 1	1,295,000
4	Investor 2	
5	Debt	3,500,000
6	Tranche 1	3,000,000
7	Tranche 2	500,000
8		
9	**Uses**	4,852,499
10	Land & Related	1,000,000
11	General & Adminis	170,000
12	Dev Chrgs & Mun C	110,000
13	Hard Cost Construc	3,401,500
14	Building Operation	60,000
15	Financing Expense	110,999
16		
17		
18		
19	**Surplus/(Deficit)**	(57,499)

Figure 8.16

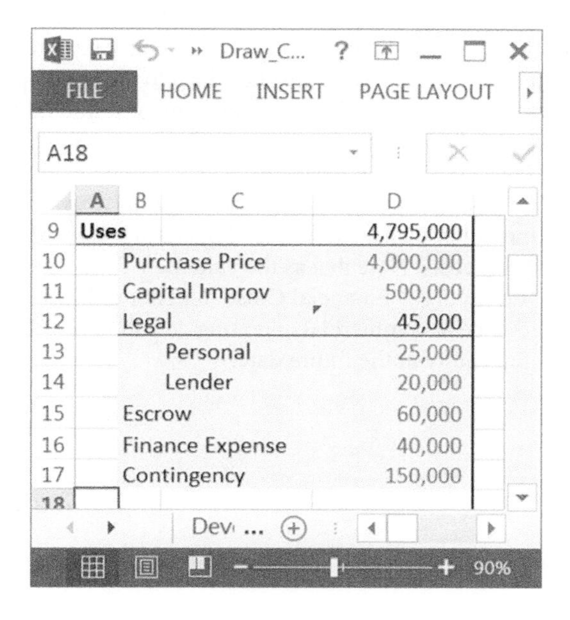

	A B C	D
9	**Uses**	4,795,000
10	Purchase Price	4,000,000
11	Capital Improv	500,000
12	Legal	45,000
13	Personal	25,000
14	Lender	20,000
15	Escrow	60,000
16	Finance Expense	40,000
17	Contingency	150,000
18		

Figure 8.17

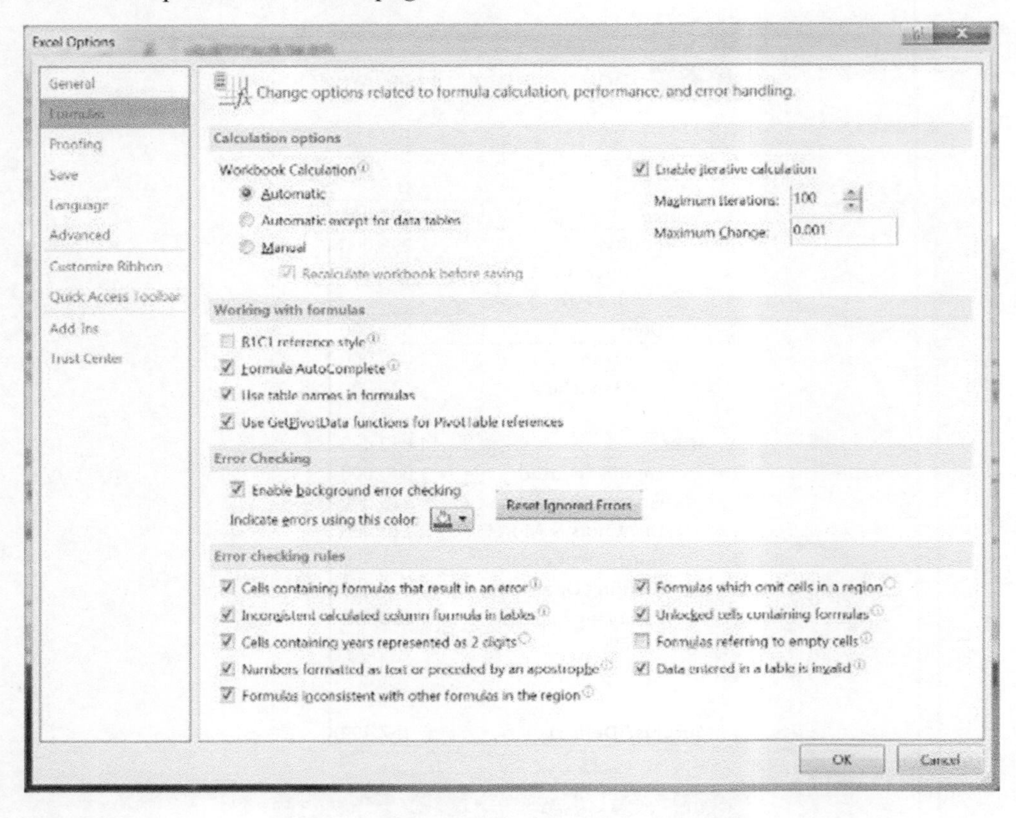

Figure 8.18

Options' will be displayed. Select 'Formulas' along the left (see Figure 8.18). Under 'Calculation Options' select 'Manual' and then check 'Enable iterative calculation.'

In order to satisfy the (Deficit), $56,000 must be added to the equity portion. Place 56,000 in cell D4, which is Investor 2 equity. The Final adjustment to the model is located on the Pro_Forma page. In cell F45 in the Valuation section, T = 0 must be adjusted. Cell D45 is no longer "=EOMONTH(G45,-1)" but is adjusted to "=Dev_Start". Note that the project return was reduced by over 500 basis points because the project is now valued a year earlier, at the beginning of construction rather than at commercial operating date. The section is now completed with the construction costs. Note that as the valuation T = 0 was adjusted to be Development Start, row 96, 'Equity FV * Financial Close' was not utilized. This row is utilized if the valuation is intended to be at commercial operating date to future value the equity, which calculates the value of the equity at the future date.

9 Waterfall

The waterfall is the method of distributing profits among partners within a transaction according to a joint venture agreement. These distributions are often not pari passu, that is proportional to the percentage of capital provided in the investment. In more complex transactions, profits do not follow an even, pari passu distribution.

A graphic depiction of a general waterfall structure is shown in Figure 9.1. Please note that rates discussed are internal rates of return (i.e. accrual rates on capital) and not simple interest rates. The distribution of profits by investor can be different at each 'tier' and is negotiated from project onset.

In this chapter a waterfall with two tiers and an "overflow" will be modelled. It will correspond to the joint venture agreement at the end of the text. The waterfall structure with splits is as follows:

1. Tier 1 (Preference Rate): Up to 15.00% (IRR)
    a. Sponsor: 10%
    b. Investor: 90%
2. Tier 2 (Secondary): 15.00–20.00% (IRR)
    a. Sponsor: 20%
    b. Investor: 80%
3. Tier 3 (Overflow): Over 20.00% (IRR)
    a. Sponsor (Promote): 40.00%
    b. Remaining: 60.00%
        i. Sponsor: 10.00%
        ii. Investor: 90.00%

The intention of the waterfall is to bifurcate cash flows to best model investor/sponsor preferences for investing. Another way to consider the waterfall is the diagram in Figure 9.2, which is meant to be viewed left to right.

The intention is to match investor and sponsor desired cash flow and risk tolerances within a particular project. In the example being modelled here, the percentage of cash flows to the sponsor increases as the project return increases. Thus the sponsor is being incentivized financially for a very profitable project. The investor's preferences are lower risk and Return of Capital. One point of note is that it is not uncommon in the first tier for the investor to receive 100.0% of the cash flows as a preference to ensure a baseline. Then the percentage

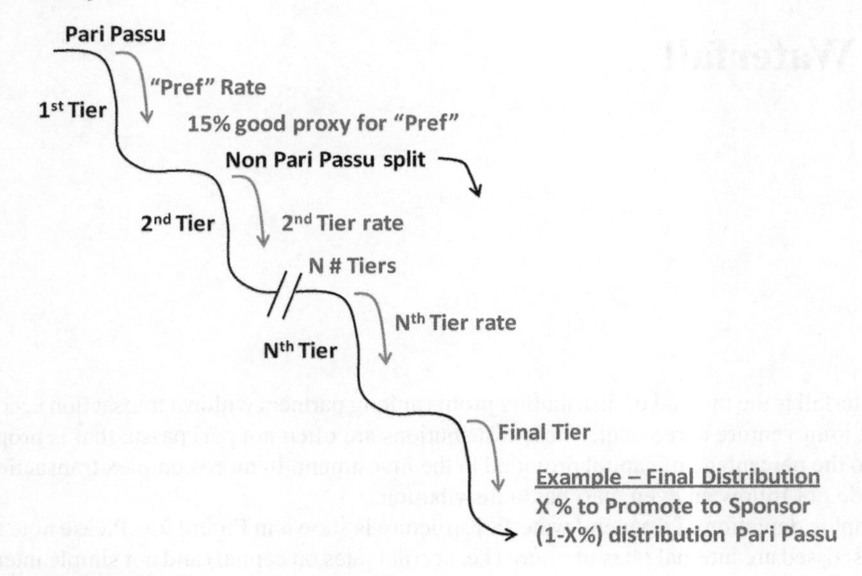

Figure 9.1 Waterfall

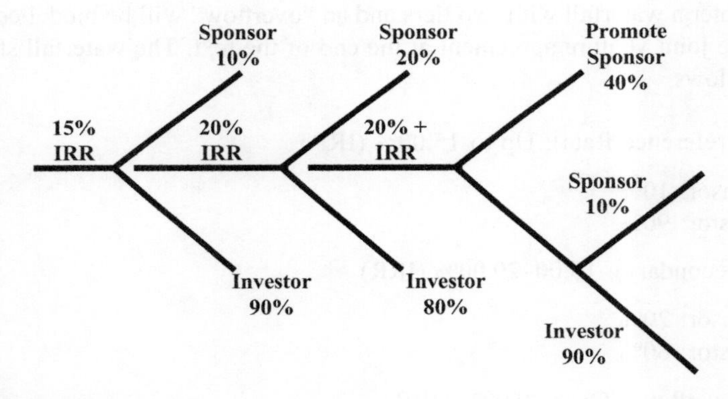

Figure 9.2 Waterfall Adjusted

to the sponsor increases with project performance. Note that the sponsor also, generally, received a development/construction/management fee, so even though the allocation for the Preference Rate may be zero, the fee is still received and earned.

While the focus of this chapter is the construction of the waterfall, it is important to demonstrate that it can be easily incorporated into – i.e. "bolted on" to – an existing spreadsheet model. In this chapter the initial waterfall to be constructed will be an 11-period cash flow model. The first period will represent the initial cash outflow. The remaining periods, 2–11, will be the 10 periods of cash flow for the project. This will then be "bolted on" to the 5-unit model constructed earlier. After successfully bolting on to the 5-unit, the waterfall will be expanded and bolted on to the N-unit construction model, thus demonstrating the scalability and modularity of both models. Finally, a discussion of how the waterfall can adjust and change investment characteristics will be demonstrated towards the end of the chapter.

A general overview of the waterfall construction steps is shared here.

Step 1: quantify equity contributions

Equity contributions are quantified within the project pro forma. For the 5-unit pro forma these are located entirely on the Inputs_Summary page. However, for the N-unit construction model, the values are modelled on the Pro_Forma page. The separation of cash flows will be as follows:

- Sponsor
- Investor(s)

Note: additional investors can be included but for the purpose of this text, only two are modelled.

Step 2: quantify project cash flows

These are the quantified cash flows from the project pro forma. This is not the project 'income', which is an accounting number. On a "Staiger" pro forma, these values will be quantified by major classification on the Inflow_Outflow page.

Step 3: quantify project characteristics

This quantification for the project characteristics determines if the project is viable prior to investor distributions. The short list of project metrics which determine viability includes the following:

- Total Profit (nominal)
- Internal Rate of Return (IRR)
- Multiple (Profit/Invested Capital)

Step 4: calculate the first waterfall ("Pref" rate)

The recommended structure for the "Pref" rate in Tier 1 uses the following line items:

- Beginning of Period Balance (BoP)
- Total equity contribution/period
- Accrual (calculate daily rate to use XIRR)
- Paydown (distributed capital at stated rate)
- End of Period Balance (EoP)
- Remaining cash
- IRR Check

An example of the structure from Excel follows in Figure 9.3.

BoP balance

The BoP (beginning of period) balance is the cash balance from the previous period. It is also the End of Period balance from the previous period.

Tier 1		
BoP Balance		
Equity Contributions		(200.0)
Accrual	15.00%	(19.2)
Paydown		219.2
EoP Balance		
Cash left for distribution		30.8
IRR Check	*15.00%*	*19.2*

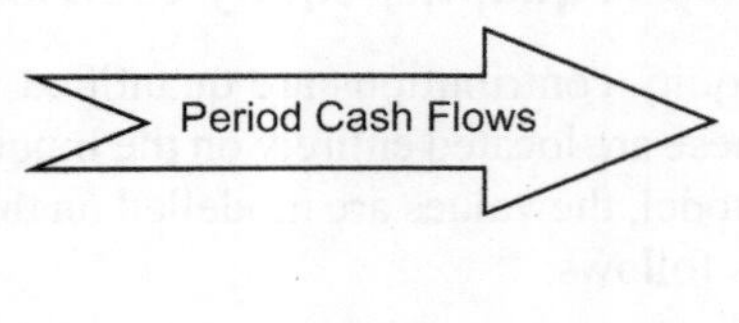

Figure 9.3 Tier 1 Waterfall

Equity contributions

This is the total funding requirement for the period. The Equity Contributions value comes directly from the summary atop the waterfall. See also 'Step 1 – Quantify Equity Contributions'.

Accrual

'Accrual' calculates the amount to achieve a total distribution equal to the specific rate for the quantified tier. For the first tier this is the "Pref" rate.

The rate is as specified in the agreement between the sponsor and investor(s). Generally the documents will state this rate as an annual rate. To complete the accrual correctly, the annual rate must first be converted to a daily rate. This allows the use of the XIRR function, which quantifies the rate on a daily basis (see Staiger Memorandum on XIRR function for additional information).

To convert an annual rate to a daily rate, the following relationship must be remembered:

$$DailyRate = \left[(1+EAY)^{1/m}\right] - 1$$

where
EAY: Equivalent Annual Yield
m: Periods (365)

For example, a 15% rate with annual compounding is equivalent to a 13.979% rate compounded daily. See the following equation.

$$ENAR = \left[(1+0.15)^{1/365}\right] - 1 = 0.000383$$

Note the daily accrual rate is 0.00038299 (0.0383%). This conversion is critical because the XIRR function in Excel discounts on a daily basis. This also provides a more accurate distribution of cash flow.

Once the daily rate is calculated, it is then used to calculate the accrual amount. The accrual is calculated as follows:

$$Accrual = \left[\left[(1+daily\_rate)^{days/period}\right] - 1\right] \times BoP$$

Paydown

The paydown is the amount distributed to achieve the rate for an individual waterfall tier. It is calculated as the lesser of the cash available/period and the absolute summation of BoP, Equity Contribution, and Accrual. It is the lesser of the two quantities given that the goal is to quantify the exact dollar amount to achieve the tier rate only. The tier rate is the IRR as stated for the particular waterfall.

EoP balance

The EoP balance is calculated as the BoP + Equity Contribution + Accrual − Paydown. Be careful regarding sign convention when building the model.

Cash left for distribution

The cash left for distribution is the difference between the cash available to distribute and the paydown amount. This is the amount to distribute above what is needed to achieve the specific tier amount.

IRR check

The cash flows are a combination of equity contribution and paydown. These cash flows are the actual investor cash flows for the tier (i.e. investor inflow and outflow attributable).

Note: repeat this step, Step 4, for the number of tiers (waterfalls) within the agreement. In addition, the tiers are all cumulative. That is, Tier 2 includes the cash flow from Tier 1. Therefore, distributions for the second tier must be Tier 2 minus Tier 1 cash flows (the delta).

Step 5: separate cash flows by tier by investor

This step proportions the cash flows by tier by investor – in other words, it separates cash flows according to division. The percentage allocation is specified in the agreement between the equity/sponsor and investors. Note that Tier 1 and Tier 2 require the delta to be calculated given that in each tier the cash flows are cumulative.

Step 6: analyze investor cash flows

Aggregate the cash flow by investor to analyze individual returns.

Example:

The Excel example depicts a $200m project funded with 10% developer/sponsor equity and 90% by an equity partner. The project provides $50m profit over the $200m invested capital. This represents a project IRR of 38.96% and a 1.25x multiple (i.e. 250/200).

The partner agreement for the waterfall is as follows:

- Pari passu to a 15% Pref Rate (IRR)
    - Up to 15%, profits are distributed proportional to invested capital
        - Developer/Sponsor receives 10%
        - Equity Partner receives 90%

- Second tier – 20% IRR
  - Profits are split unequally
    - Developer/Sponsor receives 20%
    - Equity Partner receives 80%
- Third/Final tier
  - 40% Promote
    - Developer/Sponsor receives 40% of remaining cash flow
  - 60% remaining distributed pari passu
    - Developer/Sponsor receives 10%
    - Equity Partner receives 90%

Summary

The previous structure for the stated example provides the following returns delineated by capital provider:

- Developer/Sponsor
  - IRR = 112.03%
- Equity Partner
  - IRR = 30.83%

Note the difference in return for each capital investor. The waterfall structure benefits each investor differently. This is similar to a structured finance investment.

Excel waterfall example

Excel functions utilized in the example in Figure 9.4:

- =EOMONTH()
- =COUNT()
- =SUM()
- =XIRR()
- =ABS()
- =MIN()
- =DAY()

The structure of the pro forma with the waterfall appears in Figure 9.5.

Section 1: project summary

As with all the models, the logical structure is begun in a blank spreadsheet, renaming Sheet1 'Waterfall' (Figure 9.6). Cell A1 is also labeled 'Waterfall'. Cell A3 is 'Project Cash Flows'.

Waterfall Sample		Period	1	2	3	4	5	6	7	8	9	10	11	12
($MMs)		Total	1/31/2010	2/28/2010	3/31/2010	4/30/2010	5/31/2010	6/30/2010	7/31/2010	8/31/2010	9/30/2010	10/31/2010	11/30/2010	12/31/2010
PROJECT FUNDING														
Equity Contributions			(125.0)	(25.0)	(20.0)	(15.0)	(15.0)	0.0	0.0	0.0	0.0	0.0	0.0	0.0
Sponsor	10%	(20.0)	(12.5)	(2.5)	(2.0)	(1.5)	(1.5)	0.0	0.0	0.0	0.0	0.0	0.0	0.0
Equity Partner	90%	(180.0)	(112.5)	(22.5)	(18.0)	(13.5)	(13.5)	0.0	0.0	0.0	0.0	0.0	0.0	0.0
Total	100%	(200.0)	(125.0)	(25.0)	(20.0)	(15.0)	(15.0)	0.0	0.0	0.0	0.0	0.0	0.0	0.0
Cash Proceeds for Distribution		250.0	0.0	0.0	0.0	0.0	0.0	0.0	0.0	50.0	50.0	50.0	50.0	50.0
Project Cash Flow		**50.0**	**(125.0)**	**(25.0)**	**(20.0)**	**(15.0)**	**(15.0)**	**0.0**	**0.0**	**50.0**	**50.0**	**50.0**	**50.0**	**50.0**
Profit	**$50.0**													
IRR	**38.96%**													
Multiple	**1.25x**													
STRUCTURE														
Tier 1														
BoP Balance			0.0	(125.0)	(151.3)	(173.2)	(190.2)	(207.4)	(209.8)	(212.3)	(164.9)	(116.8)	(68.2)	(18.9)
Equity Contributions		(200.0)	(125.0)	(25.0)	(20.0)	(15.0)	(15.0)	0.0	0.0	0.0	0.0	0.0	0.0	0.0
Accrual	15.00%	(19.2)	0.0	(1.3)	(1.8)	(2.0)	(2.3)	(2.4)	(2.5)	(2.5)	(1.9)	(1.4)	(0.8)	(0.2)
Paydown		219.2	0.0	0.0	0.0	0.0	0.0	0.0	0.0	50.0	50.0	50.0	50.0	19.2
EoP Balance			(125.0)	(151.3)	(173.2)	(190.2)	(207.4)	(209.8)	(212.3)	(164.9)	(116.8)	(68.2)	(18.9)	0.0
Cash left for distribution		30.8	0.0	0.0	0.0	0.0	0.0	0.0	0.0	0.0	0.0	0.0	0.0	30.8
IRR Check	*15.00%*	*19.2*	*(125.0)*	*(25.0)*	*(20.0)*	*(15.0)*	*(15.0)*	*0.0*	*0.0*	*50.0*	*50.0*	*50.0*	*50.0*	*19.2*
Tier 2														
Starting Balance			0.0	(125.0)	(151.8)	(174.1)	(191.8)	(209.8)	(212.9)	(216.2)	(169.6)	(122.2)	(74.1)	(25.2)
Equity Contributions		(200.0)	(125.0)	(25.0)	(20.0)	(15.0)	(15.0)	0.0	0.0	0.0	0.0	0.0	0.0	0.0
Accrual	20.00%		0.0	(1.8)	(2.4)	(2.6)	(3.0)	(3.2)	(3.3)	(3.4)	(2.6)	(1.9)	(1.1)	(0.4)
Paydown		225.6	0.0	0.0	0.0	0.0	0.0	0.0	0.0	50.0	50.0	50.0	50.0	25.6
Balance			(125.0)	(151.8)	(174.1)	(191.8)	(209.8)	(212.9)	(216.2)	(169.6)	(122.2)	(74.1)	(25.2)	0.0
Cash left for distribution		24.4	0.0	0.0	0.0	0.0	0.0	0.0	0.0	0.0	0.0	0.0	0.0	24.4
IRR Check	*20.0%*	*25.6*	*(125.0)*	*(25.0)*	*(20.0)*	*(15.0)*	*(15.0)*	*0.0*	*0.0*	*50.0*	*50.0*	*50.0*	*50.0*	*25.6*

Figure 9.4 Waterfall Completed

Cash Flows to each tranche:														
I. Pari Passu to an IRR of	15.00%	219.2	0.0	0.0	0.0	0.0	0.0	0.0	0.0	50.0	50.0	50.0	50.0	19.2
Sponsor	10.00%	21.9	0.0	0.0	0.0	0.0	0.0	0.0	0.0	5.0	5.0	5.0	5.0	1.9
Equity Partner	90.00%	197.3	0.0	0.0	0.0	0.0	0.0	0.0	0.0	45.0	45.0	45.0	45.0	17.3
II. Splits up to an IRR of	20.00%	6.4	0.0	0.0	0.0	0.0	0.0	0.0	0.0	0.0	0.0	0.0	0.0	6.4
Sponsor	20.00%	1.3	0.0	0.0	0.0	0.0	0.0	0.0	0.0	0.0	0.0	0.0	0.0	1.3
Equity Partner	80.00%	5.1	0.0	0.0	0.0	0.0	0.0	0.0	0.0	0.0	0.0	0.0	0.0	5.1
Cash left for distribution		24.4	0.0	0.0	0.0	0.0	0.0	0.0	0.0	0.0	0.0	0.0	0.0	24.4
III. Sponsor Promote	40.00%	9.8	0.0	0.0	0.0	0.0	0.0	0.0	0.0	0.0	0.0	0.0	0.0	9.8
Cash to Equity	60.00%	14.6	0.0	0.0	0.0	0.0	0.0	0.0	0.0	0.0	0.0	0.0	0.0	14.6
Sponsor	10.00%	1.5	0.0	0.0	0.0	0.0	0.0	0.0	0.0	0.0	0.0	0.0	0.0	1.5
Equity Partner	90.00%	13.2	0.0	0.0	0.0	0.0	0.0	0.0	0.0	0.0	0.0	0.0	0.0	13.2
INVESTOR CASH FLOWS														
SPONSOR														
Equity Investment		(20.0)	(12.5)	(2.5)	(2.0)	(1.5)	(1.5)	0.0	0.0	0.0	0.0	0.0	0.0	0.0
Proceeds		34.4	0.0	0.0	0.0	0.0	0.0	0.0	0.0	5.0	5.0	5.0	5.0	14.4
CF		14.4	(12.5)	(2.5)	(2.0)	(1.5)	(1.5)	0.0	0.0	5.0	5.0	5.0	5.0	14.4
Profit	**$14.4**													
% of Total Profit	**28.9%**													
IRR	**112.03%**													
Multiple	**1.72x**													
EQUITY PARTNER														
Equity Investment		(180.0)	(112.5)	(22.5)	(18.0)	(13.5)	(13.5)	0.0	0.0	0.0	0.0	0.0	0.0	0.0
Proceeds		215.6	0.0	0.0	0.0	0.0	0.0	0.0	0.0	45.0	45.0	45.0	45.0	35.6
CF		35.6	(112.5)	(22.5)	(18.0)	(13.5)	(13.5)	0.0	0.0	45.0	45.0	45.0	45.0	35.6
Profit	**$35.6**													
% of Total Profit	**71.1%**													
IRR	**30.83%**													
Multiple	**1.20x**													
Check	*TRUE*													

Figure 9.4 (Continued)

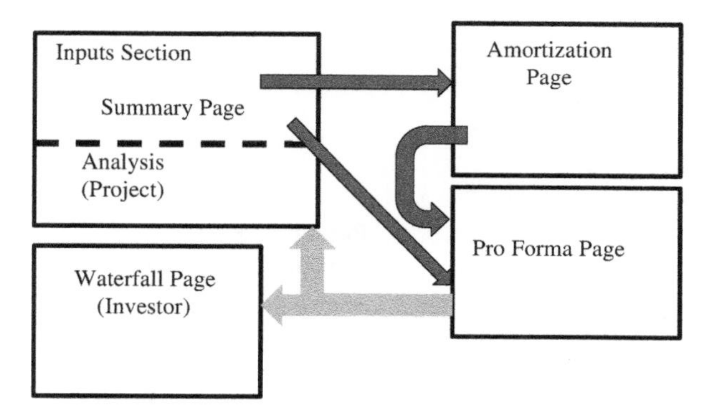

Figure 9.5 Waterfall Addition

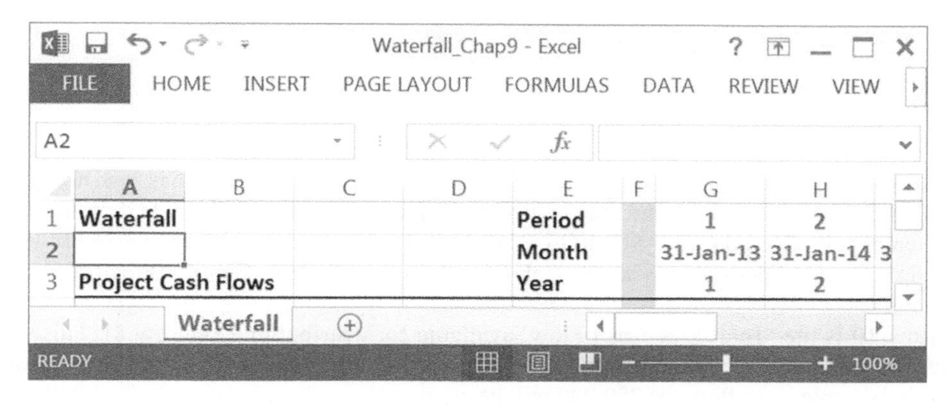

Figure 9.6

For consistency, in cell G1 place a 1, the period, and cell H1 will be "=G1+1". Then copy cell H1 across to column Q. Place the date, 31 January 2013, in cell G2. Then cell H2 is "=EOMONTH(G2,12)". Remember this is a placeholder and the dates are utilized to construct the waterfall model. Both Periods and Months will be linked at a later date when the waterfall is "bolted on" to a pro forma model. Cell H2 is copied to column Q. Column F is filled with grey as a separator and E1 and E2 are 'Period' and 'Month', respectively.

For consistency with all earlier models, cell E3 is 'Year' and cell G3 is "=G1". Note that cells G1:Q3 should be coloured blue. These will later be linked to the pro forma model.

The first section of the waterfall, as discussed, is the project characteristics (Figure 9.7). Cell A4 is 'Total Equity'. Cells A6 and A7 are 'Sponsor' and 'Investor 2', respectively. Cell A8 is 'Total Equity' again, which will be the sum of the sponsor and investor 2 equity. Cell D6 is the percentage of capital contributed by the sponsor. Cell D7 is the remaining percentage contributed by the investor. The amount contributed by the sponsor is 10% and the remaining for the investor, cell D7, is "=1-D6". Cell D8 is the summation of contribution, "=SUM(D6:D7)".

Returning to row 4, Total Equity, this is the capital, by period, invested in the project. In this example it is assumed that $1,000,000 is invested in the first year, so cell G4 is -1,000,000 (it is a project outflow so it is negative). All equity is invested at the beginning of the project so cells H4:Q4 remain zero. These ought to be blue because they are inputs from the pro forma.

	A	B	C	D	E	F	G	H	I
1	**Waterfall**				**Period**		1	2	3
2					**Month**		31-Jan-13	31-Jan-14	31-Ja
3	**Project Cash Flows**				**Year**		1	2	3
4	**Total Equity**						(1,000,000)	-	
5									
6	Sponsor			10%	(100,000)		(100,000)	-	
7	Investor 2			90%	(900,000)		(900,000)	-	
8	**Total Equity**			**100%**	**(1,000,000)**		**(1,000,000)**	-	

Figure 9.7

Rows 6 and 7 are the percentage of capital contributed by period. Cell G6 is "=$D6*G$4". Cell G6 is copied to cell G7 and then copied down the row to column Q. Cell G8 is the summation of cells G6 and G7: "=SUM(G6:G7)". Cells E6:E8 are the summation of the rows to the right, the total periodic payments. Cell E6 is "=SUM(G6:Q6)". Cell E6 is then copied to cell E8.

Row 10 is the 'Total Free Cash Flow' available for distribution (Figure 9.8). This will eventually be linked to the pro forma and is therefore to be coloured blue. For the purposes of this example, cash flows of 250,000 will be assumed in periods 2–10 (cells H10:P10) and 750,000 in cell Q10 (period 11). Cell E10 is the summation of the Total Free Cash Flow: "=SUM(G10:Q10)".

Row 12 is net cash flow for the project. Cell A12 is 'Net Cash Flow' and cell G12 is "=SUM(G8,G10)". Copy cell G12 to H12:Q12. Cell E12 is then the net cash flow received, that is "=SUM(G12:Q12)". Finally, Profit and Internal Rate of Return are quantified for the project. Cell A14 is 'Net Profit' and cell A15 'IRR'. Cell E14 is "=E12" and cell E15 "=XIRR(G12:Q12:G2:Q2)". The project then has a Net Profit of 2,000,000 and a 23.36% IRR. This concludes the first section of the waterfall, the Project Summary section.

Section 2: tier structure calculation(s)

This section quantifies the Forward Accrual for the waterfall. It quantifies the cash flows, on a forward basis, required to meet each hurdle rate as specified in the joint venture agreement. In the example being modelled here, Tier 1 IRR is 15.00% and Tier 2 IRR is 20.00%. The Structure of each Tier, i.e. row headings, is as follows:

Tier 1: Tier title for the forward accrual.

BoP Balance: Beginning of period for the project with the paydown to meet the Tier included.

Equity Contribution: This is the total project equity.

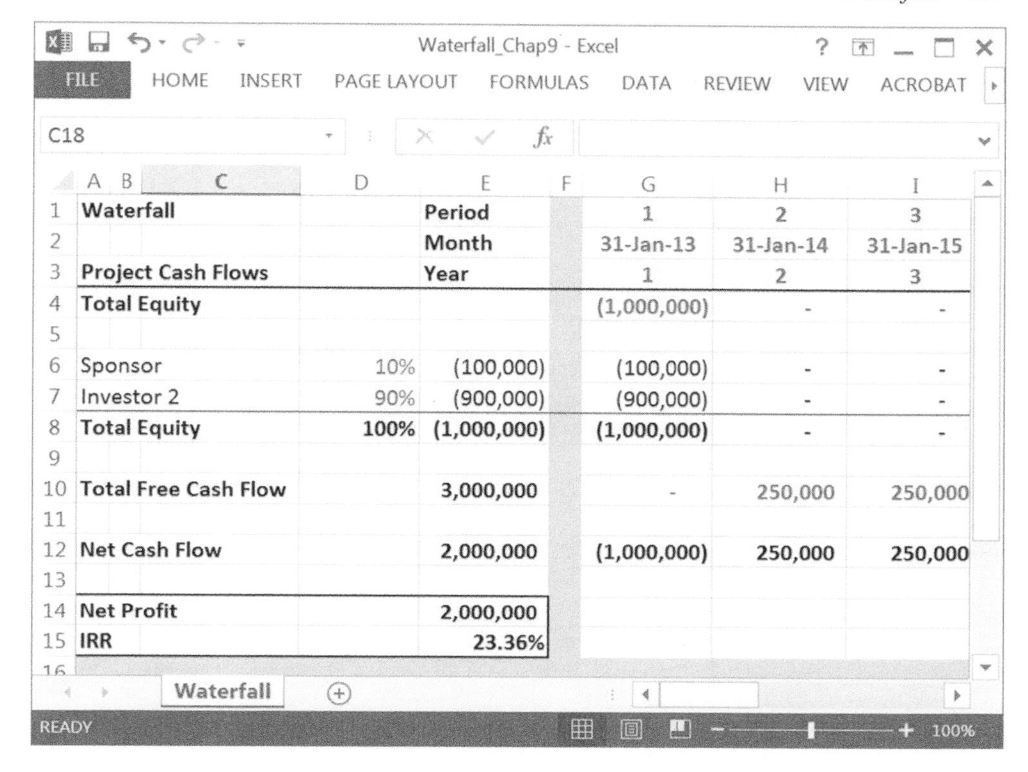

	A B C	D	E	F	G	H	I
1	**Waterfall**		**Period**		1	2	3
2			**Month**		31-Jan-13	31-Jan-14	31-Jan-15
3	**Project Cash Flows**		**Year**		1	2	3
4	**Total Equity**				(1,000,000)	-	-
5							
6	Sponsor	10%	(100,000)		(100,000)	-	-
7	Investor 2	90%	(900,000)		(900,000)	-	-
8	**Total Equity**	**100%**	**(1,000,000)**		**(1,000,000)**	**-**	**-**
9							
10	**Total Free Cash Flow**		**3,000,000**		-	250,000	250,000
11							
12	**Net Cash Flow**		**2,000,000**		**(1,000,000)**	**250,000**	**250,000**
13							
14	**Net Profit**		**2,000,000**				
15	**IRR**		**23.36%**				

Figure 9.8

Accrual (FV): The forward periodic amount to ensure the Tier return is met, i.e. compounded interest.
Paydown: Amount available to be paid down from Project Free Cash Flow.
EoP Balance: End of Period for the project with paydown of accrued and interest.
Overflow: Cash flow remaining (overflowed) after Tier is paid off.
IRR: Internal Rate of Return for Tier, i.e. verifying that Tier IRR target is met (analyst confirmation).

Cells A17:A33 should be labeled as described (Figure 9.9). Cells E20 and E29, Accrual (FV) for the respective Tiers, are 15.00% and 20.00%. These are inputs and referenced from the joint venture agreement governing this waterfall. It is also recommended that cells E20 and E29 be named "Pref_Rate" and "Second_Tier_Rate", respectively.

Cell E17 is the daily rate – i.e. the conversion – for the Preference Rate (15.00%). (See Figure 9.10.) Using the methodology described here, the annual rate must be translated to a daily rate.

$$DailyRate = \left[(1 + EAY)^{1/m}\right] - 1$$

$$DailyRate = \left[(1 + 0.15)^{1/365}\right] - 1 = 0.000383$$

$$DailyRate = \left[(1 + \text{pref\_rate})^{1/365}\right] - 1 = 0.000383$$

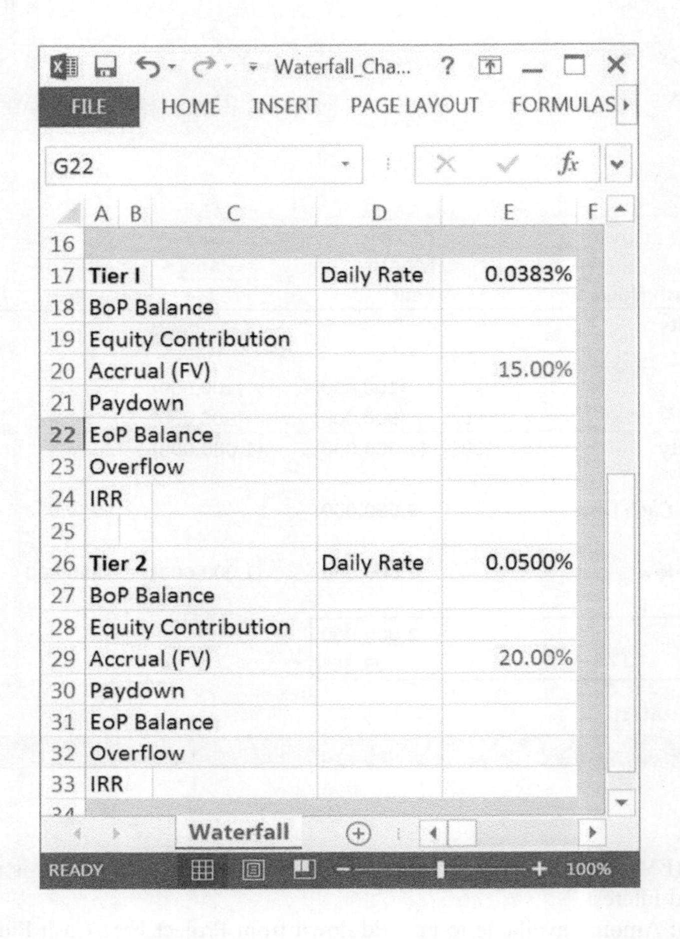

	C	D	E
16			
17	**Tier I**	Daily Rate	0.0383%
18	BoP Balance		
19	Equity Contribution		
20	Accrual (FV)		15.00%
21	Paydown		
22	EoP Balance		
23	Overflow		
24	IRR		
25			
26	**Tier 2**	Daily Rate	0.0500%
27	BoP Balance		
28	Equity Contribution		
29	Accrual (FV)		20.00%
30	Paydown		
31	EoP Balance		
32	Overflow		
33	IRR		

Figure 9.9

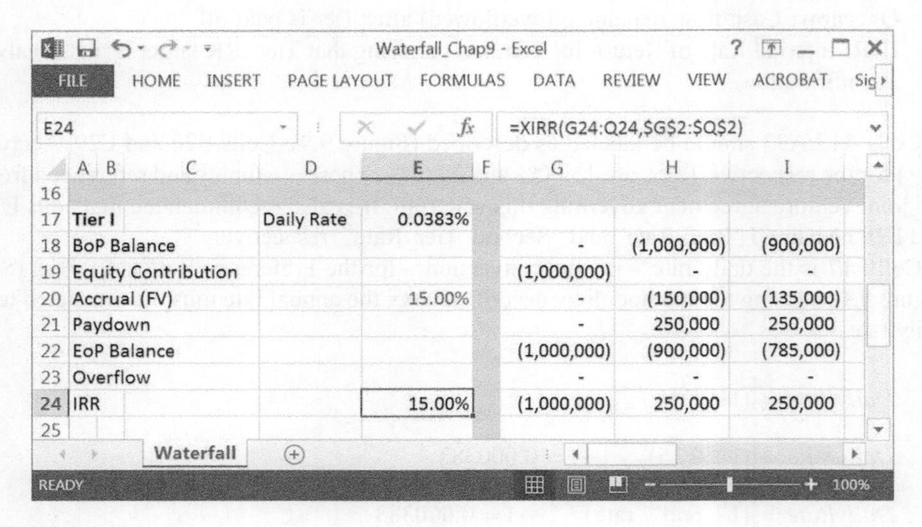

	C	D	E	G	H	I
16						
17	**Tier I**	Daily Rate	0.0383%			
18	BoP Balance			-	(1,000,000)	(900,000)
19	Equity Contribution			(1,000,000)	-	-
20	Accrual (FV)		15.00%	-	(150,000)	(135,000)
21	Paydown			-	250,000	250,000
22	EoP Balance			(1,000,000)	(900,000)	(785,000)
23	Overflow			-	-	-
24	IRR		15.00%	(1,000,000)	250,000	250,000
25						

Figure 9.10

Therefore, cell E17 is "=((1+Pref_Rate)^(1/365)-1)". Following the same logic, cell E26 is "=((1+Second_Tier_Rate)^(1/365)-1)". Now the accrual structures are constructed for each Tier. Note that by utilizing absolute and relative references for Tier 1 correctly, it will enable an easy copy of logic to Tier 2 with the adjustment only for the Second_Tier_Rate – i.e. replacing Pref_Rate with Second_Tier_Rate.

Cell G18, BoP Balance for Tier 1, is "=F22". Column F is grey and therefore all the cells are blank. This was done intentionally to allow consistency of formulas in this section. Row 19, Equity Contribution, references the total contribution by both partners: "=D$5". Row 20, Accrual (FV), follows the logic:

$$Accrual = \left(\left(1 + DailyRate\right)^{\Delta t} - 1\right) \times BoP$$

And this quantifies the accrual per period. The delta t is the number of days between time periods. The formula for cell G20 is "=((1+$E17)^(G$2-F$2)-1)*G18". Row 21 is the Paydown, which is quantified using a Minimum formula and compares cash flow available for distribution to the sum of beginning of balance, current contribution, and quantified accrual. The Paydown is the minimum of available capital to distribute and what is required to distribute to retire the Tier. Cell G21 is therefore "=MIN(G$10,-SUM(G18:G20))". The negative prior to the summation allows comparisons of cash inflow is positive but required, i.e. the summation, is negative. Row 22 is End of Period Balance and is the summation of Beginning of Period and the corresponding changes in position. Cell G22 is thus "=SUM(G18:G20)". Row 23 is the Overflow, what is left over: it is row 10 (total free cash flow) less row 21 (the paydown per period). The formula for cell 23 is "=G$10-G21". Cells G18:G23 are then copied to Q18:Q23.

Row 24 is Internal Rate of Return. It is the cash flows attributable to the Tier. Cell G24 is "=SUM(G19,G21)". This row is utilized to ensure that the Tier structure is working. Cell E24 is a logic function that checks the cash flows for the tier. The IRR for row 24 must be the Preference Rate for the Tier. Therefore, cell E24 is "=XIRR(G24:Q24,$G$2:$Q$2)".

For Tier 2, copy cells G18:Q24 and paste in cell G27 (Figure 9.11). Cell E24 is copied to E33 as well. The second tier is completed without adjustments and using simple copying, as relative and absolute references were utilized.

Waterfall_Chap9 - Excel

FILE HOME INSERT PAGE LAYOUT FORMULAS DATA REVIEW VIEW ACROBAT

A32 Overflow

	A B C	D	E	F	G	H	I	J	K	L	M	N	O	P	Q
17	**Tier I**	Daily Rate	0.0383%												
18	BoP Balance				-	(1,000,000)	(900,000)	(785,000)	(652,750)	(500,950)	(326,092)	(125,006)	-	-	-
19	Equity Contribution				(1,000,000)	-	-	-	-	-	-	-	-	-	-
20	Accrual (FV)		15.00%		-	(150,000)	(135,000)	(117,750)	(98,200)	(75,142)	(48,914)	(18,751)	-	-	-
21	Paydown				-	250,000	250,000	250,000	250,000	250,000	250,000	143,757	-	-	-
22	EoP Balance				(1,000,000)	(900,000)	(785,000)	(652,750)	(500,950)	(326,092)	(125,006)	-	-	-	-
23	Overflow				-	-	-	-	-	-	-	106,243	250,000	250,000	750,000
24	IRR		15.00%		(1,000,000)	250,000	250,000	250,000	250,000	250,000	250,000	143,757	-	-	-
25															
26	**Tier 2**	Daily Rate	0.0500%												
27	BoP Balance				-	(1,000,000)	(950,000)	(890,000)	(818,000)	(732,090)	(628,509)	(504,210)	(355,052)	(176,276)	-
28	Equity Contribution				(1,000,000)	-	-	-	-	-	-	-	-	-	-
29	Accrual (FV)		20.00%		-	(200,000)	(190,000)	(178,000)	(164,090)	(146,418)	(125,702)	(100,842)	(71,223)	(35,255)	-
30	Paydown				-	250,000	250,000	250,000	250,000	250,000	250,000	250,000	250,000	211,531	-
31	EoP Balance				(1,000,000)	(950,000)	(890,000)	(818,000)	(732,090)	(628,509)	(504,210)	(355,052)	(176,276)	-	-
32	Overflow				-	-	-	-	-	-	-	-	-	38,469	750,000
33	IRR		20.00%		(1,000,000)	250,000	250,000	250,000	250,000	250,000	250,000	250,000	250,000	211,531	-
34															

Waterfall

READY

Figure 9.11

Notice that Tier 2 overflow, row 32, is two periods less than Tier 1. This is a demonstration of the waterfall.

Section 3: separate cash flows by tier by investor

The third section (Figure 9.12) separates the cash flows by Tier (as previously) but also segregates these cash flows by investor. Cell A35 is 'Cash Flows to Each Tier'. Rows 36:38 are Tier 1 Preference Rate and sponsor/investor splits. Cell A36 is '1. Split', cell B37 is 'Sponsor', and cell B38 is 'Investor'. Column D in this section references the rate for the Tier and the percentage splits. Cell D36 is "=Pref_Rate", and D37 and D38 are "=D6" and "=D7", respectively. The joint venture agreement has the first tier being split pari passu (i.e. the same percentage as invested capital).

Rows 40:42 follow the same logic as rows 36:38 did for Tier 1. Cell A40 is '2. Tier 2 Split', cell B41 is 'Sponsor', and cell B42 is 'Investor'. Column D in this section, as above, references the rate for the Tier and the percentage splits. Cell D40 is "=Second_Tier_Rate". D41 is '20.00%', as stated in the joint venture agreement, and coloured blue as an input; D42 is "=1-D41".

Row 44 is the overflow after Tier 2. Cell A44 is 'Tier 2 Overflow'. Row 46 is the sponsor promote from the joint venture agreement at 40.00%. Cell A46 is '3. Sponsor Promote' and

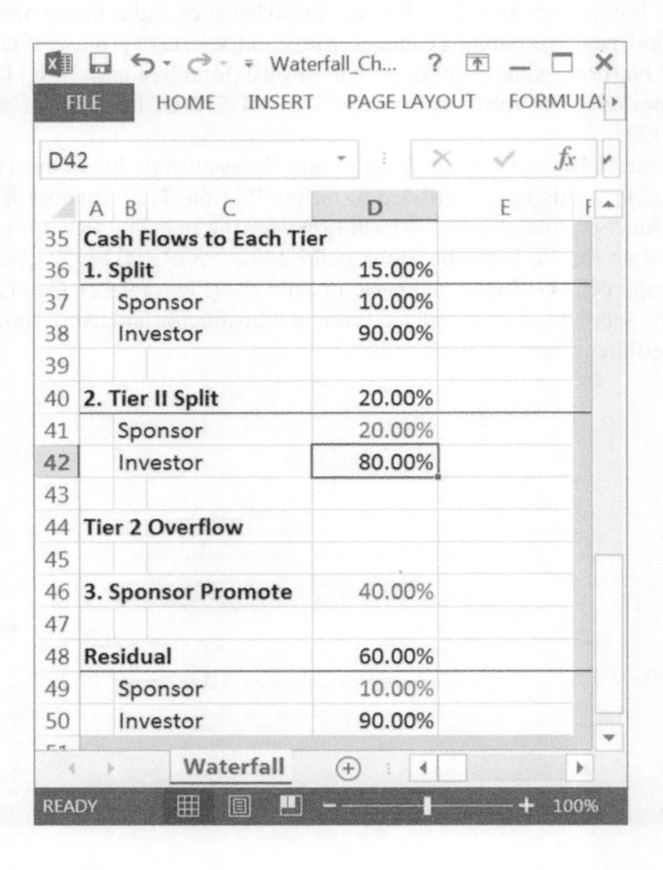

Figure 9.12

cell D46 is '40.00%'. Rows 48:50 are the residual overflow after the sponsor promote (40%). Cell A48 is 'Residual', and cells B49 and B50 are 'Sponsor' and 'Investor', respectively. Cell D48 is "=1-D46". Cell D49 is '10.00%', an input from the joint venture agreement. Cell D50 is "=1-D49".

Row 36 is the payment for Tier 1, which references row 21 from Section 2. Cell G36 is therefore "=G21". Cell G36 is then copied along row 36 to column Q (Figure 9.13). Row 40 is the payment for Tier 2, which is the difference between Tier 2 Paydown and Tier 1 Paydown. Cell G40 is therefore "=G30-G21". Cell G40 is then copied to H40:Q40. Finally, row 44 is 'Tier 2 Overflow', and cell G44 is "=G10-G30". Notice the waterfall between the two Tiers and the overflow.

Now that each of the Tier's cash flows have been modelled, these cash flows must be separated by sponsor and investor (Figure 9.14). This is simply a percentage split based upon the percentages in column D. Cell G37 is therefore "=$D37*G$36". This is then copied to cell G38 and cells G37:G38 are copied to column Q.

The separation of cash flows for Tiers 2 and the overflow are now quantified. Cell G41 is "=$D41*G$40" and is copied to row 42 and then cells G41:G42 are copied to column Q (Figure 9.15). Row 46 is the sponsor promote and is 40% of the Tier 2 Overflow (i.e. row 44).

Waterfall_Chap9 - Excel

A43

	A B C	D	E	F	G	H	I	J	K	L	M	N	O	P	Q
34															
35	**Cash Flows to Each Tier**														
36	**1. Split**	15.00%			-	250,000	250,000	250,000	250,000	250,000	250,000	143,757	-	-	-
37	Sponsor	10.00%													
38	Investor	90.00%													
39															
40	**2. Tier II Split**	20.00%			-	-	-	-	-	-	-	106,243	250,000	211,531	-
41	Sponsor	20.00%													
42	Investor	80.00%													
43															
44	**Tier 2 Overflow**				-	-	-	-	-	-	-	-	-	38,469	750,000
45															
46	**3. Sponsor Promote**	40.00%													
47															
48	**Residual**	60.00%													
49	Sponsor	10.00%													
50	Investor	90.00%													

Waterfall

Figure 9.13

Waterfall_Chap9 - Excel

G37 =$D37*G$36

	A B C	D	E	F	G	H	I	J	K	L	M	N	O	P	Q
34															
35	**Cash Flows to Each Tier**														
36	**1. Split**	15.00%			-	250,000	250,000	250,000	250,000	250,000	250,000	143,757	-	-	-
37	Sponsor	10.00%			-	25,000.00	25,000.00	25,000.00	25,000.00	25,000.00	25,000.00	14,375.73	-	-	-
38	Investor	90.00%			-	225,000.00	225,000.00	225,000.00	225,000.00	225,000.00	225,000.00	129,381.59	-	-	-
39															
40	**2. Tier II Split**	20.00%			-	-	-	-	-	-	-	106,243	250,000	211,531	-
41	Sponsor	20.00%													
42	Investor	80.00%													
43															
44	**Tier 2 Overflow**				-	-	-	-	-	-	-	-	-	38,469	750,000
45															
46	**3. Sponsor Promote**	40.00%													
47															
48	**Residual**	60.00%													
49	Sponsor	10.00%													
50	Investor	90.00%													

Waterfall

Figure 9.14

	A–C	D	E	G	H	I	J	K	L	M	N	O	P	Q
34														
35	**Cash Flows to Each Tier**													
36	**1. Split**	15.00%	1,643,757	-	250,000	250,000	250,000	250,000	250,000	250,000	143,757	-	-	-
37	Sponsor	10.00%	164,376	-	25,000	25,000	25,000	25,000	25,000	25,000	14,376	-	-	-
38	Investor	90.00%	1,479,382	-	225,000	225,000	225,000	225,000	225,000	225,000	129,382	-	-	-
39														
40	**2. Tier II Split**	20.00%	567,773	-	-	-	-	-	-	-	106,243	250,000	211,531	-
41	Sponsor	20.00%	113,555	-	-	-	-	-	-	-	21,249	50,000	42,306	-
42	Investor	80.00%	454,219	-	-	-	-	-	-	-	84,994	200,000	169,225	-
43														
44	**Tier 2 Overflow**		788,469	-	-	-	-	-	-	-	-	-	38,469	750,000
45														
46	**3. Sponsor Promote**	40.00%	315,388	-	-	-	-	-	-	-	-	-	15,388	300,000
47														
48	**Residual**	60.00%	473,082	-	-	-	-	-	-	-	-	-	23,082	450,000
49	Sponsor	10.00%	47,308	-	-	-	-	-	-	-	-	-	2,308	45,000
50	Investor	90.00%	425,773	-	-	-	-	-	-	-	-	-	20,773	405,000

Figure 9.15

Cell G46 is "=$D46*G$44" and is copied to column Q. Row 48 is the residual, which is after the promote and split pari passu. Cell G48 is "=$D48*G$44" and copied to column Q. Lastly, rows 49 and 50 are sponsor and investor splits for the residual. Cell G49 is "=$D49*G$48" and is copied to cell G50. Then cells G49 and G50 are copied to column Q.

The final addition to this section is column E, the summary of the payments in each period for the respective row. Cell E36 is "=SUM(G36:Q36)". Cell E36 is then copied down column E to rows 37:38, 40:42, 44, 46, and 48:50, using paste special formulas. The use of paste special formulas ensures the original formatting in each cell is retained. Section 3, cash flows to each tier, is now completed.

Section 4: analyze investor cash flows

In this final section, the cash flows for both the sponsor and investor are analyzed and valued (Figure 9.16). Cell A52 is 'Investor Cash Flows' and cell A53 is 'Sponsor'. Cells B54 and B55 are 'Equity Investment(s)' and 'Distribution(s)', respectively. Cell C56 is 'Total Cash Flow' for the investor type.

The valuation is then completed for the investor. Cell A58 is 'Total Profits', cell A59 is '% of Total Profits', and cell A60 is 'IRR'. The same headers are completed for the investor starting in row 62.

Row 54, Equity Investment(s), references the capital inflow for the sponsor (Figure 9.17). Therefore cell G54 is "=G6". Cell G54 is then copied to column Q. Row 55 is the Distributions for the sponsor. Cell G55 is "=SUM(G37,G41,G46,G49)". These reference the rows for Tier 1 Sponsor, Tier 2 Sponsor, Promote, and Overflow Sponsor. Cell G55 is copied to column Q. Row 56 is the Total Cash Flow to the Sponsor and is the summation of the Equity Investment(s) and Distribution(s); thus cell G56 is "=SUM(G54:G55)". Cell G56 is copied to column Q. To complete rows 54:56, column E represents the summation of the periodic payments for each. Cell E54 is "=SUM(G54:Q54)" and is copied to cells E55 and E56.

Cell E58 is Total Profits for the Sponsor. Cell E58 is "=E56". Cell E59, % of Total Profits, is "=E58/E14". Note that cell E14 is total project profit. The sponsor's IRR, cell E60, is "=XIRR(G56:Q56,G2:Q2)".

The equivalent process is completed for the investor cash flows and analysis. Cell G63 is "=G7". Cell 64, distributions, is "=SUM(G38,G42,G50)". These represent Tier 1, Tier 2, and

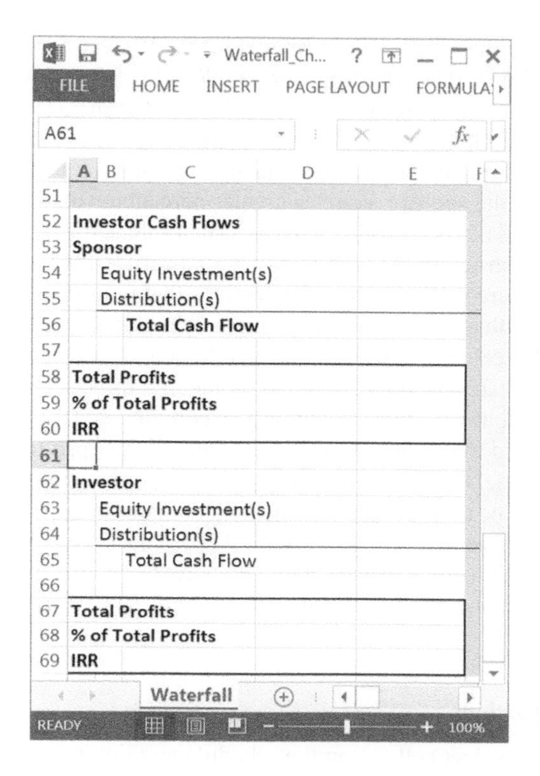

Figure 9.16

Row	A–C	E	G	H	I	J	K	L	M	N	O	P	Q
52	**Investor Cash Flows**												
53	**Sponsor**												
54	Equity Investment(s)	(100,000)	(100,000)	-	-	-	-	-	-	-	-	-	-
55	Distribution(s)	640,626	-	25,000	25,000	25,000	25,000	25,000	25,000	35,624	50,000	60,002	345,000
56	**Total Cash Flow**	**540,626**	**(100,000)**	**25,000**	**25,000**	**25,000**	**25,000**	**25,000**	**25,000**	**35,624**	**50,000**	**60,002**	**345,000**
57													
58	**Total Profits**	**540,626**											
59	**% of Total Profits**	**27.03%**											
60	**IRR**	**32.06%**											
61													
62	**Investor**												
63	Equity Investment(s)	(900,000)	(900,000)	-	-	-	-	-	-	-	-	-	-
64	Distribution(s)	2,359,374	-	225,000	225,000	225,000	225,000	225,000	225,000	214,376	200,000	189,998	405,000
65	**Total Cash Flow**	**1,459,374**	**(900,000)**	**225,000**	**225,000**	**225,000**	**225,000**	**225,000**	**225,000**	**214,376**	**200,000**	**189,998**	**405,000**
66													
67	**Total Profits**	**1,459,374**											
68	**% of Total Profits**	**72.97%**											
69	**IRR**	**21.78%**											

Figure 9.17

Overflow distribution to the investor. Note that the promote was provided only to the sponsor, according to the joint venture agreement. Cell G65 is Total Cash Flow for the investor and is therefore "=SUM(G63:G64)". Cells G63:G65 are copied to column Q. Cells E63:E65 are the summation of columns G:Q.

Cell E67, Total Profits for investor, is "=E65". Cell E68, % of Total Profits for the investor, is "=E67/E12". The Internal Rate of Return for the investor, cell E69, is "=XIRR(G65:Q65,G2:Q2)". The waterfall is now complete!

Of particular note are the differences in returns for the Project, Sponsor, and Investor. The original capital was provided as 10% Sponsor and 90% Investor. However, the final returns as quantified by Internal Rate of Return are 32.06% Sponsor and 21.78% Investor. The joint venture agreement, through the use of a waterfall structure, redistributes profits. In the structure that was modelled, it incentivizes the sponsor to pursue a strategy of higher, and potentially riskier, returns as the sponsor shares a greater percentage of the higher project return.

Note that the waterfall has been constructed as a stand-alone file – i.e. it is not incorporated into a larger pro forma structure. As was demonstrated earlier, the waterfall must be "bolted on" to a pro forma (Figure 9.18). Each of the sections that require external pro forma inputs were to be indicated in blue. In short these were the Schedule, Equity, Total Free Cash Flow, Tier Splits from JV Agreement, and Preference and Second Tier rates, the Tier rates for the waterfall (Figure 9.19). These inputs are found on the first three sections of the waterfall. In addition, the waterfall that was constructed only accommodated 12 periods. The pro forma will need to be expanded to the right as well to fit into a larger pro forma.

Returning to the model completed in the previous chapter with the Construction/Draw page, this waterfall will be added to the model. With both models open, right click the Waterfall sheet and select 'Move or Copy. . .' (Figure 9.20). The 'Move or Copy' menu appears; select the new file. Select '(move to end)' and then 'Create a copy'. Select 'OK' and this will copy the Waterfall sheet and add it to the new pro forma. The incorporation of the waterfall will take several steps.

Step 1: link project portions

On the Waterfall sheet, select cell G1 on the schedule and link this to the Pro_Forma sheet. Cell G1 is then "=Pro_Forma!G1" (Figure 9.21). Copy cell G1 and paste to G2:G3, then copy cells G1:G3 and paste to EH1:EH3, period 132 (11 years). It is recommended that copying G1 to G2 and G3 be done with copy and then 'paste special'. Copy only the formula to ensure that formatting is unchanged. While G1:EH3 are selected, the cells should be coloured black; they are no longer inputs given that they have been incorporated into the Pro Forma.

Total Equity, row 4 on the waterfall, must be linked. This requires attention as the Pro_ Forma Valuation section has the initial equity in column F, which does not parallel the schedule starting in column G. Therefore the equity from the Pro_Forma page must be escalated to the first period of the waterfall, which is commercial operating date. Note that the sponsor/investor valuations will be valued as of the commercial operating date, whereas the

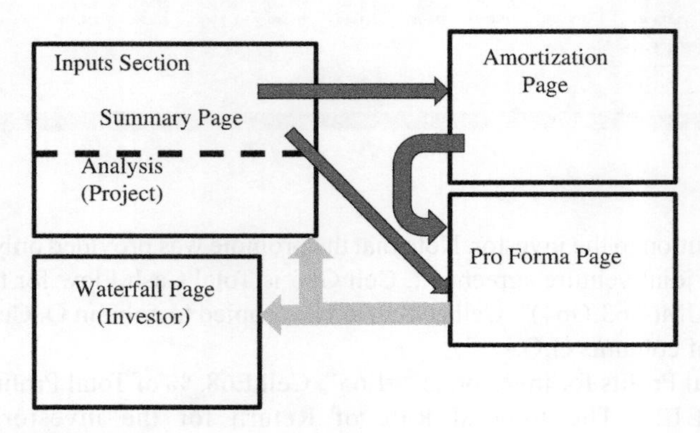

Figure 9.18 Waterfall Addition

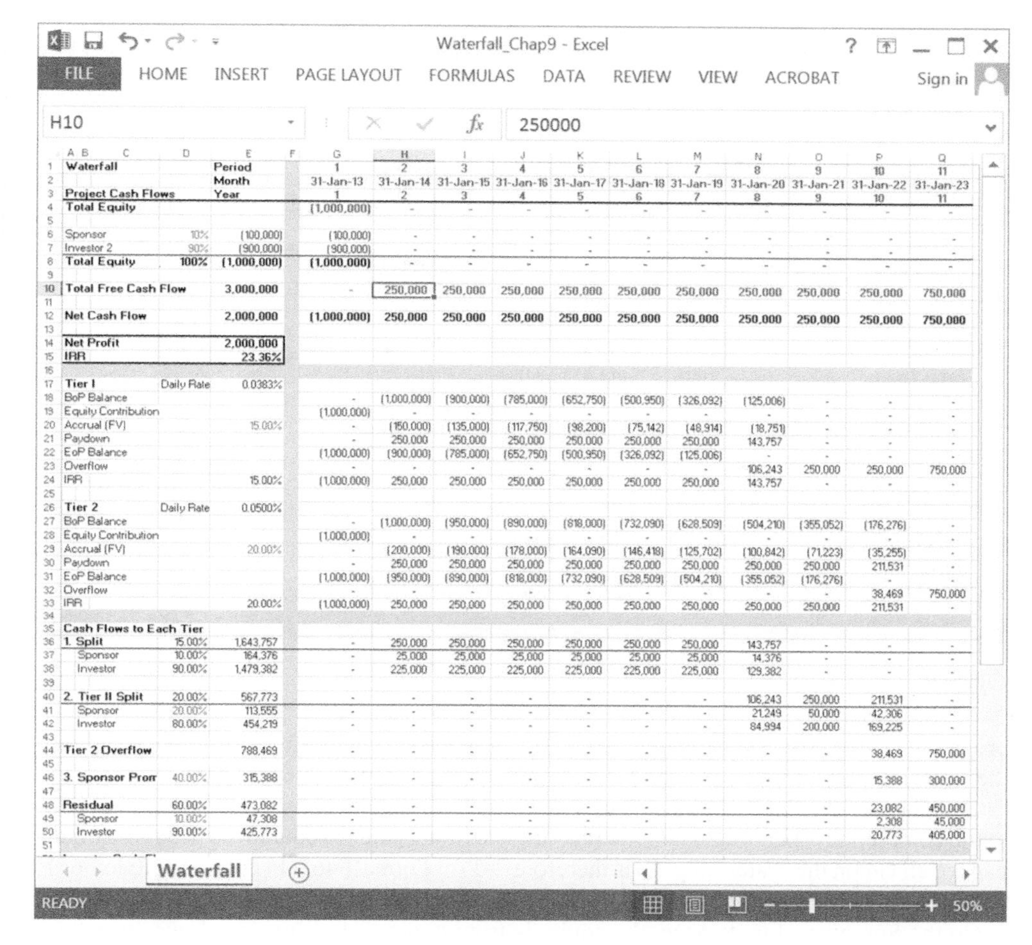

Waterfall_Chap9 - Excel

FILE HOME INSERT PAGE LAYOUT FORMULAS DATA REVIEW VIEW ACROBAT Sign in

H10 250000

A B C	D	E	F	G	H	I	J	K	L	M	N	O	P	Q
Waterfall		Period		1	2	3	4	5	6	7	8	9	10	11
		Month		31-Jan-13	31-Jan-14	31-Jan-15	31-Jan-16	31-Jan-17	31-Jan-18	31-Jan-19	31-Jan-20	31-Jan-21	31-Jan-22	31-Jan-23
Project Cash Flows		Year		1	2	3	4	5	6	7	8	9	10	11
Total Equity				(1,000,000)	-	-	-	-	-	-	-	-	-	-
Sponsor	10%	(100,000)		(100,000)	-	-	-	-	-	-	-	-	-	-
Investor 2	90%	(900,000)		(900,000)	-	-	-	-	-	-	-	-	-	-
Total Equity	100%	(1,000,000)		(1,000,000)	-	-	-	-	-	-	-	-	-	-
Total Free Cash Flow		3,000,000		-	250,000	250,000	250,000	250,000	250,000	250,000	250,000	250,000	250,000	750,000
Net Cash Flow		2,000,000		(1,000,000)	250,000	250,000	250,000	250,000	250,000	250,000	250,000	250,000	250,000	750,000
Net Profit		2,000,000												
IRR		23.36%												
Tier I	Daily Rate	0.0383%												
BoP Balance				-	(1,000,000)	(900,000)	(785,000)	(652,750)	(500,950)	(326,092)	(125,006)	-	-	-
Equity Contribution				(1,000,000)	-	-	-	-	-	-	-	-	-	-
Accrual (FV)		15.00%		-	(150,000)	(135,000)	(117,750)	(98,200)	(75,142)	(48,914)	(18,751)	-	-	-
Paydown				-	250,000	250,000	250,000	250,000	250,000	250,000	143,757	-	-	-
EoP Balance				(1,000,000)	(900,000)	(785,000)	(652,750)	(500,950)	(326,092)	(125,006)	-	-	-	-
Overflow				-	-	-	-	-	-	-	106,243	250,000	250,000	750,000
IRR		15.00%		(1,000,000)	250,000	250,000	250,000	250,000	250,000	250,000	143,757	-	-	-
Tier 2	Daily Rate	0.0500%												
BoP Balance				-	(1,000,000)	(950,000)	(890,000)	(818,000)	(732,090)	(628,509)	(504,210)	(355,052)	(176,276)	-
Equity Contribution				(1,000,000)	-	-	-	-	-	-	-	-	-	-
Accrual (FV)		20.00%		-	(200,000)	(190,000)	(178,000)	(164,090)	(146,418)	(125,702)	(100,842)	(71,223)	(35,255)	-
Paydown				-	250,000	250,000	250,000	250,000	250,000	250,000	250,000	250,000	211,531	-
EoP Balance				(1,000,000)	(950,000)	(890,000)	(818,000)	(732,090)	(628,509)	(504,210)	(355,052)	(176,276)	-	-
Overflow				-	-	-	-	-	-	-	-	-	38,469	750,000
IRR		20.00%		(1,000,000)	250,000	250,000	250,000	250,000	250,000	250,000	250,000	250,000	211,531	-
Cash Flows to Each Tier														
1. Split	15.00%	1,643,757		-	250,000	250,000	250,000	250,000	250,000	250,000	143,757	-	-	-
Sponsor	10.00%	164,376		-	25,000	25,000	25,000	25,000	25,000	25,000	14,376	-	-	-
Investor	90.00%	1,479,382		-	225,000	225,000	225,000	225,000	225,000	225,000	129,382	-	-	-
2. Tier II Split	20.00%	567,773		-	-	-	-	-	-	-	106,243	250,000	211,531	-
Sponsor	20.00%	113,555		-	-	-	-	-	-	-	21,249	50,000	42,306	-
Investor	80.00%	454,219		-	-	-	-	-	-	-	84,994	200,000	169,225	-
Tier 2 Overflow		788,469		-	-	-	-	-	-	-	-	-	38,469	750,000
3. Sponsor Prom	40.00%	315,388		-	-	-	-	-	-	-	-	-	15,388	300,000
Residual	60.00%	473,082		-	-	-	-	-	-	-	-	-	23,082	450,000
Sponsor	10.00%	47,308		-	-	-	-	-	-	-	-	-	2,308	45,000
Investor	90.00%	425,773		-	-	-	-	-	-	-	-	-	20,773	405,000

Waterfall

READY 50%

Figure 9.19

Move or Copy

Move selected sheets

To book:

Draw_Waterfall_Chap9.xlsx

Before sheet:

Development
Unit_Characteristics
Summary_Inputs
Debt
Pro_Forma
Steps_RR
Steps_Draw
(move to end)

Create a copy

OK Cancel

Figure 9.20

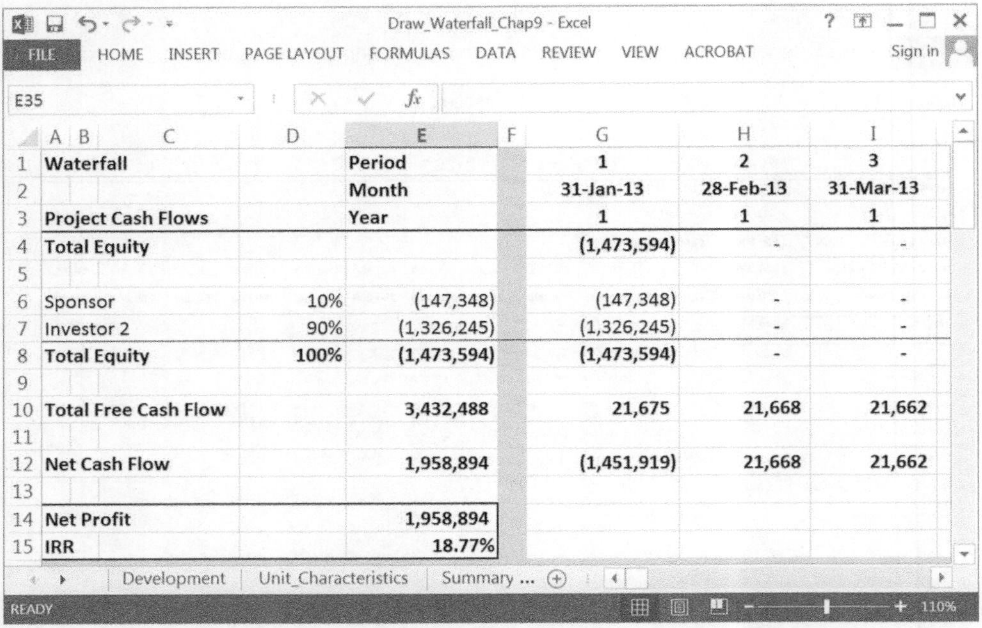

Draw_Waterfall_Chap9 - Excel

	A B C	D	E	F	G	H	I
1	**Waterfall**		**Period**		**1**	**2**	**3**
2			**Month**		**31-Jan-13**	**28-Feb-13**	**31-Mar-13**
3	**Project Cash Flows**		**Year**		**1**	**1**	**1**
4	**Total Equity**				**(1,473,594)**	**-**	**-**
5							
6	Sponsor	10%	(147,348)		(147,348)	-	-
7	Investor 2	90%	(1,326,245)		(1,326,245)	-	-
8	**Total Equity**	**100%**	**(1,473,594)**		**(1,473,594)**	-	-
9							
10	**Total Free Cash Flow**		**3,432,488**		**21,675**	**21,668**	**21,662**
11							
12	**Net Cash Flow**		**1,958,894**		**(1,451,919)**	**21,668**	**21,662**
13							
14	**Net Profit**		**1,958,894**				
15	**IRR**		**18.77%**				

Figure 9.21

project is valued as of Development Start. This will also account for the difference between the waterfall project value and the project value on the Summary_Inputs page.

Cell G4 uses the Future Value equation to escalate the equity from the Development Start to the commercial operating date. The rate of escalation is the Discount Rate from the Inputs page. Cell G4 on the Waterfall page is therefore "=Pro_Forma!F46*(1+Discount_Rate/365)^(Pro_Forma!G45-Pro_Forma!F45)".

On the Summary_Inputs page, the equity is from two sources: Investor 1 and Investor 2 with equity of $1,295,000 and $56,000, respectively. As the total equity is $1,351,000, the investors will be adjusted; cell C3 is 'Sponsor' and C4 is 'Investor'. The sponsor's equity is replaced with $135,100. The investor's equity is replaced with $1,216,000. Returning to the Waterfall page (Figure 9.21), cell D6 (sponsor equity percentage) is "=Summary_Inputs!D3/Summary_Inputs!D2". Cells G6:G8 are then copied to column DV (period 120), as it is a 10-year pro forma. Also, cells E6:E8 must be adjusted to summarize to column DV as well.

On the Waterfall page, row 10, Total Free Cash Flow, is then linked to the Pro_Forma page. Cell G10 is "=Pro_Forma!G46" and is then copied across to column DV. Cell E10 is adjusted to summarize to column DV. Row 12, Net Cash Flow, is also copied to column DV. The project IRR on the Waterfall page, cell E15, needs to be adjusted to column DV. Cell E15 is therefore "=XIRR(G12:DV12,G2:DV2)". As completed in cells above, cell E12 is adjusted to summarize through column DV.

Step 2: link the tier rates

Only two cells must be adjusted for this section of the waterfall (Figure 9.22). The Internal Rate of Return checks, cells E24 and E33, are copied to column DV, the 120th period. Cells Q16:Q34 are then copied through rows 16:34 to column DV to expand the section.

	A B C	D	E	G	H	I
16						
17	**Tier I**	Daily Rate	0.0383%			
18	BoP Balance			-	(1,451,919)	(1,445,901)
19	Equity Contribution			(1,473,594)	-	-
20	Accrual (FV)		15.00%	-	(15,650)	(17,265)
21	Paydown			21,675	21,668	21,662
22	EoP Balance			(1,451,919)	(1,445,901)	(1,441,504)
23	Overflow			-	-	-
24	IRR		15.00%	(1,451,919)	21,668	21,662
25						
26	**Tier 2**	Daily Rate	0.0500%			
27	BoP Balance			-	(1,451,919)	(1,450,700)
28	Equity Contribution			(1,473,594)	-	-
29	Accrual (FV)		20.00%	-	(20,450)	(22,639)
30	Paydown			21,675	21,668	21,662
31	EoP Balance			(1,451,919)	(1,450,700)	(1,451,677)
32	Overflow			-	-	-
33	IRR		18.77%	(1,451,919)	21,668	21,662

Figure 9.22

Note that cell E33 is not 20.00% as the second tier rate but rather 18.77%. This is because the project's maximum return is 18.77% and the second tier rate is not met. Note that the inputs for the Preference Rate and Second Tier Rate remain inputs in this section and thus are to be coloured blue. This is done deliberately to separate the page for the sponsor and investor.

Step 3: link cash flows to each tier

In order to link cash flows to each tier (see Figure 9.23), cells Q35:Q51 are selected and then copied to column DV. In addition, cells E36:E50 are adjusted to summarize through column DV.

Step 4: adjust investor cash flows

Cells Q54:Q65 are now copied to column DV to adjust the cash flows for the investor. Cells E54:E56 and E63:E65 are adjusted to summarize through column DV (Figure 9.24). The sponsor and investor IRR cells, E60 and E69, are adjusted as well through column DV.

Note that the sponsor and investor IRRs are 21.50% and 18.42%, respectively. The project return, as stated in Section 1 of the waterfall, is 18.77%. The returns for the sponsor and investor have been adjusted according to the joint venture agreement.

Draw_Waterfall_Chap9 - Excel

	A B C	D	E	F	G	H	I
34							
35	**Cash Flows to Each Tier**						
36	**1. Split**	15.00%	2,606,805		21,675	21,668	21,662
37	Sponsor	10.00%	260,661		2,167	2,167	2,166
38	Investor	90.00%	2,346,144		19,508	19,502	19,496
39							
40	**2. Tier II Split**	20.00%	825,682		-	-	-
41	Sponsor	20.00%	165,136		-	-	-
42	Investor	80.00%	660,546		-	-	-
43							
44	**Tier 2 Overflow**		-		-	-	-
45							
46	**3. Sponsor Promote**	40.00%	-		-	-	-
47							
48	**Residual**	60.00%	-		-	-	-
49	Sponsor	10.00%	-		-	-	-
50	Investor	90.00%	-		-	-	-
51							

Figure 9.23

Draw_Waterfall_Chap9 - Excel

	A B C	D	E	F	G	H	I
52	**Investor Cash Flows**						
53	**Sponsor**						
54	Equity Investment(s)		(147,348)		(147,348)	-	-
55	Distribution(s)		425,798		2,167	2,167	2,166
56	**Total Cash Flow**		**278,449**		**(145,181)**	**2,167**	**2,166**
57							
58	**Total Profits**		**278,449**				
59	**% of Total Profits**		**14.21%**				
60	**IRR**		**21.50%**				
61							
62	**Investor**						
63	Equity Investment(s)		(1,326,245)		(1,326,245)	-	-
64	Distribution(s)		3,006,690		19,508	19,502	19,496
65	**Total Cash Flow**		**1,680,445**		**(1,306,737)**	**19,502**	**19,496**
66							
67	**Total Profits**		**1,680,445**				
68	**% of Total Profits**		**85.79%**				
69	**IRR**		**18.42%**				

Figure 9.24

Further, the project return on the Summary_Inputs page is 16.55%. The values are different as a result of times of valuation and because the equity was escalated at the Discount Rate on the waterfall.

Joint venture agreement example

[date]

[Sponsor]

[Address]

Re: Equity Investment – [property]

Dear [Principal]:

This letter outlines the basic terms and conditions upon which Stage Capital, LLC (with its successors, "Investor"), or its designee and/or affiliate (Investor, or such designee or affiliate being sometimes referred to in this letter as "STAGE"), will consider entering into a transaction (the "Equity Investment") pertaining to the property described in Paragraph A below (the "Property") whose owner will be limited liability company ("Owner"), an entity whose managing member will be [Sponsor], or its designee or affiliate ("Sponsor").

A. Property description

Project Name:	
Property:	Acquisition of approximately [____] square feet of land and development of [____] square feet of *[____]*. Owner shall be the fee simple owner of the Property.
Location:	The Property is located in [____].

B. Documentation

The parties' obligations will be set forth in definitive documentation ("*Definitive Documents*") in a manner mutually satisfactory to STAGE and Sponsor. STAGE's counsel will begin preparation of definitive documentation upon the Approval Date, and the parties anticipate that final Definitive Documents will be ready for execution within thirty (30) days after the Approval Date.

C. Equity investment terms

The proposed Equity Investment terms are as outlined on *Exhibit A* attached hereto.

D. Due diligence

For a period (the "*Due Diligence Period*") of thirty (30) days commencing on the date (the "*Commencement Date*") STAGE receives a copy or original of this LOI executed by you, you will provide STAGE with access to the Property and all information regarding the Property for the purpose of performing its due diligence with respect to the Equity Investment to its sole satisfaction. At any time prior to the end of the Due Diligence Period (the "*Approval Date*"), STAGE may give notice as to whether (i) STAGE desires to proceed with the Equity Investment subject to satisfaction of the closing conditions specified in any Definitive Documentation or (ii) STAGE desires to terminate this letter (which STAGE may elect to do in STAGE's sole and absolute discretion). If STAGE elects to proceed, STAGE shall provide Sponsor written notice of such election no later than the Approval Date. If STAGE does not elect in writing to proceed by the close of business (Eastern Standard Time) on the Approval Date, this letter automatically shall terminate. In connection with STAGE's due diligence, Sponsor will, not later than two (2) business days following the Commencement Date, prepare and deliver to STAGE (to the extent reasonably available to Sponsor and not previously provided to STAGE) and to STAGE's representatives the due diligence materials identified on *Exhibit B* attached hereto. Sponsor will also make available other documentation, in a timely manner, as reasonably requested during the course of the due diligence process. Sponsor shall execute and deliver in conjunction with the Due Diligence materials, the Credit Report Authorization and Release (*Exhibit C*) for each Principal herein defined. Without limiting the preceding provisions of this paragraph, STAGE shall have the right in its sole discretion, upon written notice given to Sponsor, to terminate this letter at any time prior to the expiration of the Due Diligence Period.

E. Conditions to closing for the benefit of STAGE

The conditions to the closing of the Equity Investment by STAGE include, but are not limited to:

STAGE giving notice that it intends to proceed on or before the conclusion of the Due Diligence Period;

The approval of STAGE's investment committee, which will be sought during the Due Diligence Period;

Equity Investment documentation satisfactory to STAGE and satisfaction of closing conditions set forth in such documentation, including, without limitation, opinions of counsel satisfactory to STAGE and joint venture documentation satisfactory to STAGE;

Receipt of third-party estoppels, subordination, nondisturbance and attornment agreements, consents and will-serve undertakings, and other deliveries as required by STAGE;

Such property or financial conditions as may be specified by STAGE; and

That there has not occurred a material adverse change that affects in any way Sponsor or the Property or the Equity Investment.

F. Costs

As part of the project budget, Owner shall be responsible for and shall pay all transfer, mortgage, note, intangible and similar taxes, escrow, title, lien and violations search, survey, recording and filing fees, and all other fees and expenses (including without limitation all of STAGE's out-of-pocket expenses, reasonable legal fees, due diligence expenses, consultants' expenses, accountants' fees and expenses, and printing costs and any other third-party fees) associated with the Equity Investment and the negotiation and documentation of the governing documents of Owner;

provided that, upon closing the Equity Investment, all of such costs, fees and expenses will be payable by Owner, in part from Equity Investment proceeds as may be approved by STAGE. Owner and Sponsor shall be solely responsible for the payment of advisory or brokerage fees to all parties that it has engaged.

G. Exclusivity

In consideration of STAGE's effort and expense in analyzing this transaction, Sponsor agrees that as long as STAGE is proceeding in good faith to underwrite the Equity Investment and/or negotiate formal written Equity Investment agreements with regard to the Equity Investment, at all times after the date hereof but prior to the termination of this letter (which letter may be terminated in accordance with Paragraph D above), Sponsor, for itself and on behalf of each of their respective affiliates and their respective representatives, agents and employees, will not directly or indirectly make, accept, negotiate, entertain or otherwise pursue any offers to either sell the Property or any interest therein or to engage in any financing or other capital transaction regarding the Property, other than the investment contemplated hereby with STAGE.

____

Initials of Sponsor

H. Confidentiality

By execution of this LOI, Sponsor agrees to maintain the confidentiality of STAGE's involvement in this possible transaction and the structure and pricing thereof. Prior to Closing, Sponsor further agrees not to disclose any information regarding STAGE's involvement in this transaction to any person or entity other than, on an as-needed basis, with its advisors, agents and consultants who will assist Sponsor in its transaction with STAGE, and Sponsor will inform each of them of the confidentiality requirements of this letter and their duty to comply with its terms. The foregoing shall not apply to any disclosures required by law or disclosures consented to by STAGE.

STAGE, by execution of this letter, agrees to maintain prior to Closing the confidentiality of the information contained in the documents provided by Sponsor to STAGE pursuant to Paragraph A above. STAGE further agrees not to disclose any such information to any person or entity other than, on an as-needed basis, with its advisors, agents and consultants who will assist STAGE in its transaction with Sponsor and STAGE will inform each of them of the confidentiality requirements of this letter and their duty to comply with its terms. The foregoing shall not apply to any disclosures required by law or disclosures consented to by Sponsor.

I. Brokers

Each party represents and warrants to the other that such party has not engaged any person to whom a commission or finders' fee may be owing by reason of the transactions contemplated by this LOI. Sponsor agrees to defend, indemnify and hold STAGE, its successors, assigns, trustees, shareholders, directors and officers harmless from and against any claims of any persons or entities claiming a fee or commission by reason of the Equity Investment or the transactions contemplated by this LOI resulting from the actions of the indemnifying party. STAGE agrees to defend, indemnify and hold Sponsor, its successors, assigns, trustees, shareholders, directors and officers harmless from and against any claims of any persons or entities claiming a fee or commission by reason of the Equity Investment or the transactions contemplated by this letter.

J. Not binding agreement

This letter represents a statement of the parties' general intent only, except that the provisions of Paragraph F, G, H, I and this Paragraph J are intended by the parties to be and shall be binding. Notwithstanding Paragraph D, none of the parties hereto will have any legal obligation under this letter unless and until subsequent formal written documentation (an Equity Investment Agreement and other documents required by STAGE to govern, evidence and secure the Equity Investment) is executed and delivered by STAGE and Owner/Sponsor. This letter shall be governed by, and construed in accordance with, the internal laws of the Commonwealth of Virginia. To the fullest extent permitted by law, each party to this letter expressly waives all rights to trial by jury in any litigation relating to this letter and all rights to punitive, consequential or special damages on account of this letter. The prevailing party in any litigation relating to this letter shall be entitled to recover its actual reasonable attorneys' fees and disbursements, expert witness fees and expenses and court costs from the non-prevailing party in such litigation.

K. Survival

The provisions of Paragraph F, G, H, I and J of this letter shall survive a termination of this letter.

Subject to the foregoing, if this letter is acceptable to you, please execute a copy of this letter in the space provided for below and return same to us on or before 5:00 p.m. EDT, [Month day, year]. If this letter is not so executed and returned to us by such time and date then the proposal set forth herein shall be deemed withdrawn and this letter shall be of no further force or effect whatsoever.

We look forward to working with you on this transaction. If you have any questions, please call me at (XXX) XXX-XXXX.

Best regards,

STAGE Fund I, LP
By: Stage Capital, LLC
By:______________________
Managing Director Principal

Agreed to and Accepted this [ ] day of [ ] [year].
By:______________________
Name:
Title:
cc: Oversight Principal
General Counsel

Exhibit A

Equity investment terms

Owner:	A to-be-formed limited liability company, whose Members will be Sponsor and STAGE to be governed by an operating agreement (the "LLC Agreement").
Principals:	
Purpose:	Approval, entitlement, development and sale of [______] in accordance with the Business Plan.
Estimated Capitalization (Total Project):	$
Equity Investment Amount:	[90]% of the equity requirement. Investor will never be obligated to fund more than $[____].
Sponsor Equity:	[10]% of the equity requirement.
Funding:	Upon acquisition of the Property.
Closing and Documentation:	The parties, in good faith, will attempt to agree on the Operating LLC Agreement and Business Plan (including operating projections) within thirty (30) days after the Approval Date.
Management:	Sponsor will have day-to-day operational control of Owner and will manage Owner in accordance with a fiduciary standard of care consistent with industry standards as further defined in the LLC Agreement, and pursuant to a Business Plan to be prepared by Sponsor and approved by STAGE. STAGE shall have approval of all major decisions and matters consistent with prevailing custom and practice.
Sponsor Development Management Fee:	Sponsor and its affiliates shall receive no compensation except for a Development Management Fee equal to [__]% of total hard and soft costs in the aggregate, to be paid to Sponsor over the projected course of development and stabilization. Sponsor shall receive reimbursement for project related costs and expenses to be approved by Investor (see Approval).
STAGE Development Management Fee:	Investor shall be entitled to a Development Management Fee equal to 1.0% of total hard and soft costs in the aggregate, to be paid to Investor over the projected course of development and stabilization. The payment of the Development Management Fee to the Investor shall create no obligation, liability or duty from the Investor to the Owner, the Sponsor, any other Member of Owner, or any affiliate, principal, agent or employee of the foregoing.
Financing Guarantees:	Sponsor and, to the extent necessary, the Principals shall provide any required guarantees (such as a completion guaranty) necessary to obtain construction financing.
Sponsor and Principal Liability:	Sponsor and each Principal is obligated to Investor for breach of material representation or fraud, gross negligence, misappropriation or misapplication of funds, the transfer of Sponsor's and/or Principals' interests in Owner or of the Project or any portion thereof without Investor's consent, and any intentional failure of either Owner or Sponsor to cure any default beyond the applicable cure period following notice and an opportunity to cure, as well as other acts of malfeasance and omissions customarily constituting malicious breach of conduct or duty by a managing member, manager or fiduciary.

Cost Overrun Guaranty: Principals shall be responsible for controllable cost completion overruns and the completion of the construction of the Project's improvements.

Expenses: As part of the approved Business Plan, Owner will be responsible for all expenses associated with the Equity Investment, including but not limited to closing costs, recording and filing fees, Sponsor and Investor's counsel fees and expenses, the costs of third-party reports, and Sponsor and Investor's legal, due diligence, administrative and other expenses, provided that if Closing occurs, such costs will be paid from Equity Investment proceeds as approved by Investor.

Additional Capital: In the event Additional Capital is required, and it cannot be funded from Construction Loan proceeds (including as a result of the inability to obtain the Construction Loan), Investor shall have the right, but not the obligation, to make additional advances *pari passu* with Sponsor on a [50]% (Investor) / [50]% (Sponsor) basis, failing which such additional capital (if due to controllable cost overruns) will be required to be contributed by Sponsor.

Member Loans: If either Member fails to fund all or a portion of such required equity contribution, the other Member may elect to fund the unfunded amount (the "Contributing Member") as a Member Loan. In addition, any delinquent contributions shall result in 200% penalty dilution of the defaulting Member's interest in Owner and, if applicable, Sponsor's Promote.

Distributions: All available cash flow and capital distributions (after repayment of the Construction Loan) will be distributed to Investor and Sponsor in the following manner:

1. First, *pari passu* to any providers of Additional Equity (pro rata) until such provider has received a [20]% IRR (including the return of all outstanding Additional Equity contributions) on such Additional Equity;
2. Second, to the Investor and Sponsor *pari passu*, until each has received a [15]% IRR (compounded monthly) on required invested capital;
3. Third, to the Investor [80%] and Sponsor [20%] until each has received a [20]% IRR (compounded monthly) on required invested capital;
4. Fourth, to the Members, [40]% to Sponsor (the "Promote") and [60]% to equity, to be split pro rata between Sponsor and Investor based on their respective equity contributions.

[NOTE IN THIS EXAMPLE, A 90/10 DEAL, SPONSOR GETS 40% OVER THE PREF OF 20% PLUS ITS PRO RATA RATE 10% OF THE REMAINING 60% OR ANOTHER 6%, SO THE SPLIT OVER 20% IRR IS ESSENTIALLY 46% SPONSOR/54% INVESTOR.]

Approvals Sponsor will manage the Property according to an approved Business Plan; provided, however, that any material deviation from the Business Plan will require Investor's approval. As part of the LLC Agreement, Investor will approve a Business Plan, which will include development costs, financing, all budgets, ADR and occupancy assumptions, and sales parameters and costs, and a timing schedule to take the Property through pre-development, development, stabilization and sale. Investor will have the right to require Sponsor to update the Business Plan should any material circumstances or assumptions change. Investor shall have approval (as well as certain control) rights with respect to affiliate transactions, capital expenditures, capital improvements, leasing and sales parameters and costs, financings or refinancings, transfer of membership interests or additions of new Members, all extraordinary company actions, material changes in the Business Plan and any other prior approvals of the Investor, material contracts and agreements, the retention and dismissal of professionals and other material expenditures and acts (collectively, "Major Decisions"). Any approval, consent, exercise of judgment or other determination to be made by the Investor, or the exercise of any option by the Investor, may be made, given, withheld or conditioned in the sole, but good faith, interest of the Investor.

Covenants: Sponsor will make customary and complete entity-level and Property-level representations and warranties and covenants consistent with the scope of the transaction. Owner documents will include such covenants as Investor requires including, but not limited to, financial covenants, financial reporting covenants, insurance covenants, covenants limiting distributions, prohibitions on Sponsor transfer and encumbrance, covenants restricting the incurring of debt, and covenants regarding the SPE status of Owner. Sponsor will also covenant to manage and operate the Property in the ordinary and usual manner, and, in addition, after the expiration of the Due Diligence Period, and pending approval of a Business Plan, not to enter into any lease or occupancy agreement, or any service, construction or other contract (or to extend, modify or terminate any of the same), except as may be approved by STAGE in its reasonable discretion.

Buy-Sell The Definitive Documents will contain a buy-sell provision that will only be effective (i) upon the Parties being unable to agree on a Major Decision pertaining to that asset, (ii) the Parties being unable to agree on a Development Budget, the Business Plan, or change to the Business Plan on that asset after negotiation thereof for at least three (3) months, or (iii) at any time after 12 months. The Party exercising the Buy-Sell will provide written notice including an Exercise Price at which it will buy or sell. The other Party will have 60 days to respond. Once the Buying Party is determined, it must post a 2% non-refundable deposit into escrow within 5 business days and close within 120 days. If the Buying Party defaults, the other Party will have the opportunity to buy at 95% of the original Exercise Price and keep Buying Party's 2% deposit.

Right to Cause Sale: Investor shall have the right to cause a sale of the Project (or any phase) (i) at any time after 36 months after the Closing, but not during the period after construction has commenced until receipt of a certificate of occupancy with respect to any individual phase and (ii) at any time that a "Cause Event" has occurred in respect of Sponsor or any Principal.

Exhibit B

Items to be furnished to STAGE

1. Audited financial statements for Sponsor for the preceding three calendar years and unaudited financial statements for the current year (to the most recent quarter).
2. Financial statements for each of the Principals for the preceding three calendar years and for the current year (to the most recent quarter).
3. Complete litigation and defaulted loan history for Sponsor and each of the Principals for the past 10 years, including a Credit Report Authorization and Release (see Exhibit C).
4. True and complete copies of all loan applications and documentation related to Sponsor and the Property (including all documentation relating to the Senior Loan), leases, ground leases, purchase and sale agreements, collective bargaining agreements, pension and benefit plan documentation, and other material agreements to which Sponsor is a party or by which Sponsor, Sponsor or the Property is or will be bound.
5. Phase I environmental report and, if required by STAGE, a Phase II environmental report.
6. Property Condition and Physical Inspection report.
7. Proposed capital and operating budgets for the Project.
8. Proposed plans and specifications for the Project.
9. Proposed pre-development and development budgets for the Project.
10. Existing surveys, site plans, title insurance policies and commitment, evidence of zoning and legal compliance, evidence of necessary entitlements, CC&Rs (covenants, conditions and restrictions), association related reports, appraisals, marketing studies, engineering reports, soils reports, ADA compliance studies, environmental site assessments, asbestos studies, traffic reports, and other material agreements and documents affecting the Property (as available).
11. Historical tax bills for the current year and the three preceding calendar years.
12. Licenses, permits, authorizations, approvals, certificates of occupancy, and certificates of insurance (as available).
13. Documentation regarding Sponsor's and Sponsor's capital structure and true and complete copies of organizational documents for Sponsor, its direct and indirect owners, Sponsor, and other entities reasonably required by STAGE.
14. A comprehensive project life cycle budget and strategic redevelopment and operating/marketing plan (the "Business Plan") prepared by Sponsor, which sets forth Sponsor's objectives and business plan, on a qualitative and quantitative basis, with respect to the Project and the entitlement, pre-development, development, marketing, and sale thereof, all as applicable, and setting forth all anticipated income, operating, entitlement, development expenses, and capital and other costs and expenses of Sponsor, together with projected monthly/annual capital returns and aggregate IRRs to Sponsor and STAGE.

Exhibit C

Credit report authorization and release

Authorization is hereby granted to Stage Capital, LLC to obtain a credit report through a credit reporting agency chosen by Stage Capital, LLC.

My signature below authorizes release to the credit reporting agency, to obtain information regarding my home, employment, savings, other deposit or money market accounts, outstanding credit accounts such as mortgage, auto, personal loans, charge cards, or credit unions accounts. Authorization is further granted to the reporting agency to use a photocopy of this authorization, if necessary, to obtain any information regarding the above mentioned information.

Any reproduction of this credit report authorization and release made by photocopy or facsimile is considered an original.

Name: ____________________

Current address: ____________________

Previous address if above 5 years or less: ____________________

____

Social Security Number: ____________________

Owner's signature: ____________________

Date: ____________________

10 Accounting statement(s)

Financial statements are critical for any real estate project requiring external financing, that is equity and/or debt. The statements provide an accounting picture of the firm's operations and financial position at present and in the future (i.e. pro forma). The statements provide a common 'template' with standard rules which demonstrate the operational performance for a real estate project. These statements all follow one of two forms: US GAAP or International Financial Reporting Standards (IFRS). The statements are the basis of most financial decisions for a project which includes equity and debt financing. The four financial statements are Balance Sheet, Income Statement, Statement of Cash Flow, and Statement of Retained Earnings. The audience for these statements is a myriad of individuals to include owners/sponsors, investors, lenders, managers, and eventual purchasers.

In general there are two accepted methods for developing financial statements: (1) Cash Basis and (2) Accrual Basis. The cash basis records revenues and expenses when received/paid. This is utilized for small companies and real estate pro formas because it most closely ties to cash. The second method, accrual, records revenues/expenses when earned/incurred. This is utilized for larger companies and intends to 'smooth' income for reporting purposes. The common account rules are established by the Financial Accounting Standards Board (FASB) and the International Accounting Standards Board (IASB).

A balance sheet is a snapshot, a point-in-time perspective, of a project or entity. The fundamental premise of a balance sheet is that it identifies three 'bins' of transactions/capital: Assets, Liabilities, and Shareholder Equity. The fundamental relationship for a balance sheet and, in reality, for all of accounting is that Assets must equal the summation of Liabilities and Shareholder Equity (Figure 10.1).

What is critical is that Assets less Liabilities is Net Worth of the project at a particular point in time. In general all account balances are ending balances and there can be a substantial amount of estimates in this statement. However, regardless of estimates, the balance sheet, by definition, must balance.

Assets are considered economic resources available to the project. They provide an ability or potential to provide future benefits to the project. The order of assets listed corresponds to their liquidity; in other words, the most liquid (cash) is the first asset listed and the least liquid – Plant, Property, and Equipment (PP&E) – is listed last. Note that Patents, Goodwill, and so forth are examples of assets that can have less liquidity than PP&E, though they are not likely to be found in a real estate pro forma. The Assets represents rights the project has acquired for future benefits.

There are four classes of Assets: (1) Current, (2) Investments, (3) Plant, Property, and Equipment, and (4) Intangibles. Current Assets are used in production during the current period. Investments are long-term investment securities of other firms (not found in a real

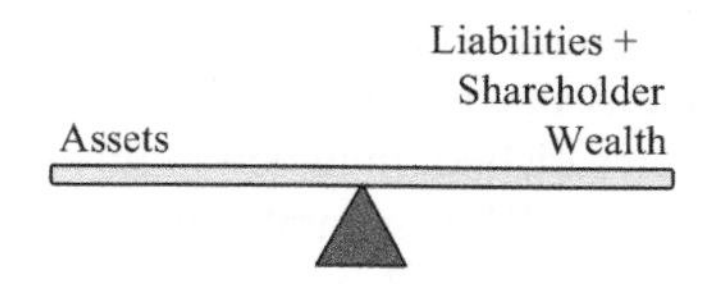

Figure 10.1 Balance Sheet

estate pro forma). Plant, Property, and Equipment are tangible assets used in the project's operations (the focus of a real estate pro forma). Examples of Intangibles are Patents, Licenses, and Trademarks, which most likely will not be represented in a real estate pro forma model.

Additional examples of Asset titles appear in the following image. However, many of these are not applicable to a real estate pro forma and are included only for information.

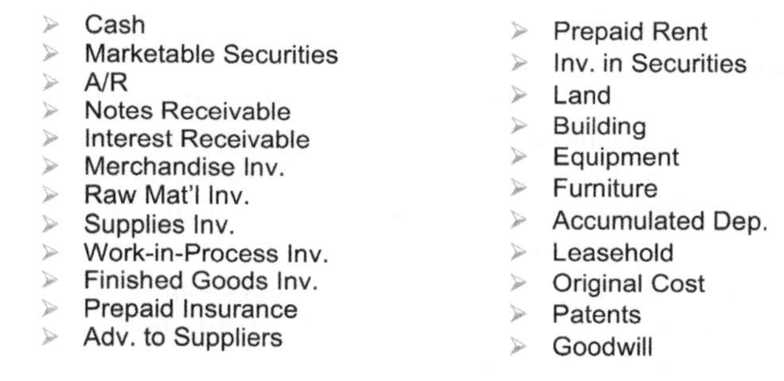

Liabilities are a creditor's claim on the assets of a project. They are essentially funding from creditors and are an obligation of the project or claims from creditors on assets. Liabilities arise when benefit is received from a promise to pay.

There are three classes of Liabilities: (1) Current, (2) Long-Term Debt, and (3) Other Long-Term Debt. Current Liabilities are utilized in operations over the next reporting period, and Long-Term Debt is debt that matures outside of the reporting period, typically one year. Other Long-Term Debts are liabilities not fitting the two previous categories, a category generally not found in a real estate project. However, examples of Other Long-Term Debt are deferred income tax and pension liability.

Additional examples of Liability titles follow in the figure. However, many of these are not applicable to a real estate pro forma and are included only for the reader's information.

- A/P
- Notes Payable
- Interest Payable
- Income Taxes Payable
- Adv. from Customers
- Adv. from Tenants
- Mortgage Payable
- Bonds Payable
- Convertible Bonds Payable
- Capitalized Lease Obligations
- Deferred Income Tax

The final section of the balance sheet is Shareholder Equity (i.e. Net Worth). Noting that owners have a residual claim on assets, it represents the owner's claim on the assets of the firm after all liabilities have been satisfied. Included in this section is also the Statement of Retained Earnings (SRE). While the SRE is separated in most firm accounting, for modelling purposes, it is embedded at the base of the balance sheet and modelled in Shareholder Equity.

The portions of Shareholder Equity are Contributed Capital (Common Shares and Additional Paid-in Capital) and Retained Earnings. The Contributed Capital reflects funds invested by shareholders (investors/sponsors) for ownership interests. Retained Earnings are earnings realized by the project in excess of dividends distributed to shareholders. Assets reinvested by the project sponsor for the benefit of the investors are retained earnings as well.

Shareholder Equity titles are Common Stock, Preferred Stock, Capital Contribution in Excess of Par, Retained Earnings, and Treasury Shares. Again, most of these classifications will not be utilized for real estate projects and are included here for context.

What is unique about the balance sheet is the dual effect of any account. The balance sheet, again, must balance. If Assets decrease, Liabilities/Shareholder Equity must decrease. If Assets decrease, another asset must decrease in an equal amount. The balance sheet is also the most manipulated accounting statement but provides several useful metrics to be discussed later in the chapter.

The income statement reports sources and amounts of a project's revenue and the nature of the project's exposure that summarize to earnings in a particular period. The income statement links the beginning and ending period for a balance sheet. The statement measures income across a period of time (i.e. a periodic statement). In short, it summarizes transactions for the project across a given period.

In general it is a stair step of expenses. As discussed in Chapter 9, the income statement is a waterfall of income for a project (Figure 10.2). The top portion is revenue (monies earned), and the next is direct expense (monies paid).

Revenues are simply a measure of inflows of assets – i.e. liability reduction – from renting space (selling goods) or providing services. Expenses measure outflow of assets – i.e. increased liability – used in generating revenues. Net Operating Income is revenue less expense.

Note that in corporate financing there are additional nuances, and a corporate finance text or accounting text must be referenced. An example of the stair step of expenses for a manufacturing or typical firm are included in Figure 10.3. However, this is simplified for a real estate pro forma analysis. Notice that for a corporate income statement, (1) Revenues are equivalent to a real estate pro forma; however, Expenses (2) is direct expenses used in production. The net is then Gross Income/Profit or Gross Margin. Operating Expenses (3 and 4) are indirect expenses and generally considered overhead. Selling Expenses (3) are largely variable expenses while General & Administrative (G&A) Expense (4) is overhead –fixed

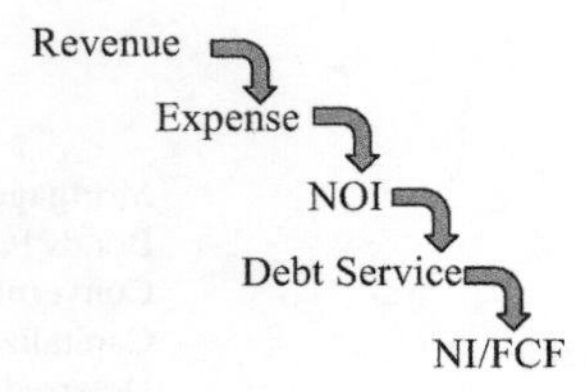

Figure 10.2 Pro Forma "Waterfall"

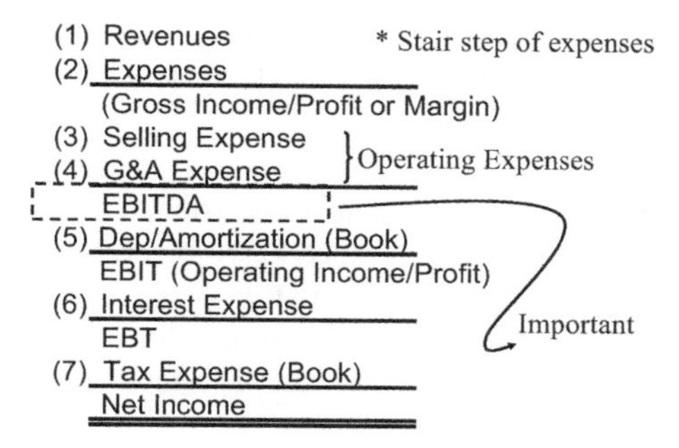

Figure 10.3 Income Statement Flow

expense which is somewhat independent of sales. The net of Selling Expense and G&A Expense is Earnings Before Interest, Taxes, Depreciation, and Amortization (EBITDA).

EBITDA is very important in corporate finance and valuation. It represents the last number in the stair step that is not accounting adjusted (sort of). EBITDA also maps to Net Operating Income (NOI) in a real estate pro forma. When valuing a corporation, a typical method is an EBITDA multiplier. An investment bank analyzes recent or comparable transactions in the capital markets and boils down the value of the company to so many times EBITDA. This is almost identical to a real estate valuation, which uses a capitalization rate. The difference is actually mathematical in that corporate finance uses a multiplier and real estate uses a divider. The resultant is the same: expected value.

Following EBITDA is depreciation/amortizations (D/A). This is a non-cash expense utilized to place the 'cost' of large capital items in the period for which it is incurred. For instance, assuming a 30-year depreciation schedule for a building, 1/30th of the cost is allocated per year. EBITDA less D/A results in Earnings Before Interest, Taxes (EBIT). Interest Expense (6) is the interest paid for any capital utilized in the capital stack for a project. Note that Principal is not included. While Principal is a cash outflow, under US GAAP, it is not considered an expense item as it is, essentially, a buildup of equity rather than a negative outflow. The principal payment is accounted for on the Statement of Cash Flow, discussed later. EBIT less Interest Expense (6) is Earnings Before Tax (EBT).

Finally Book taxes are subtracted as Tax Expense (Book) (7) from Earnings Before Tax. As noted these are 'Book' taxes and not cash taxes. There will be a page added to the pro forma to accommodate depreciation/amortization and taxes, both book and cash. The resultant of EBT less Tax Expense (Book) (7) is Net Income. Note that Net Income is not a cash number but an accounting value. Because it was derived using accounting treatment, it is meant to represent theoretical cash available if revenues and expenses were all incurred in the period of reporting.

One aspect that many analysts and financial statement forecasters miss is the interrelationship of the statements; in other words, these are not stand-alone statements but rather link to financially demonstrate how a project works. As an example of this consider the income statement and balance sheet relationship. Notice the basic equation for a balance sheet and then notice the contribution to Shareholder Equity attributable to the income statement across the period:

Assets = Liability + Shareholder Equity
Assets = Liability + Contributed Capital + Retained Earnings (RE)

$$\text{Assets} = \text{Liability} + \text{Contributed Capital} + RE_{BoP} + \text{Net Income} - \text{Dividends}_{Period}$$

$$\text{Assets} = \text{Liability} + \text{Contributed Capital} + RE_{BoP} + \text{Revenue} - \text{Expense} - \text{Dividends}_{Period}$$

The Statement of Retained Earnings, which is constructed as part of the balance sheet in this text, is the quantified amount of the project's earnings retained rather than paid out in distributions (i.e. dividends). This is a representation of a claim against assets and not the actual assets. A larger project will hold 'retain' earnings for additional capital expenditures and may not hold cash above a minimum amount. The retained earnings represent capital that is not available for distribution.

The final statement, Statement of Cash Flow, is the key statement as it demonstrates all cash. The statement also nets the differences between periods on the balance sheet. Because it relates to cash, it can clearly expose the financial strategy of a project or business and identifies if the project is a net producer or user of cash. As with the income statement and Statement of Retained Earnings, the Statement of Cash Flow is a periodic statement that reports cash flow in three segments: Operations, Investment, and Financing. It shows the principal inflows/outflows of cash from these three activities.

The relationship between the Statement of Cash Flow and income statement is one where actual cash flows are shown on the cash flow statement while net income is calculated using the accrual method of accounting. Also, projects can receive cash from sources not related directly to the project, though this is rare in the case of a real estate project.

There are two types of Statements of Cash Flow: Direct and Indirect. A direct cash flow statement calculates cash flow from operations subtracting cash disbursements. An indirect cash flow statement calculates cash flow from operations by adjusting Net Income for non-cash revenue and expenses.

The Operations section of the cash flow (the core of the statement) identifies cash from the core business. It is the excess cash received over paying direct costs. Excess cash from operations is then used for distributions or to adjust the capital stack through decreasing leverage.

The Investment section of the cash flow identifies payments to maintain or increase current operating levels. It is for the acquisition of additional real estate for the project – e.g. capital expenditures. Could be cash obtained through the sale of existing assets.

The Financing section of the cash flow identifies financing activities to support operating and investing. A project obtains cash from short- and long-term obligations. A project can also adjust ownership structure through payouts to investors according to joint venture and operating agreements.

The Statement of Cash Flow relates to the balance sheet and income statement in three main ways:

1 Explains change in cash between periods
2 Identifies and captures cash for major investing and financing activities
3 Demonstrates how operations affected cash flow for a period from the income statement.

Please note how the statements fit together and interlock. Note again that the balance sheet is a point-in-time statement while the remaining three – income statement, Statement of Cash Flow, and Statement of Retained Earnings – are periodic statements. It is these relationships that will be modelled when the accounting statements are added to the real estate pro forma model. Once the model is adjusted to accommodate the financial statements, a discussion of Financial Statement Analysis will follow.

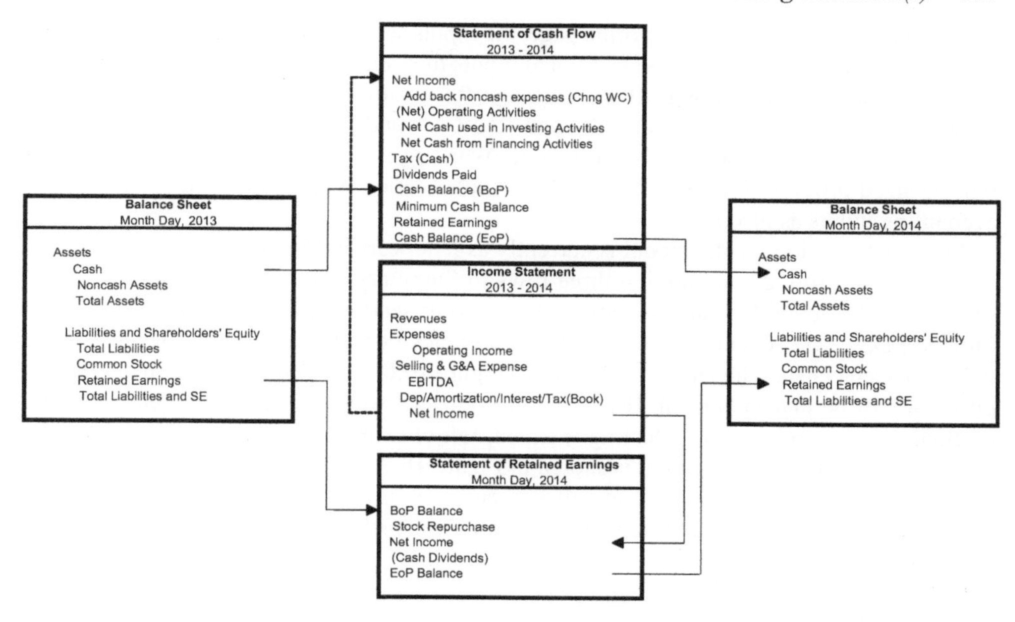

Beginning with the model as constructed with the Rent Roll and Construction/Draw pages, the addition of accounting statements can be considered an adjustment of the pro forma page on the basic model. Essentially the pro forma sheet is expanded to five sheets: Income Statement, Balance Sheet, Statement of Cash Flow, Depreciation, and Tax. The model graphically is as in Figure 10.4.

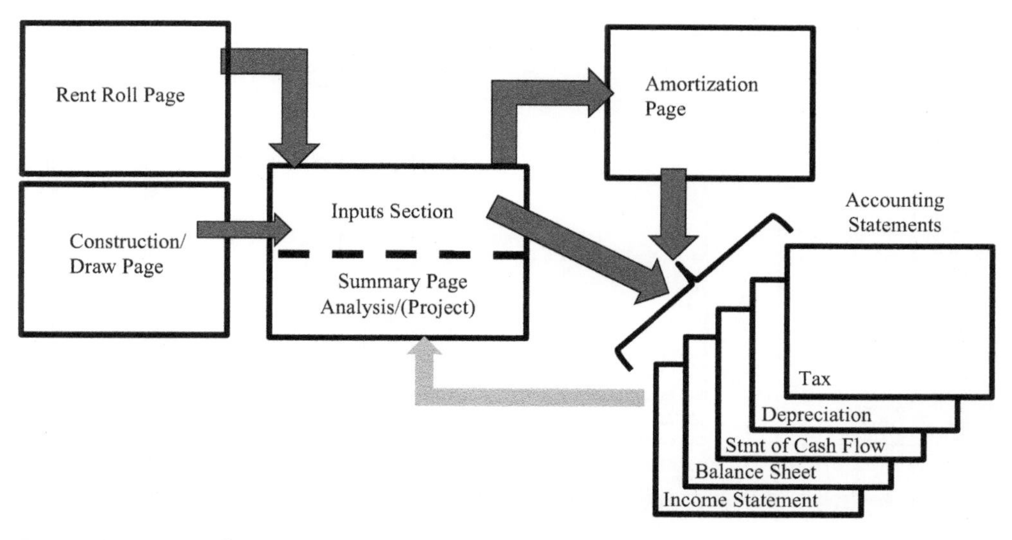

Figure 10.4 Accounting Statement Additions

In addition to adjusting the pro forma page, the Inputs section of the Summary_Inputs page must be adjusted to accommodate additional inputs that relate to the accounting statements. The adjustments for accounting will include Days Receivable, Days Payable, Minimum Cash Balance, Depreciation (Cash and Book), and Tax (Cash and Book). Working Capital will be sized to 3 months' Principal and Interest. Note: Working Capital was modelled earlier, but it will be sized differently in this chapter. The Inputs section will be adjusted to include the accounting inputs as discussed.

The Inputs section for the model is currently as it appears in Figure 10.5.

By selecting row 27 and adding three rows to Summary_Inputs, the accounting inputs section can be added beneath Valuation. The current section underneath 'Valuation' with Purchase/Dev Start, Comm Operating Date, and Sales Period is labeled as 'Schedule' and will be placed beneath the Operations section (Figure 10.6). The DSCR Minimum and Cash-on-Cash will be placed beneath IRR underneath Sources/Uses. The rearrangement of the sections is accomplished by selecting the area, left clicking the section, and dragging the entire section. The rearrangement is shown in Figure 10.7.

The title 'Schedule' was added in cell F25 and 'I/P' in I25. It is always critical that, as much as possible, formatting be kept and remain transparent. Cell K21 is titled 'Accounting'

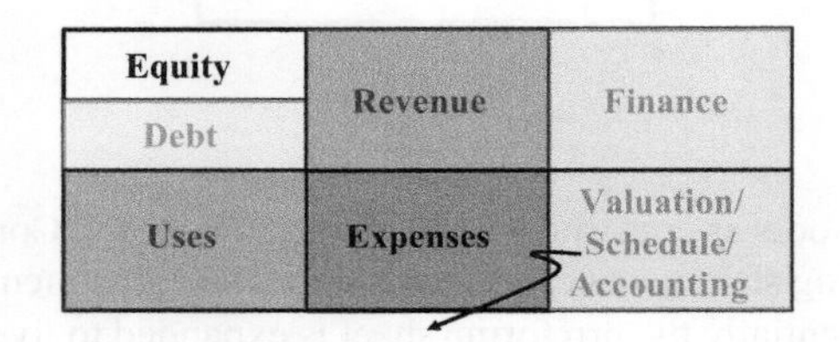

Figure 10.5 Diagram

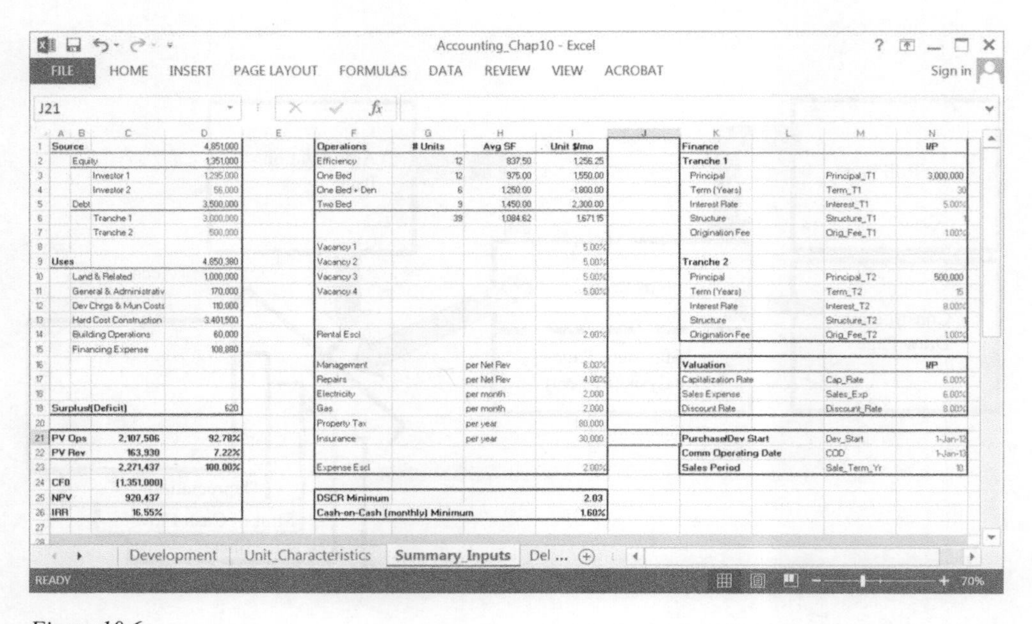

Row	A–B	C	D
1	Source		4,851,000
2	Equity		1,351,000
3		Investor 1	1,295,000
4		Investor 2	56,000
5	Debt		3,500,000
6		Tranche 1	3,000,000
7		Tranche 2	500,000
8			
9	Uses		4,850,380
10	Land & Related		1,000,000
11	General & Administrativ		170,000
12	Dev Chrgs & Mun Costs		110,000
13	Hard Cost Construction		3,401,500
14	Building Operations		60,000
15	Financing Expense		108,880
16			
17			
18			
19	Surplus/(Deficit)		620
20			
21	PV Ops	2,107,506	92.78%
22	PV Rev	163,930	7.22%
23		2,271,437	100.00%
24	CF0	(1,351,000)	
25	NPV	920,437	
26	IRR	16.55%	

Row	F	G	H	I
1	Operations	# Units	Avg SF	Unit $/mo
2	Efficiency	12	837.50	1,256.25
3	One Bed	12	975.00	1,550.00
4	One Bed + Den	6	1,250.00	1,800.00
5	Two Bed	9	1,450.00	2,300.00
6		39	1,084.62	1,671.15
7				
8	Vacancy 1			5.00%
9	Vacancy 2			5.00%
10	Vacancy 3			5.00%
11	Vacancy 4			5.00%
12				
13				
14	Rental Escl			2.00%
15				
16	Management		per Net Rev	6.00%
17	Repairs		per Net Rev	4.00%
18	Electricity		per month	2,000
19	Gas		per month	2,000
20	Property Tax		per year	80,000
21	Insurance		per year	30,000
22				
23	Expense Escl			2.00%
24				
25	DSCR Minimum			2.03
26	Cash-on-Cash (monthly) Minimum			1.60%

Row	K	M	N
1	Finance		I/P
2	Tranche 1		
3	Principal	Principal_T1	3,000,000
4	Term (Years)	Term_T1	30
5	Interest Rate	Interest_T1	5.00%
6	Structure	Structure_T1	1
7	Origination Fee	Orig_Fee_T1	1.00%
8			
9	Tranche 2		
10	Principal	Principal_T2	500,000
11	Term (Years)	Term_T2	15
12	Interest Rate	Interest_T2	8.00%
13	Structure	Structure_T2	1
14	Origination Fee	Orig_Fee_T2	1.00%
15			
16	Valuation		I/P
17	Capitalization Rate	Cap_Rate	6.00%
18	Sales Expense	Sales_Exp	6.00%
19	Discount Rate	Discount_Rate	8.00%
20			
21	Purchase/Dev Start	Dev_Start	1-Jan-12
22	Comm Operating Date	COD	1-Jan-13
23	Sales Period	Sale_Term_Yr	10

Figure 10.6

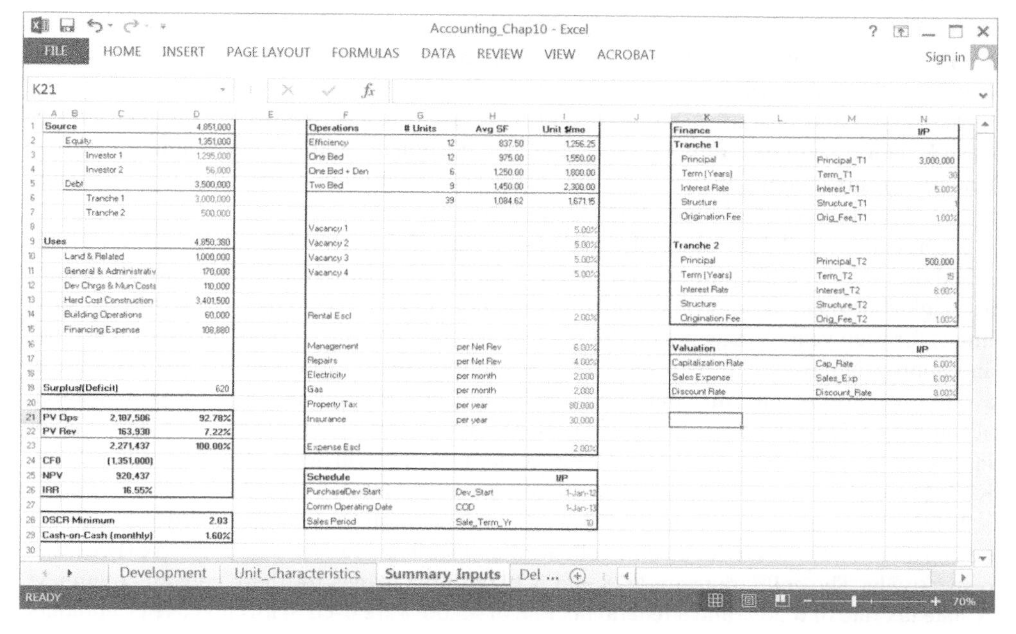

Figure 10.7

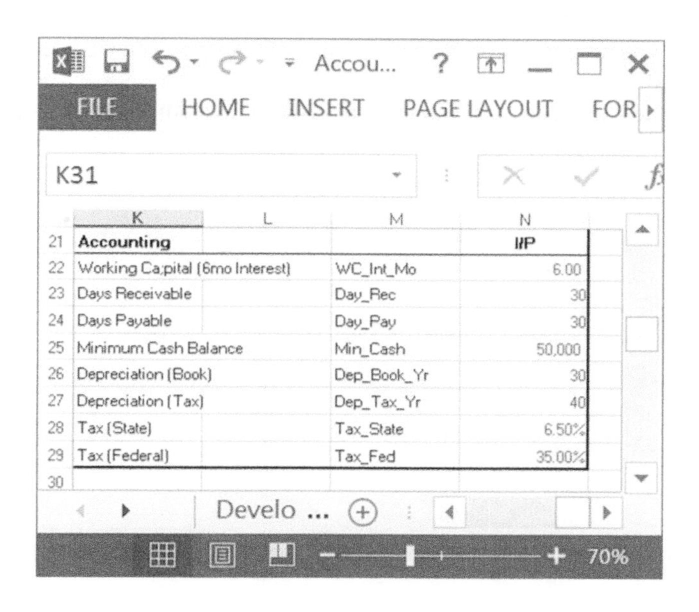

Figure 10.8

and cell N21 'I/P' (Figure 10.8). Beginning in K22 and continuing downward through K29 the following titles are placed: Working Capital (6mo Interest), Days Receivable, Days Payable, Minimum Cash Balance, Depreciation (Book), Depreciation (Tax), Tax (State), and Tax (Federal).

Working Capital was sized for 6 months of interest charges for the debt. The name for cell N22 is 'WC_Int_Mo', which is placed in cell M22, beside the actual input cell. For Days

Receivable and Payable, 30 days is chosen as the input. Later Days Receivable may be eliminated (i.e. placed as zero), as rents are generally paid in advance rather than arrears. However, the model will be constructed to provide ability for payment in arrears. The names for cells N23 and N24 are 'Day_Rec' and 'Day_Pay', respectively. Cell K25, 'Minimum Cash Balance', will generally be set by the loan documents and is a negotiating point with lenders. In this model $50,000 is set and represents the minimum cash that must be maintained in the project at all times, that is, not distributed to the partners.

Depreciation and Tax follow the Minimum Cash Balance. There are two Depreciation schedules, one for book purposes and another for tax purposes. Both will be modelled using a straight-line methodology – i.e. applying the amount of 1/n ("n" is the year in the respective input) for each year. As this is a modelling text, a detailed explanation of different scheduled and appropriate uses will be reserved for accounting texts. However, the analyst must be aware that various schedules exist and will be utilized for larger projects. Please consult an accounting text or the accounting department within the respective firm. The depreciation inputs are 'Depreciation (Book)' and 'Depreciation (Tax)', which are placed in cells K26 and K27. The cell names and inputs are in the respective columns, M and N.

While there can also be county/city tax, for this text only the state and federal rates will be considered. These are located in cells K28 and K29 as 'Tax (State)' and 'Tax (Federal)', respectively. The cell inputs are N28 and N29 with the names for these cells in column M. A state tax rate of 6.50% and a federal tax rate of 35.00% are used in this text as placeholders. The analyst should check with the respective tax department prior to presenting the project. Cells N22:N29 are named by selecting cells M22:N29 and choosing the Formulas tab from the Ribbon, then selecting 'Create from Selection' in the 'Defined Names' section and choosing 'Left Column'.

At this point the Inputs section has been adjusted to accommodate the addition of the Accounting inputs. The Inputs section is as in Figure 10.9.

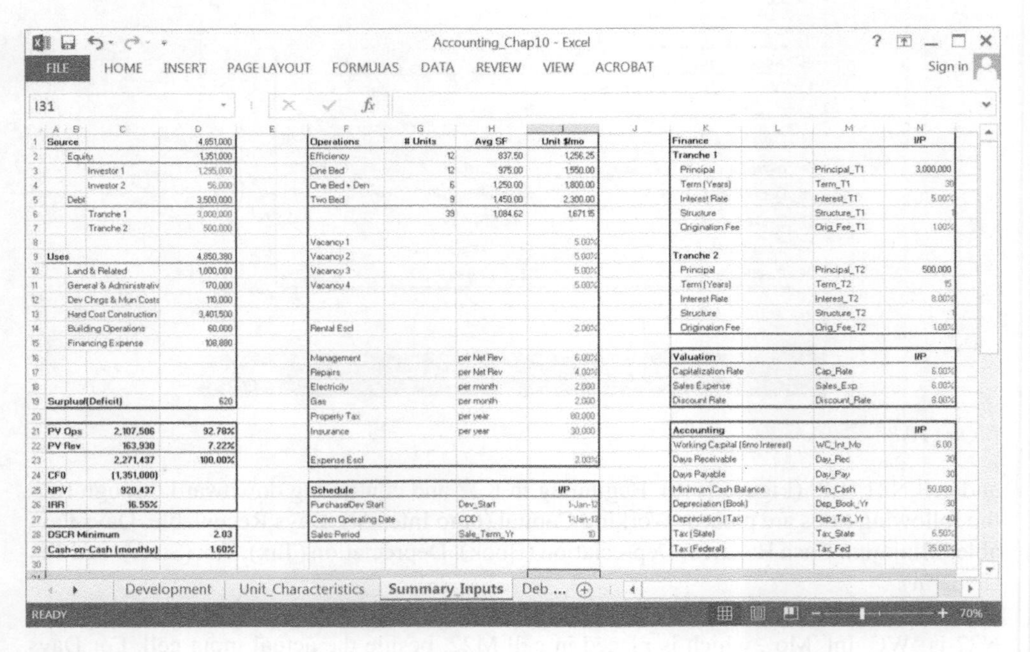

Figure 10.9

Income statement – the structure

The next step is to adjust the Pro_Forma page to be an income statement. The first step in this process is to rename 'Pro_Forma' as 'Income'. Double click on the sheet tab and type 'Income' and the sheet name will be replaced (Figure 10.10). Next the expenses must be separated into fixed and variable. Select row 19 and add a row; this will be for the Variable expense header. Then select row 22, electricity (prior to row addition, this was row 21), and add a row as well. This will be for Fixed expenses. Select cells B20:B26 and move these cells, together, to column C. This will visually indent the expense items. In cell B19 type 'Variable' and in B22 type 'Fixed'. As these are summary rows, cells G19 and G22 will be "=Subtotal()" formulas. Cell G19 is therefore "=Subtotal(9,G20:G21)" and cell G22 is "=Subtotal(9,G23:G26)". As there are now subtotals within the expenses, the formula in cell G27 must be adjusted to accommodate: this is "=Subtotal(9,G19:G26)". Cells G19, G22, and G27 are then copied to the corresponding rows through column EH.

Cell A29, 'Net Operating Income', is then relabeled 'EBITDA (NOI)'. After selecting row 30, four rows are added to accommodate the Depreciation/Amortization Schedule. In cell A31, 'Depreciation/Amortization' is typed. In cell A33 type 'EBIT' (Earnings Before Interest, Taxes). Cell G33 is therefore "=G29-G31". Cell G33 is then copied to H33:EH33. Note that row 31, Depreciation/Amortization, has been left blank. This will be completed after the Dep/Amort page has been added and completed.

Cell A35, Debt, is adjusted to be Other Income/Expense. Cells B36 and B37 are adjusted to Tranche 1 – Interest and Tranche 2 – Interest, respectively. In cell B36 type 'Interest Expense'. Cell C38, Total Debt, is adjusted to 'Total Other Inc/Exp'. The formula for Tranche 1 and Tranche 2 in this section was total payment (principal and interest). However, now that this is modelled as an income statement, only interest is presented on the income statement. Interest is the only periodic expense given that principal is essentially an equity payment; while principal is a real cash outlay captured on the Statement of Cash Flow, it is not captured on the income statement. Therefore, cells G36 and G37 must have the "=Vlookup()" formula adjusted to reference column 4 rather than column 5. Column 5 on the amortization table is total Principal and Interest, while column 4 is simply interest. After cells G36 and G37 are adjusted, these are then copied to EH36:EH37. Cell C38 is then replaced with 'Total Interest Expense' and cell G38 is "=Subtotal(9,G36:G37)". Cell G38 is then copied to column EH.

Rows 39 and 40 are deleted next. Cell B41, 'Principal Repay – Tranche 1', is replaced with 'Book Value'. Note that 'Book Value' will be referenced from the balance sheet when completed. Row 42, 'Principal Repay – Tranche 2', is deleted and cell C42, 'Net Sales Proceeds', is replaced with 'Gain on Sale'. Cells G41:EH41 are deleted and left blank and, as stated, will be linked to the Balance Sheet.

Select row 43 and add 11 rows. In cells B43:B44 type 'Accounts Receivable' and 'Accounts Payable', respectively. In cell C45 type 'Net Working Capital'. Rows 43 and 44 will be completed later, but cell G45 is "=subtotal(9,G43:G44)". Then copy cell G45 to column EH. Cell B46 is 'Return of Cash to Partners' and will reference the balance sheet when completed later. Cell C47 is then titled 'Total Other Income/Expense'. Cell G47 is "=subtotal(9,G36:G46)", which is then copied to column EH.

In cell A49 type 'Earnings Before Taxes (EBT)'. In cell G49 the formula is "=G33-DV38+DV42". Cell G49 is then copied to column EH. In cell A51 type "Federal + State Tax". Row 51 will remain empty until later in this chapter when taxes are calculated on the Tax page. In cell A53 type 'Net Income'. Cell G53 is then "=G49-G51" and is copied to column EH. Rows 55:57, Operating, Reversion, and Total FCF, will remain at present. These rows will later be deleted.

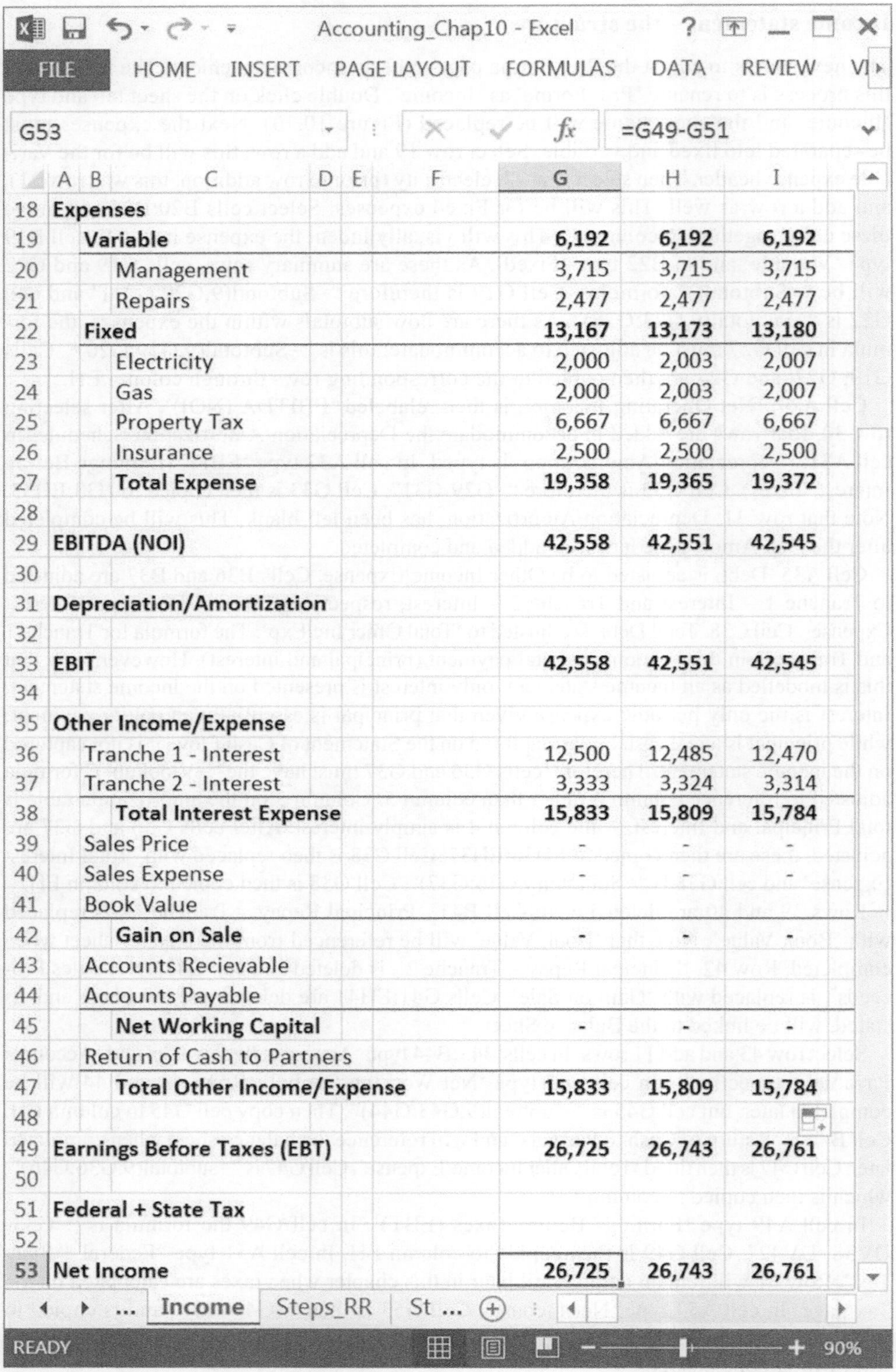

Accounting_Chap10 - Excel

FILE HOME INSERT PAGE LAYOUT FORMULAS DATA REVIEW VI

G53 | f_x | =G49-G51

	A B C D E F	G	H	I
18	**Expenses**			
19	**Variable**	**6,192**	**6,192**	**6,192**
20	Management	3,715	3,715	3,715
21	Repairs	2,477	2,477	2,477
22	**Fixed**	**13,167**	**13,173**	**13,180**
23	Electricity	2,000	2,003	2,007
24	Gas	2,000	2,003	2,007
25	Property Tax	6,667	6,667	6,667
26	Insurance	2,500	2,500	2,500
27	**Total Expense**	**19,358**	**19,365**	**19,372**
28				
29	**EBITDA (NOI)**	**42,558**	**42,551**	**42,545**
30				
31	**Depreciation/Amortization**			
32				
33	**EBIT**	**42,558**	**42,551**	**42,545**
34				
35	**Other Income/Expense**			
36	Tranche 1 - Interest	12,500	12,485	12,470
37	Tranche 2 - Interest	3,333	3,324	3,314
38	**Total Interest Expense**	**15,833**	**15,809**	**15,784**
39	Sales Price	-	-	-
40	Sales Expense	-	-	-
41	Book Value			
42	**Gain on Sale**	**-**	**-**	**-**
43	Accounts Recievable			
44	Accounts Payable			
45	**Net Working Capital**	**-**	**-**	**-**
46	Return of Cash to Partners			
47	**Total Other Income/Expense**	**15,833**	**15,809**	**15,784**
48				
49	**Earnings Before Taxes (EBT)**	**26,725**	**26,743**	**26,761**
50				
51	**Federal + State Tax**			
52				
53	**Net Income**	**26,725**	**26,743**	**26,761**

... Income | Steps_RR | St ...

READY 90%

Figure 10.10

Balance sheet – the structure

A new sheet must be added for the balance sheet and labeled 'Balance'. The sheet's header is to be identical to that of the Income Sheet (Figure 10.11). Cells E1:E3 are 'Periods', 'Months', and 'Year', respectively. Cells G1:G3, as on the Income Sheet, are linked to the Unit_Characteristics page. Cells G1:G3 are then copied to column EH. Column F represents T = 0 or COD date and therefore represents the balance sheet at completion (i.e. start-up). Cell A3 is 'Balance'.

Prior to beginning to model the balance sheet, the line items, or accounts, must be labeled (Figure 10.12). Cell A4 is 'Current Assets', which have the following categories beginning

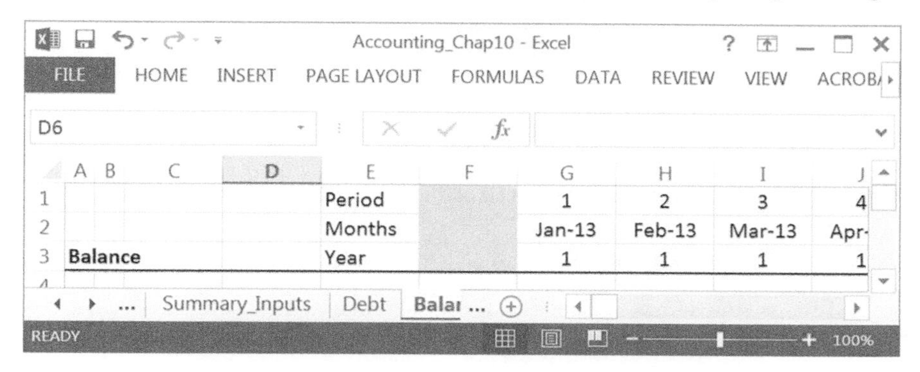

Figure 10.11

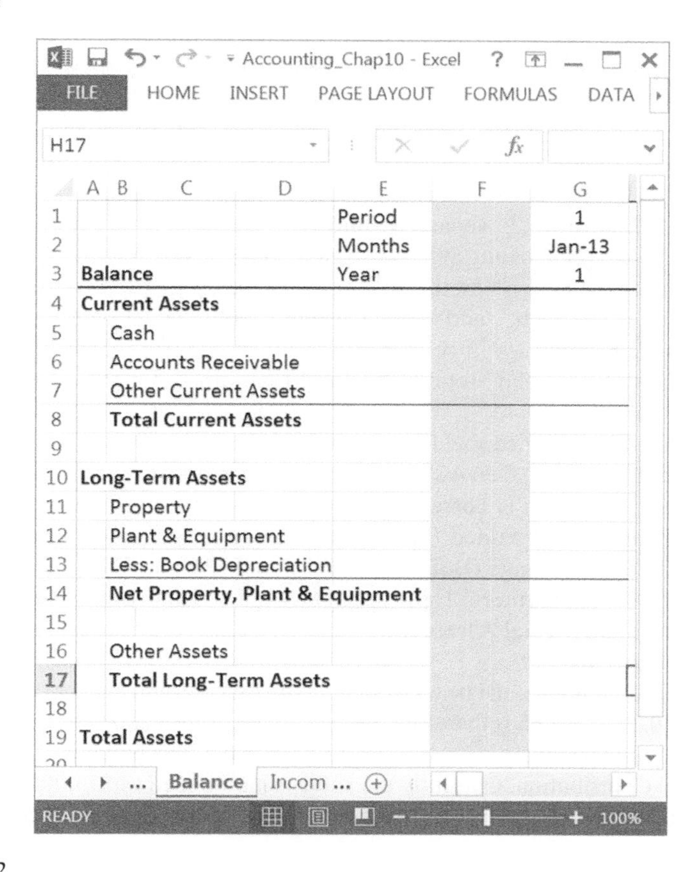

Figure 10.12

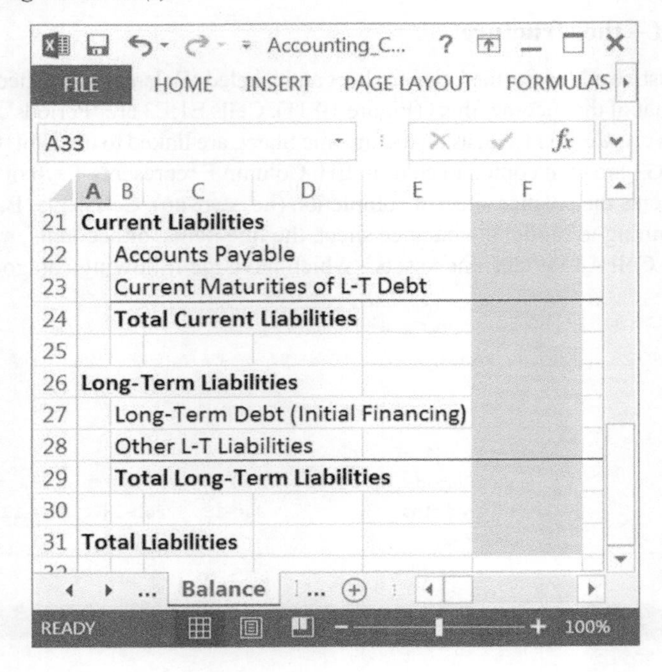

Figure 10.13

in B5:B8: Cash, Accounts Receivable, Other Current Assets, and Total Current Assets. 'Long-Term Assets' is placed in cell A10. The subcategories in cells B11:B14 are 'Property', 'Plant & Equipment', 'Less: Book Depreciation', and 'Net Property, Plant & Equipment'. Finishing Long-Term Assets are in cell B16 'Other Assets' and cell B17 'Total Long-Term Assets'. Finally in cell A19 'Total Assets' is placed.

As with Assets, Liabilities is separated into two sections: Current and Long-Term (Figure 10.13). The account listing will be similar to Assets. In cell A21 type 'Current Liabilities'. In cells B22:B24 place the following category titles: 'Accounts Payable', 'Current Maturities of L-T Debt', and 'Total Current Liabilities'. 'Long-Term Liabilities' is in cell A26. The subcategories in cells B27:B29 are 'Long-Term Debt (Initial Financing)', 'Other L-T Liabilities', and 'Total Long-Term Liabilities'. Finally, 'Total Liabilities' is placed in cell B31.

The final section of the balance sheet is shareholder equity, or in the case of a real estate project, Equity (Figure 10.14). Cell A33 is called 'Partners' Equity' and cell B34 is 'Paid-In Capital'. While the section is correctly labeled Partners' Equity, the embedded next section is the Statement of Retained Earnings. Cell C36 is 'Retained Earnings (BoP)'. Cells D37 and D38 are 'Earnings (Before Inc. Tax)' and 'Partner Distributions', respectively. Cell B41 is 'Total Partners' Equity'. Cell A43 is 'Total Liabilities & Partners' Equity', which must equal Total Assets from row 19. Cell A45 is a check of this balance and is labeled 'A = L + PE ???'.

Next the balance sheet logic must be constructed. The first account beneath Current Assets is 'Cash' (Figure 10.15). Cell F5 is therefore the cash of the project at Commercial Operating Date (COD). This amount is working capital. Returning to the 'Development' page, cell E41 is Working Capital Contribution. Currently the value is hard-coded as $35,000. However, as stated prior, Working Capital logic has been added as part of the accounting section. Working

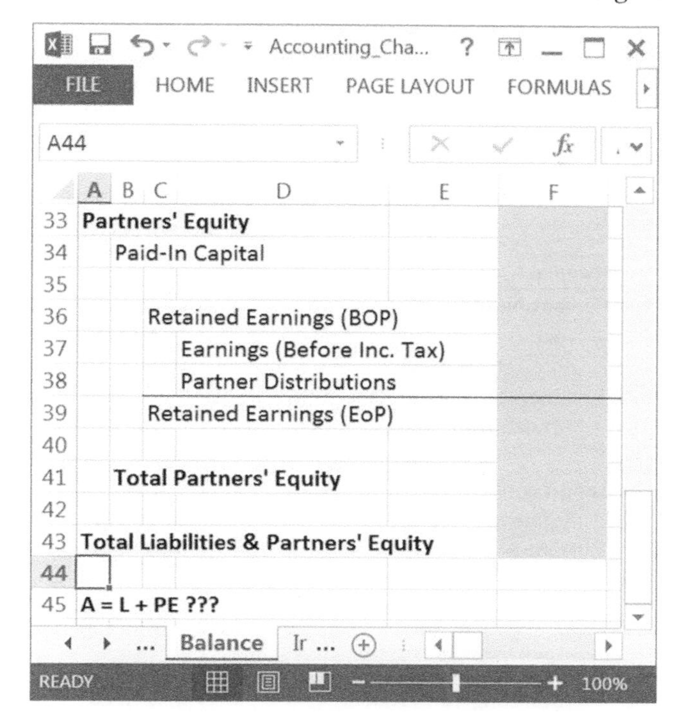

Figure 10.14

Capital is 6 months' Interest. Therefore, cell E41, Working Capital Contribution, on the Development page is as follows:

"=Sum(Principal_T1,Principal_T2)*Interest_T1/12*WC_Int_Mo".

Returning to the Summary_Inputs page, it is seen that there is now a Deficit for the Sources & Uses. Equity should be adjusted upward to accommodate the shortfall. 'Investor 2 Equity', cell D4, is adjusted upward to equate sources and uses. Therefore cell D4 is "=D9-D5-D3". By making this dynamic, it ensures the Sources/Uses section always equates.

Back on the 'Balance' page (Figure 10.15), cell F5, 'Cash' at COD, is "=Development!E41". Cells F6 and F7, 'Accounts Receivable' and 'Other Current Assets', are zero at COD. Cell F8, 'Total Current Assets', is "=Subtotal(9,F5:F7)". Cell F11, 'Property', is non-depreciable land. The formula is therefore "=Summary_Inputs!D10", which is land from the Sources section. Cell F12, 'Plant & Equipment', includes depreciable items. The formula is "=Summary_Inputs!D11:D14)-F5-F12". Notice that it excludes land (featured above), working capital (also featured above), and Interest During Construction (to be featured below). Cell F13, 'Less: Book Depreciation', is zero as it is COD. Cell F14, 'Net Property, Plant & Equipment', is the summation of the section: "=Subtotal(9,F11:F13)". Cell F16, 'Other Assets', is the Interest During Construction value, "=Development!E44". 'Total Long-Term Assets' in cell F17 is the summation of Long-Term Assets: "=Subtotal(9,F10:F16). Finally, 'Total Assets', cell F19, is "=Subtotal(9,F4:F18)". Total Assets must equal total uses from the Summary page. This should be manually checked by any modeller/analyst.

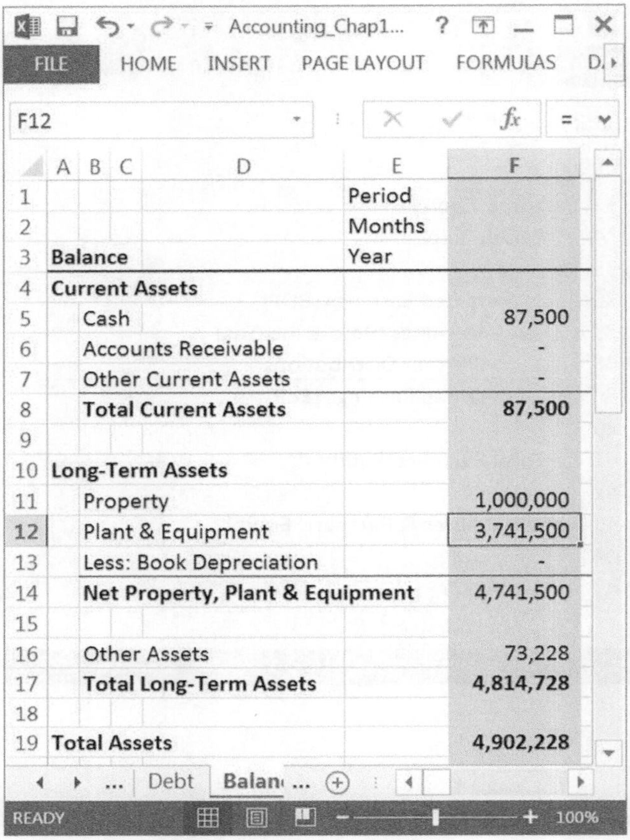

Accounting_Chap1...

FILE HOME INSERT PAGE LAYOUT FORMULAS D.

F12 fx =

	A	B	C	D	E	F
1					Period	
2					Months	
3	**Balance**				Year	
4	**Current Assets**					
5		Cash				87,500
6		Accounts Receivable				-
7		Other Current Assets				-
8		**Total Current Assets**				**87,500**
9						
10	**Long-Term Assets**					
11		Property				1,000,000
12		Plant & Equipment				3,741,500
13		Less: Book Depreciation				-
14		**Net Property, Plant & Equipment**				4,741,500
15						
16		Other Assets				73,228
17		**Total Long-Term Assets**				**4,814,728**
18						
19	**Total Assets**					**4,902,228**

... Debt Balan ... READY 100%

Figure 10.15

The liability section at COD, column F, represents the short- and long-term debt obligations of the project (Figure 10.16). At COD there are no payables and therefore 'Accounts Payable', cell F22, is zero. 'Current Maturities of L-T Debt', cell F23, is the next period maturity (i.e. principal) for the project. Cell F23 is therefore "=Vlookup(F1+1,Tranche_1,3)+ Vlookup(F1+1,Tranche_2,3)". The "F1+1" represents the next period – principal to be paid down between periods – and "3" is the principal payment for the period. 'Total Current Liabilities', cell F24, is "=Subtotal(9,F22:F23)". Cell F27, 'Long-Term Debt (Initial Financing)', follows a similar logic but represents the end of period balance for the next period: "=Vlookup(F1+1,Tranche_1,6)+Vlookup(F1+1,Tranche_2,6)". Cell F28, 'Other L-T Liabilities', is zero. 'Total Long-Term Liabilities' in cell F29 is the summation of F27:F28. 'Total Liabilities', cell F31, is "=Subtotal(9,F21:F30)". Total Liabilities must equal the total debt value on the Uses section on the Summary page.

The final section of the balance sheet is Partners' Equity (Figure 10.17). 'Paid-In Capital', cell F34, is the original equity at COD and is therefore "=Summary_Inputs!D2". Cells F36:F38 for COD are all zero. Cell F39 is "=Subtotal(9,F36:F38)". 'Total Partners' Equity' (cell F41) is "=Subtotal(9,F33:F40)". 'Total Liabilities & Partners' Equity', cell F43, is "=Subtotal(9,F21:F42)". Finally, cell F45, which is the 'check' cell, is "=F19-F43" and should be zero if column F, COD, balances.

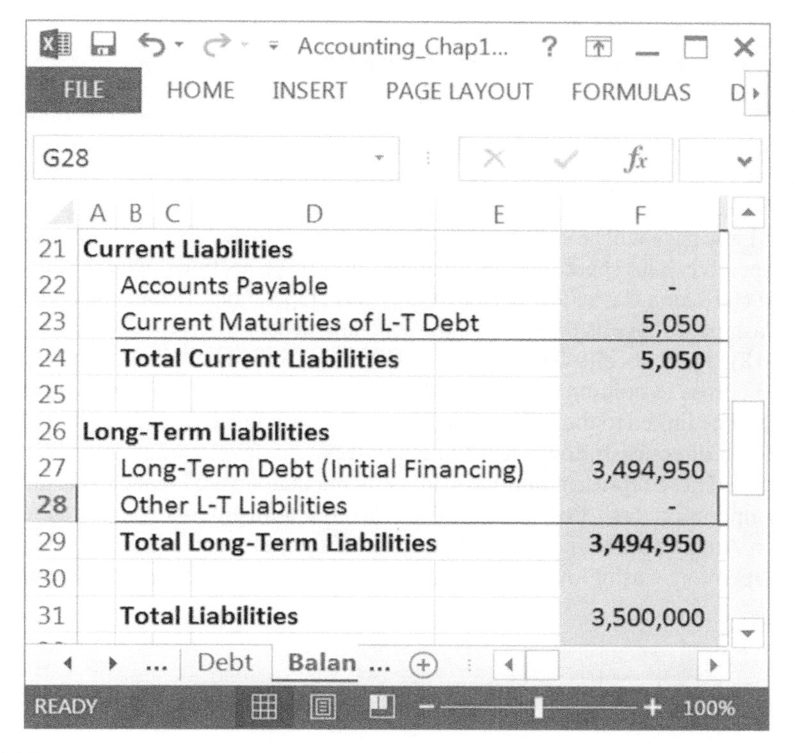

Figure 10.16

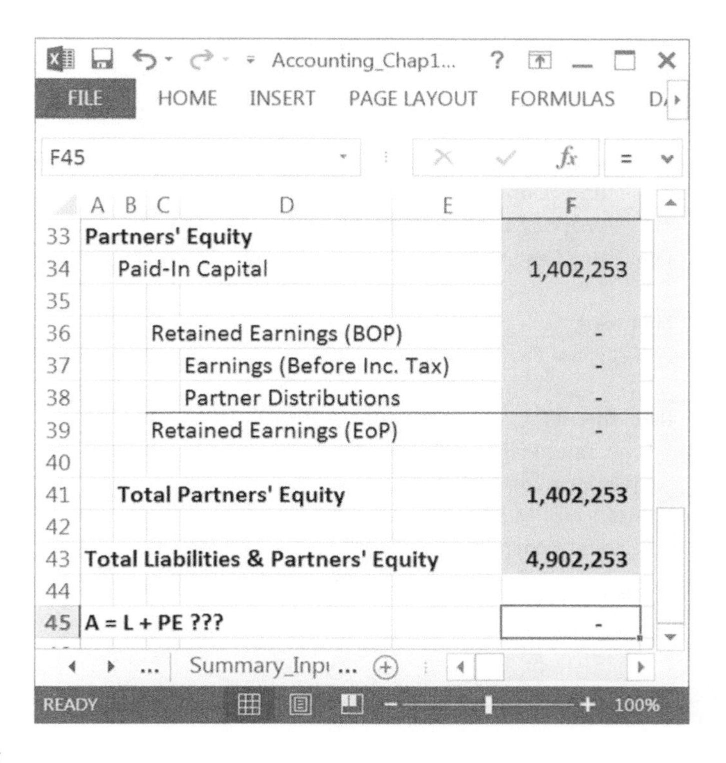

Figure 10.17

The statement of cash flow – the structure

A new sheet must be added for the Cash Flow sheet. The Cash Flow sheet is the guts and glory of a real estate project. The statement demonstrates the cash inflows and outflows. It is similar to the cash inflow/outflow sheet that was developed on the 'Development' sheet in the current model for the construction portion of this model.

The Cash Flow page will be structured similar to other pages in the model from a schedule and header perspective. The sheet will be separated into three sections, which is the generally accepted structure for a Cash Flow page by US GAAP: Operations, Financing, and Investment.

On the Cash page, in cells F1:F3, type the schedule header, 'Period', 'Months', and 'Year' (Figure 10.18). Then in cells G1:G3 link the schedule to the Unit_Characteristics page and copy G1:G3 across to column EH. In cell A3 type 'Cash'. In cell A1 type 'Earnings Before Tax'. This will be linked to the Income page later. Because this is an indirect cash flow sheet, as opposed to a direct cash flow sheet (which will not be developed in this text), cell B5 is 'Adjustments'. These represent non-cash adjustments that must be made to earnings prior to calculating operation cash flow. Cells C6:C7 are the adjustments and are labeled 'Book Depreciation/Amortization' and 'Working Capital (Increase)/Decrease', respectively. Finally, cell A8 is 'Operating Cash Flow', the guts and glory of the statement and the financial model.

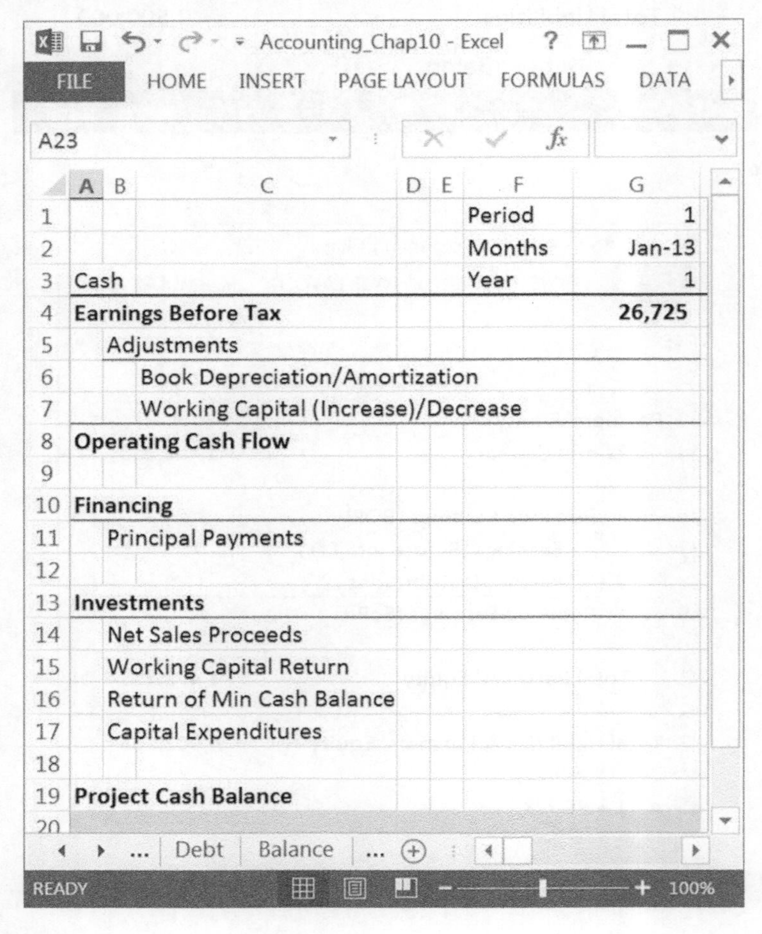

	A	B	C	D	E	F	G
1						Period	1
2						Months	Jan-13
3	Cash					Year	1
4	**Earnings Before Tax**						**26,725**
5		Adjustments					
6			Book Depreciation/Amortization				
7			Working Capital (Increase)/Decrease				
8	**Operating Cash Flow**						
9							
10	**Financing**						
11		Principal Payments					
12							
13	**Investments**						
14		Net Sales Proceeds					
15		Working Capital Return					
16		Return of Min Cash Balance					
17		Capital Expenditures					
18							
19	**Project Cash Balance**						

Figure 10.18

Cell G4 is Earnings Before Tax or EBT, which is "=Income!G33-Income!G38". This is EBIT less Interest Expense. The remaining items on the Income page – Net Sales Proceeds, Working Capital Return, and Return of Minimum Cash Balance – will be netted again on the Cash Flow page in the Investments section. Cell G4 is then copied to column EH.

Cell A10 is 'Financing' and cell B11 is 'Principal Payments'. Given that, for a real estate project, there will not be any additional transactions for financing (e.g. stock issuance), this section is relatively short.

The next and final section of the cash flow statement is Investments. In cell A13 'Investments' is labeled. Following in cells B14:B17 are 'Net Sales Proceeds', 'Working Capital Return', 'Return of Min Cash Balance', and 'Capital Expenditures', respectively. Cell A19 is 'Project Cash Balance'. Fill row 20 with a solid grey line indicating a new section.

This new section, 'Cash Balance and Distribution to Partners' (in cell A21), will demonstrate distributions and cash retainage, that is minimum cash balance (Figure 10.19). Cell A23, 'Cash Balance (BoP)', references the Beginning of Period balance of cash on the balance sheet. Cells B24 and B25 are 'Project Cash Flow' and 'Minimum Cash Balance', respectively. Cell C26 is then 'Total Cash Balance'. Beneath are the un-distributable and distributable

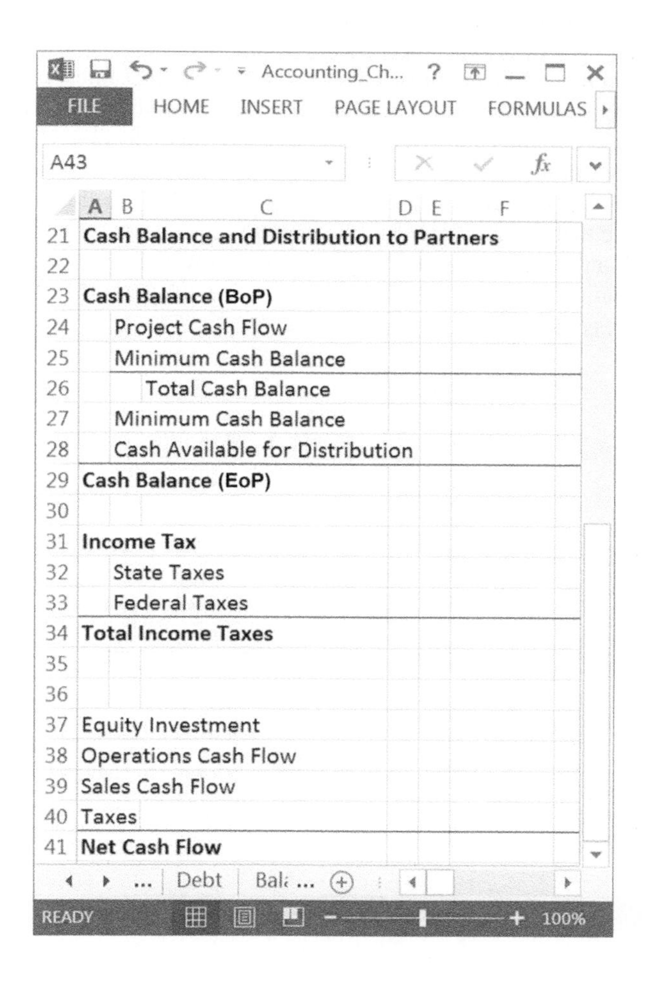

Figure 10.19

amounts. Cell B27 is again ‘Minimum Cash Balance’ and cell B28 is ‘Cash Available for Distribution’. Finally, cell A29 is ‘Cash Balance (EoP)’.

What follows is a summary of income taxes to be paid at the project level. Cell A31 is ‘Income Tax’, and B32 and B33 are ‘State Taxes’ and ‘Federal Taxes’, respectively. Finally, A34 is ‘Total Income Taxes’.

The final cash flows for valuation purposes are ‘Equity Investment’, ‘Operations Cash Flow’, ‘Sales Cash Flow’, ‘Taxes’, and ‘Net Cash Flow’, which are placed in cells A37:A41, respectively. Net Cash Flow will be the value from which a new valuation section will be built later in this chapter.

Depreciation – the structure

Depreciation is an accounting method of allocating costs to a particular period when they are utilized. For instance, when a car is purchased all cash, the expense is incurred immediately. However, the car has an economic life. Therefore, assuming a 5-year depreciation schedule, the cost of the car is allocated to 20% (straight-line depreciation) a year for 5 years. For example, a $50,000 vehicle, paid in cash, reduces cash on the Cash Flow page immediately; however, only $10,000 is ‘expensed’ per year (for 5 years) on the income statement. On the balance sheet, the value of the asset is reduced each year by the cumulative depreciable amount.

The Depreciation page, like the other pages, is structured the same. A new sheet, ‘Depr’, must be created (Figure 10.20). Cells F1:F3 are ‘Period’, ‘Months’, and ‘Year’, respectively. Cells G1:G3 are linked to the Unit_Characteristics page and then copied to column EH. Cell A3 is titled ‘Depreciation’.

Cell A4 is ‘Book Depreciation’ and B5 is “=Dep_Book_Yr”. Cell C5 is ‘’-Yr SL’ (‘year straight-line’). Note that the apostrophe at the beginning is necessary within Excel. Cell C6 is ‘Cap Cost less Land’. The next section, rows 8–10, follow the same structure for Tax. Cell A8 is ‘Tax Depreciation’. Cell B9 is “=Dep_Tax_Yr” and cell C9 is therefore ‘’-Yr SL’. Cell C10 is ‘Cap Cost less Land’.

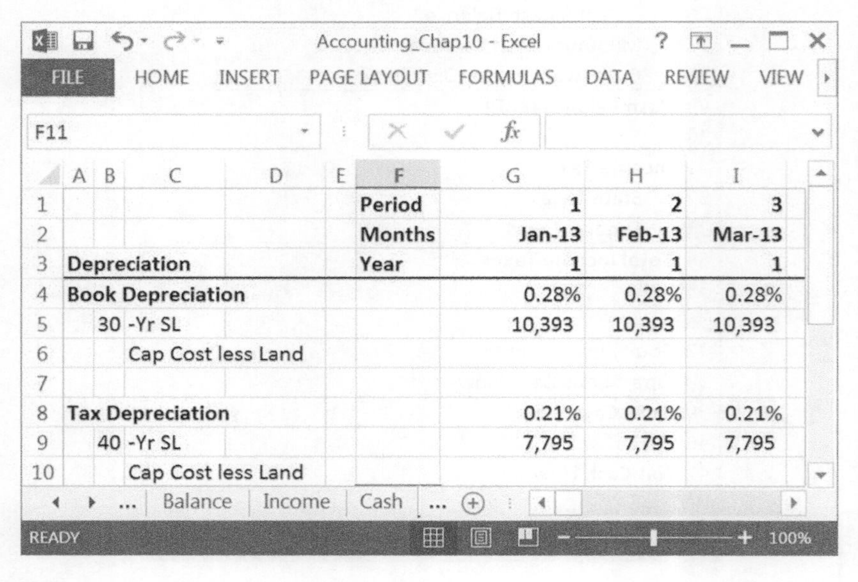

	A	B	C	D	E	F	G	H	I
1						Period	1	2	3
2						Months	Jan-13	Feb-13	Mar-13
3	Depreciation					Year	1	1	1
4	Book Depreciation						0.28%	0.28%	0.28%
5		30	-Yr SL				10,393	10,393	10,393
6			Cap Cost less Land						
7									
8	Tax Depreciation						0.21%	0.21%	0.21%
9		40	-Yr SL				7,795	7,795	7,795
10			Cap Cost less Land						

Figure 10.20

Rows 4 and 8 are the depreciation schedules for both book and tax. Cell G4 is "=IF(G3>Dep_Book_Yr,0,1/Dep_Book_Yr/12)". Cell G8 is the same logic, only 'Dep_Tax_Yr' is substituted. Cells G4 and G8 are copied to column EH. Cell G5 represents the periodic book depreciation – i.e. "=Balance!$F$12*G4". It references the balance sheet value for capital depreciation. Cell G9 follows the same logic: "=Balance!$F$12*G4". Both cells G5 and G9 are then copied to column EH. The Depreciation page is now complete.

Tax – the structure

The tax page is set up identical to all the other pages. A new sheet is added to the workbook, i.e. 'Tax' (Figure 10.21). Then cell A3 is 'Tax' and cells F1:F3 'Period', 'Months', and

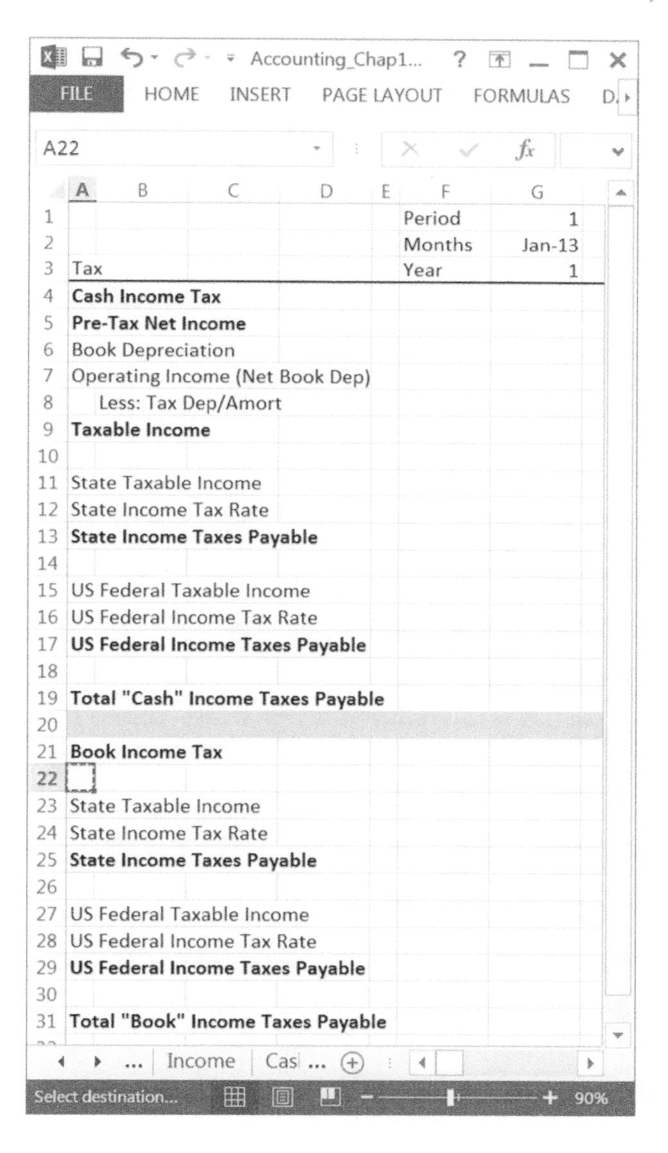

Figure 10.21

'Year', respectively. Finally, G1:G3 are linked to the Unit_Characteristics page and copied to column EH.

Cell A4 is 'Cash Income Tax' and cell A5 is 'Pre-Tax Net Income'. Cells A6 and A7 are 'Book Depreciation' and 'Operating Income (Net Book Dep)', respectively. Cell A8 is 'Less: Tax Dep/Amort' and cell A9 is 'Taxable Income'.

The next section is state tax calculation. Cells A11:A13 are labeled 'State Taxable Income', 'State Income Tax Rate', and 'State Income Taxes Payable', respectively. Following is the federal tax calculation. Cells A15:A17 are 'US Federal Taxable Income', 'US Federal Income Tax Rate', and 'US Federal Income Taxes Payable'. Cell A19 is 'Total "Cash" Income Taxes Payable'. Row 21 is solid grey to separate the next section.

The next section is the Book Income Tax. Cell A21 is 'Book Income Tax'. Then copy cells A11:A17 and paste in cell A23. Cell A31 is then 'Total "Book" Income Taxes Payable'.

Returning to the top of the Tax page, 'Cash Income Tax', cell G5, is referenced from the Income page: "=Income!G49" (Figure 10.22). Cell G6, 'Book Depreciation', is

Accounting_Chap10 - Excel

FILE | HOME | INSERT | PAGE LAYOUT | FORMULAS | DATA | REVIEW | VIE

D14

	A	F	G	H	I
1		Period	1	2	3
2		Months	Jan-13	Feb-13	Mar-13
3	Tax	Year	1	1	1
4	Cash Income Tax				
5	Pre-Tax Net Income		26,725	26,725	26,725
6	Book Depreciation		10,393	10,393	10,393
7	Operating Income (Net Book Dep)		37,118	37,118	37,118
8	Less: Tax Dep/Amort		(7,795)	(7,795)	(7,795)
9	Taxable Income		29,323	29,323	29,323
10					
11	State Taxable Income		29,323	29,323	29,323
12	State Income Tax Rate		6.50%	6.50%	6.50%
13	State Income Taxes Payable		1,906	1,906	1,906
14					
15	US Federal Taxable Income		27,417	27,417	27,417
16	US Federal Income Tax Rate		35.00%	35.00%	35.00%
17	US Federal Income Taxes Payable		9,596	9,596	9,596
18					
19	Total "Cash" Income Taxes Payable		11,502	11,502	11,502
20					
21	Book Income Tax		26,725	26,725	26,725
22					
23	State Taxable Income		26,725	26,725	26,725
24	State Income Tax Rate		6.50%	6.50%	6.50%
25	State Income Taxes Payable		1,737	1,737	1,737
26					
27	US Federal Taxable Income		24,988	24,988	24,988
28	US Federal Income Tax Rate		35.00%	35.00%	35.00%
29	US Federal Income Taxes Payable		8,746	8,746	8,746
30					
31	Total "Book" Income Taxes Payable		10,483	10,483	10,483

... | Income | Cash | Depr | ...

READY | 90%

Figure 10.22

linked to the depreciation page: "=Depr!G5". 'Operating Income (Net Book Dep)' in cell G7 is the summation of the two above – i.e. "=Subtotal(9,G5:G6)". Then row 8 subtracts tax, cash, and depreciation with the formula "=-Depr!G9". Row 9, taxable income, is therefore the summation of everything in this section; that is, G9 is "=Subtotal(9,G5:G8)".

State Taxable Income, row 11, is Taxable Income from row 9. The State Income Tax Rate, row 12, is "=Tax_State". The State Income Taxes Payable, row 13, is the product of income tax and the state rate or "=product(G11:G12)".

Row 15, US Federal Taxable Income Tax, is Taxable Income less State Income Taxes Payable. At the time of this text, state tax could be deducted from federal taxes. Cell G15 is therefore "=G9-G13". Cell G16 is the Federal Tax Rate or "=Tax_Fed". Row 17 is US Federal Income Taxes Payable and is calculated as the state quantity was calculated, "=product(G15:G16)". Total "Cash" Income Taxes Payable, row 19, is the summation of state and federal: "=sum(G13,G15)".

The next section on the Tax page is Book Income Tax. Row 21, Book Income Tax, is Earnings Before Tax, unadjusted, from the Income sheet. Cell G21 is "=Income!G49". The remainder of the sheet follows the same logic as above. Cell G23 equals cell G21. The State Income Tax Rate, cell G24, is "=Tax_State". The State Income Taxes Payable, cell G25, is the product of the two above or "=product(G23:G24)".

Row 27, US Federal Taxable Income, is Book Income Tax less State Income Taxes Payable or "=G21-G25". The US Federal Income Tax Rate is the same Federal Tax rate, so cell G28 is "=Tax_Fed". The US Federal Income Taxes Payable, G29, is "=Product(G27:G28)". Finally, row 31, Total "Book" Income Taxes Payable, is the summation of state and federal taxes; thus cell G31 is "=sum(G25,G29)". Next select cells G5:G31 and copy these onto column EH. The Tax page is now complete.

Upon completing the Tax page, the 'holes' in the accounting statements must be addressed – i.e. the sheets must be completed now that the structures have been developed. Returning to the balance sheet (Figure 10.23), row 5, Cash, must be linked to the End of Period Cash Balance on the Cash Flow. Cell G5 on the Balance page is "=Cash!G29". As the Cash page has not been completed, this value will result in zero initially. Accounts Receivable, row 6, is the days of receivables. Therefore, the respective Net Revenue divided by days per month is then multiplied by the number of days receivable: cell G6 is "=Income!G16/DAY(G2)*Day_Rec". Row 7, Other Current Assets, is left blank and row 8, Total Current Assets, is the summation: cell G8 is "=Subtotal(9,G5:G7)". Cells G5:G8 are then copied to column EH.

The Long-Term Assets section for Net Property, Plant & Equipment has three subcategories: Property, Plant & Equipment, and Less: Book Depreciation. As there are not additional capital purchases during the project, Property and Plant & Equipment initial values remain constant. Cells G11 and G12 are therefore "=G11" and "=G12", respectively. Cell G13, 'Less: Book Depreciation', is the cumulate of book depreciation for the project. This is therefore linked to the Depreciation page as "=-Sum(Depr!$G5:G5)". Cell G14 is the summation of the section: "=Subtotal(9,G11:G13)". Cells G11:G14 are then copied to column EH.

Other Assets, row 16, is Interest During Construction which is capitalized, i.e. depreciated, as part of Book Depreciation (row 13). Therefore cell G16 is "=F16". Cell G17, 'Total Long-Term Assets', is the summation of Long-Term Assets, or "=Subtotal(9,G11:G16)". In row 19, Total Assets, cell G19 is "=Subtotal(9,G5:G17)". Cells G16:G19 are then copied to column EH to complete the Assets section of the balance sheet.

Accounting_Chap10 - Excel

FILE HOME INSERT PAGE LAYOUT FORMULAS DATA REVIEW VIEW

G40

	A B C	D	E	F	G	H
1			Period		1	2
2			Months		Jan-13	Feb-13
3	**Balance**		Year		1	1
4	**Current Assets**					
5	Cash			87,500	-	-
6	Accounts Receivable			-	59,919	66,339
7	Other Current Assets			-		
8	**Total Current Assets**			**87,500**	**59,919**	**66,339**
9						
10	**Long-Term Assets**					
11	Property			1,000,000	1,000,000	1,000,000
12	Plant & Equipment			3,741,500	3,741,500	3,741,500
13	Less: Book Depreciation			-	(10,393)	(20,786)
14	**Net Property, Plant & Equipment**			4,741,500	4,731,107	4,720,714
15						
16	Other Assets			73,253	73,253	73,253
17	**Total Long-Term Assets**			**4,814,753**	**4,804,360**	**4,793,967**
18						
19	**Total Assets**			**4,902,253**	**4,864,279**	**4,860,306**
20						

Balance Income Cash

READY CALCULATE 100%

Figure 10.23

The Liabilities section is next on the balance sheet (Figure 10.24). Row 22, Accounts Payable, is the payables for the period (Expenses plus Depreciation for the Days Payable period). Cell G22 is "=Sum(Income!G27,Income!G31)/Day(G2)*Day_Pay". Row 23, Current Maturities of L-T Debt, is the same logic as from T = 0, column F. Cell G23 is "=VLOOKUP(G1+1,Tranche_1,3)+VLOOKUP(G1_1,Tranche_2,3)". Cell G24 is "=Subtotal(9,G22:G23)". Long-Term Liabilities follow the same logic; cells G27:G31 can be populated by selecting cells F27:F31, copying, and paste special only the formulas in cells G27:G31. Then cells G27:G31 are selected and copied to column EH.

The Partners' Equity is the final section. Row 34, Paid-In Capital, remains constant for the project as there is assumed to be no capital calls. Therefore G34 is "=F34", which is then copied to column EH.

Retained Earnings (BoP) in cell G36 is the end of period balance from the previous period (i.e. "=F39"). Earnings (Before Inc. Tax), row 37, references EBT from the income statement; thus Cell G37 is "=Income!G49". Finally, Partner Distributions in cell G38 references the Cash Flow page: "=-Cash!G28". Retained Earnings (EoP) is the summation of the section,

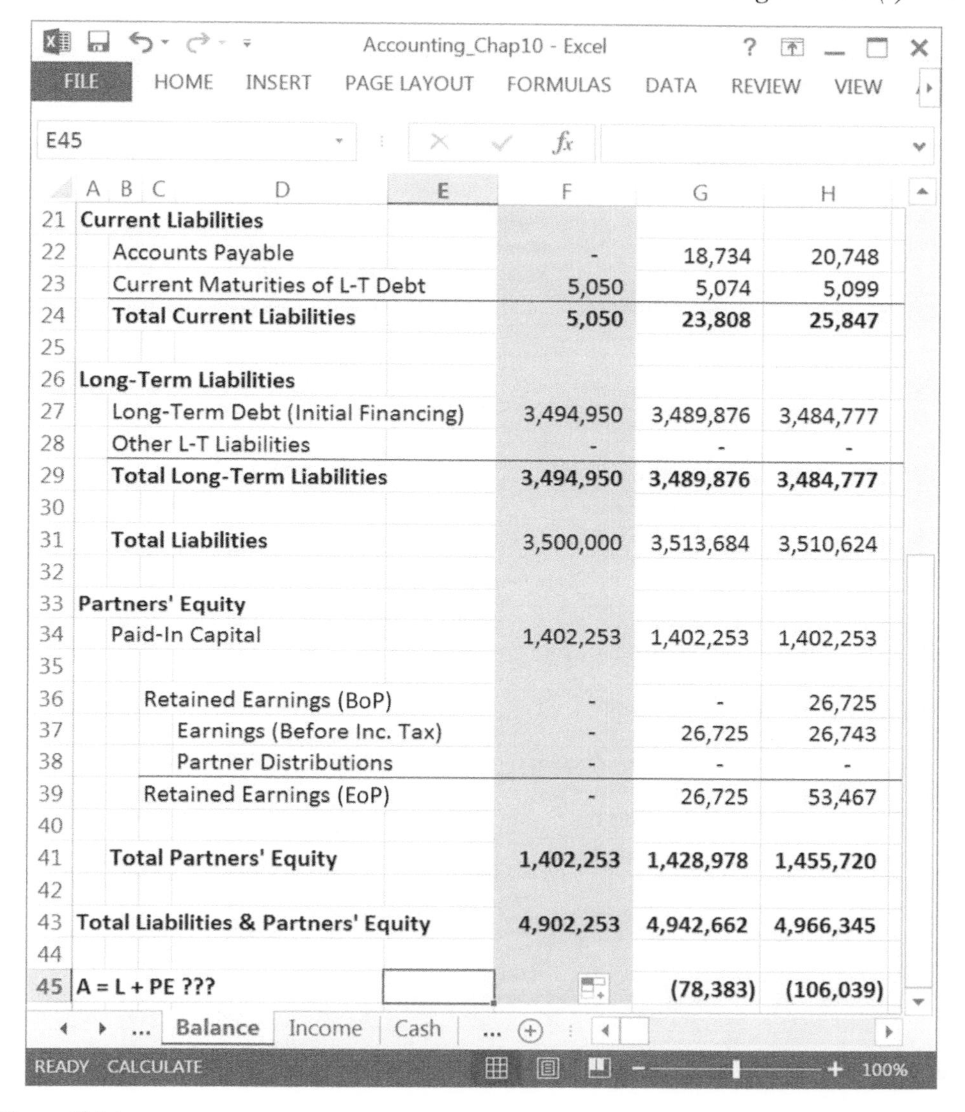

	A B C D	E	F	G	H
21	**Current Liabilities**				
22	Accounts Payable		-	18,734	20,748
23	Current Maturities of L-T Debt		5,050	5,074	5,099
24	**Total Current Liabilities**		**5,050**	**23,808**	**25,847**
25					
26	**Long-Term Liabilities**				
27	Long-Term Debt (Initial Financing)		3,494,950	3,489,876	3,484,777
28	Other L-T Liabilities		-	-	-
29	**Total Long-Term Liabilities**		**3,494,950**	**3,489,876**	**3,484,777**
30					
31	**Total Liabilities**		3,500,000	3,513,684	3,510,624
32					
33	**Partners' Equity**				
34	Paid-In Capital		1,402,253	1,402,253	1,402,253
35					
36	Retained Earnings (BoP)		-	-	26,725
37	Earnings (Before Inc. Tax)		-	26,725	26,743
38	Partner Distributions		-	-	-
39	Retained Earnings (EoP)		-	26,725	53,467
40					
41	**Total Partners' Equity**		**1,402,253**	**1,428,978**	**1,455,720**
42					
43	**Total Liabilities & Partners' Equity**		**4,902,253**	**4,942,662**	**4,966,345**
44					
45	**A = L + PE ???**			**(78,383)**	**(106,039)**

Figure 10.24

so cell G39 is "=Subtotal(9,G36:G38)". Total Partners' Equity in row 41 requires "=Subtotal(9,G34:G39)" to be placed in cell G41. Therefore, cell G43 is "=Subtotal(9,G22:G41)". Therefore, cell G45, the test cell, is "=G19-G43". Then select cells G36:G45 and copy these to the final column, EH.

Notice that the balance sheet does not balance. This will be corrected when Partnership Distributions are added in the Cash Flow sheet.

Income page

It is time to return to the Income page once again and complete the page (Figure 10.25). Row 31 is Depreciation/Amortization, which directly references the Depreciation page.

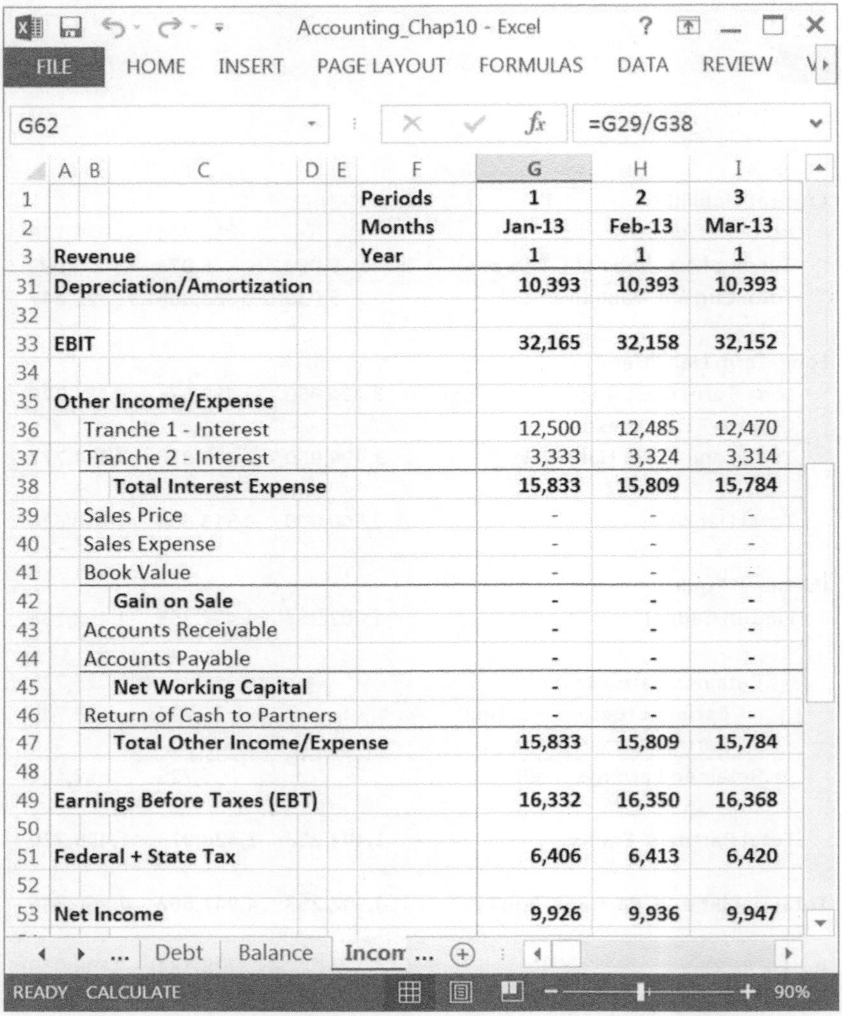

Accounting_Chap10 - Excel

FILE HOME INSERT PAGE LAYOUT FORMULAS DATA REVIEW

G62 =G29/G38

	A B C D E	F	G	H	I
1		**Periods**	**1**	**2**	**3**
2		**Months**	**Jan-13**	**Feb-13**	**Mar-13**
3	**Revenue**	**Year**	**1**	**1**	**1**
31	**Depreciation/Amortization**		**10,393**	**10,393**	**10,393**
32					
33	**EBIT**		**32,165**	**32,158**	**32,152**
34					
35	**Other Income/Expense**				
36	Tranche 1 - Interest		12,500	12,485	12,470
37	Tranche 2 - Interest		3,333	3,324	3,314
38	**Total Interest Expense**		**15,833**	**15,809**	**15,784**
39	Sales Price		-	-	-
40	Sales Expense		-	-	-
41	Book Value		-	-	-
42	**Gain on Sale**		**-**	**-**	**-**
43	Accounts Receivable		-	-	-
44	Accounts Payable		-	-	-
45	**Net Working Capital**		**-**	**-**	**-**
46	Return of Cash to Partners		-	-	-
47	**Total Other Income/Expense**		**15,833**	**15,809**	**15,784**
48					
49	**Earnings Before Taxes (EBT)**		**16,332**	**16,350**	**16,368**
50					
51	**Federal + State Tax**		**6,406**	**6,413**	**6,420**
52					
53	**Net Income**		**9,926**	**9,936**	**9,947**

... Debt Balance Incom ...

READY CALCULATE 90%

Figure 10.25

Therefore cell G31 is "=Depr!G5" and is copied to column EH. Row 41, Book Value, is the current book value of the property. Remember that the section is 'Gain on Sale' and is calculating the net return over the current book value, which has depreciated over time. Thus cell G41 is "=-If(G39–0,0,Balance!G14+Balance!F14-Income!G31)".

Rows 43 and 44, Accounts Receivable and Accounts Payable, respectively, are closed out at time of sale. Therefore, Accounts Receivable can be modelled as being returned and Accounts Payable as being paid. Note that there is no collection loss/allowance. This can be added by the individual analyst for specific projects. Cell G43 is "=IF(G39=0,0,Balance!G6)" and G44 is "=IF(G39=0,0,-Balance!G22)". Cells G43:G44 are then copied to column EH.

Row 46, Return of Cash to Partners, is the Minimum Cash Balance that is returned upon sale. As this only has a value upon sale, the IF/THEN formula must be utilized. Cell G46 is therefore "=IF(G39=0,0,Cash!G27)". Cell G46 is then copied to column EH.

Finally, row 51, Federal + State Tax, references the book tax from the Tax page. Cell G51 is "=Tax!G31", which is the Total "Book" Income Taxes Payable per period from that page. Cell G51 is then copied to column EH. The income statement is now completed. It should and must be reviewed upon the completion of the Cash Flow page for consistency.

Statement of cash flow

The cash flow statement, other than the structure, has yet to be completed and was saved for last. (Note: Some minor additions to the balance sheet remain.) As this is an Indirect Cash Flow Statement, Book Depreciation/Amortization, row 6, must be added back as part of the Operating Cash Flow. Cell G6 is therefore "=Income!G31". Row 7, Working Capital (Increase)/Decrease, is the difference in Working Capital – i.e. Accounts Receivable minus Accounts Payable – between periods. The formula for cell G7 is "=-(Balance!G6-Balance!G22)+(Balance!F6-Balance!F22)+(Balance!F16-Balance!G16)". Returning to Adjustments in row 5, cell G5 is then "=Subtotal(9,G6:G7)". Operating Cash Flow, row 8, is then the summation of above: cell G8 is "=Subtotal(9,G4:G7)". Cells G5:G8 are then copied to column EH.

The Financing section of the Cash Flow page represents the principal payment per period and principal balloon payment upon sale. Though there are two Tranches, both have been combined within this single row. However, the analyst can break them out to separate rows, as desired, for the analysis. The first part of this formula captures the principal payment and the second the balloon payment upon sale. The first part of cell G11 is "=-SUM(VLOOKUP(G1,Tranche_1,3),VLOOKUP(G1,Tranche_2,3))-. . . .

IF(G1/12=Sale_Term_Yr,Sum(VLOOKUP(G1,Tranche_1,6),VLOOKUP(G1,Tranche_2,6)),0)" (Figure 10.26). Cell G10 is "=Subtotal(9,G11)". Cells G10:G11 are then copied to column EH.

The Investments section, rows 13–17, provides cash perspective at time of sale (or purchase) of asset. Row 13, Investments, is "=Subtotal(9,G14:G17)". Row 14, Net Sales Proceeds, is the difference between Gross Sales Price and Sales Expense. For the ease of calculation, both will be combined into the single formulaic cell. Cell G14 is "=SUM(Income!G39:G40)". Row 15, Working Capital Return, is referenced from the Income page, so that cell G15 is "=Income!G45". Row 16, Return of Minimum ("Min") Cash Balance, is the cash balance in the sale period. Cell G16 therefore has a logic formula to determine sales period and then references the balance sheet Cash Balance for the period: "=IF(G1/12=Sale_Term_Yr,Balance!G5,0)". As there are no additional Capital Expenditures, row 17 is left blank. Cells G13:G17 are selected and copied to column EH.

Project Cash Balance, row 19, is the summation of the Cash Flow Sheet. Cell G19 is "=Subtotal(9,G4:G17)" and is then copied to column EH.

The next section of the Statement of Cash Flow is the Cash Balance and Distribution to Partners section (Figure 10.27). The Cash Balance (BoP), row 23, is the respective cash balance for the project from the balance sheet. Cell G23 is therefore "=Balance!F5". Project Cash Flow references the above calculation – i.e. cell G24 is "=G19". Row 25, Minimum Cash Balance, is the amount from the Input sheet that is required to remain in the project at all times. Cell G25 is "=Min_Cash". Row 26, Total Cash Balance, is the summation of Cash (BoP) (row 23) and Project Cash Flow (row 24); thus cell G26 is "=SUM(G23,G24)". Row 25 is there only as a reference. The Minimum Cash Balance is referenced again in row 27, that is, cell G27 is "=G25". Cash Available for Distribution, row 28, is Total Cash Balance (row 26) less Minimum Cash Balance (row 27), i.e. Cell G28 is "=G26-G27". Row 29, Cash Balance

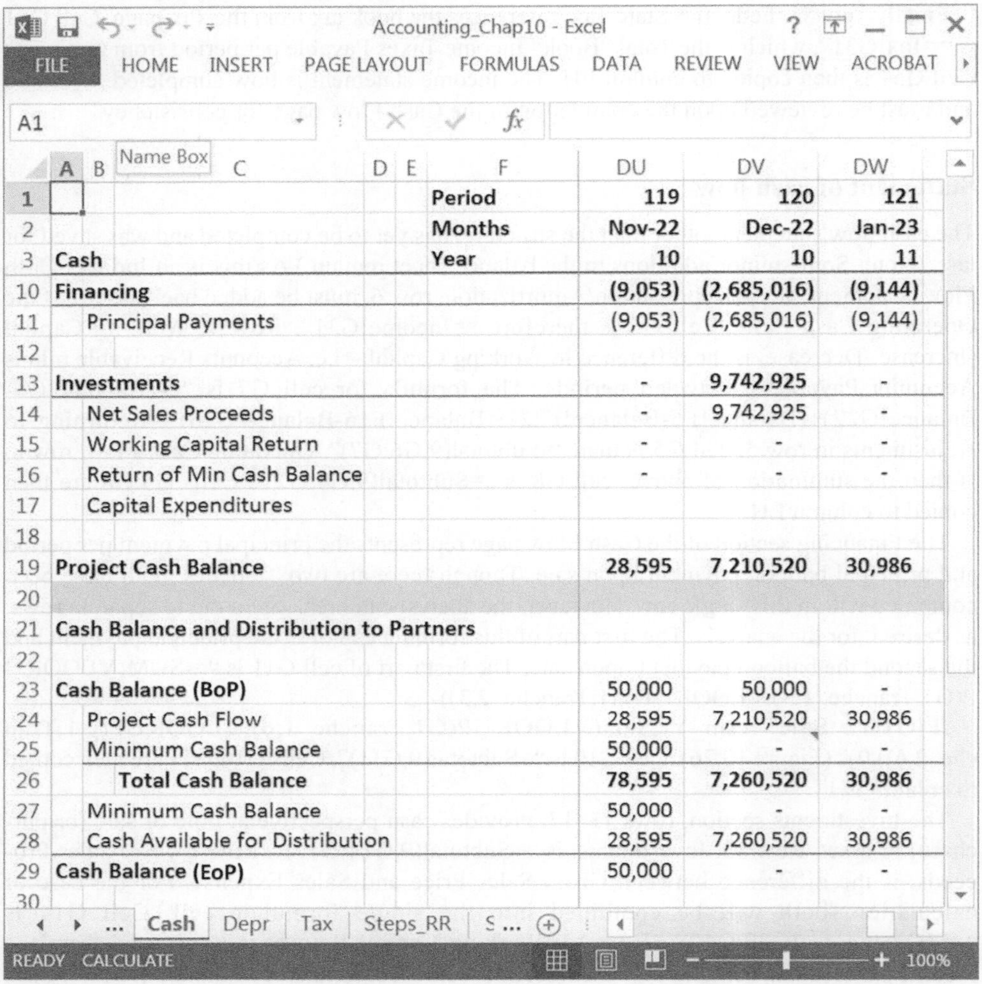

Accounting_Chap10 - Excel

FILE HOME INSERT PAGE LAYOUT FORMULAS DATA REVIEW VIEW ACROBAT

A1 | Name Box | fx

	A/B/C	F	DU	DV	DW
1		**Period**	**119**	**120**	**121**
2		**Months**	**Nov-22**	**Dec-22**	**Jan-23**
3	**Cash**	**Year**	**10**	**10**	**11**
10	**Financing**		**(9,053)**	**(2,685,016)**	**(9,144)**
11	Principal Payments		(9,053)	(2,685,016)	(9,144)
12					
13	**Investments**		**-**	**9,742,925**	**-**
14	Net Sales Proceeds		-	9,742,925	-
15	Working Capital Return		-	-	-
16	Return of Min Cash Balance		-	-	-
17	Capital Expenditures				
18					
19	**Project Cash Balance**		**28,595**	**7,210,520**	**30,986**
20					
21	**Cash Balance and Distribution to Partners**				
22					
23	**Cash Balance (BoP)**		50,000	50,000	-
24	Project Cash Flow		28,595	7,210,520	30,986
25	Minimum Cash Balance		50,000	-	-
26	**Total Cash Balance**		**78,595**	**7,260,520**	**30,986**
27	Minimum Cash Balance		50,000	-	-
28	Cash Available for Distribution		28,595	7,260,520	30,986
29	**Cash Balance (EoP)**		50,000	-	-
30					

... Cash | Depr | Tax | Steps_RR | S ...

READY CALCULATE 100%

Figure 10.26

(EoP), is Total Cash Balance (row 26) minus Cash Available for Distribution (row 28), so cell G29 is "=G26-G28". Cells G23:G29 are then selected and copied to column EH.

The Income Tax section on the Cash Flow page, rows 31:34, references the Tax page (Figure 10.28). Cell G32, 'State Taxes', is "=Tax!G13". Cell G33, 'Federal Taxes', is therefore "=Tax!(G17)". While row 34 is the summation of State and Federal Taxes, it is best to reference the Tax page where it is calculated. Remember that it is best to only complete a calculation once when modelling and then reference throughout the model. Cell G34 is "=Tax!(G19)". Then select cells G32:G34 and copy to row EH.

The final section of the Statement of Cash Flow is the respective cash flows from which to value the project. Notice there are two blank rows, rows 35 and 36, which was deliberate. While row 35 will remain blank, row 36 is to capture the dates for the respective cash flows. Cell G36 is "=G2" (see Figure 10.29). Cell F36 is the date for the equity infusion. Cell F36 is therefore "=DATE(Year(G36),Month(G36),1)". This sets the equity date as the first of the

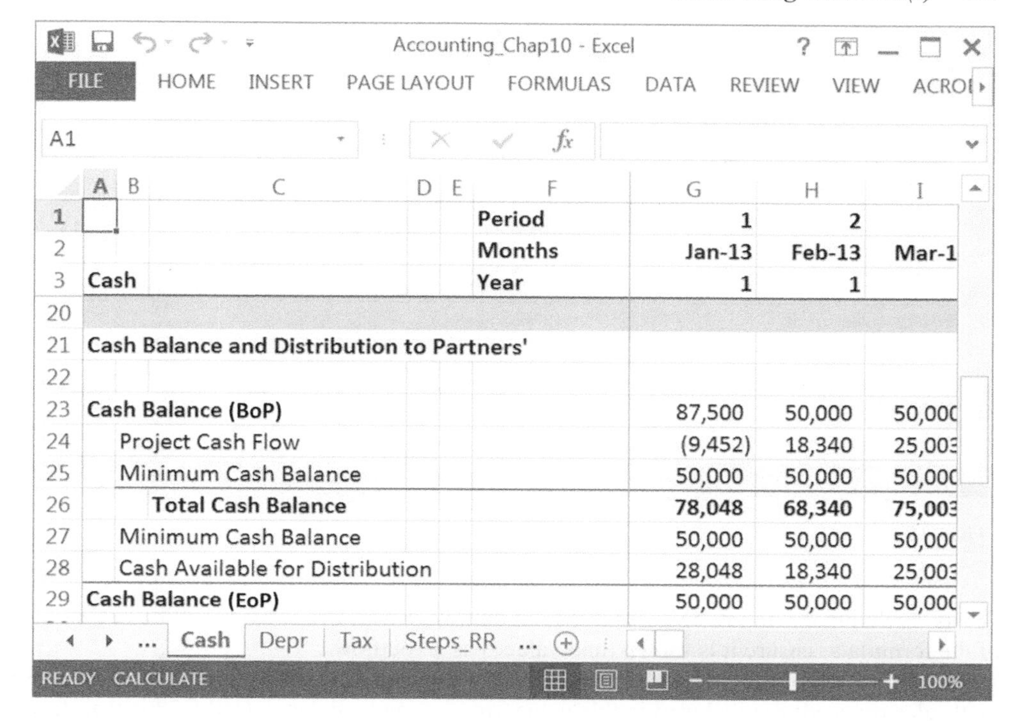

	A B C	D E F	G	H	I
1		Period	1	2	
2		Months	Jan-13	Feb-13	Mar-1
3	Cash	Year	1	1	
20					
21	Cash Balance and Distribution to Partners'				
22					
23	Cash Balance (BoP)		87,500	50,000	50,00
24	Project Cash Flow		(9,452)	18,340	25,00
25	Minimum Cash Balance		50,000	50,000	50,00
26	Total Cash Balance		78,048	68,340	75,00
27	Minimum Cash Balance		50,000	50,000	50,00
28	Cash Available for Distribution		28,048	18,340	25,00
29	Cash Balance (EoP)		50,000	50,000	50,00

Figure 10.27

Accounting_Chap10 - Excel

	A B C	D E F	G	H	I
1		Period	1	2	3
2		Months	Jan-13	Feb-13	Mar-13
3	Cash	Year	1	1	1
31	Income Tax				
32	State Taxes		1,230	1,232	1,233
33	Federal Taxes		6,195	6,201	6,207
34	Total Income Taxes		7,425	7,432	7,439

Figure 10.28

month while the period cash flow is the end of the month, 31 January in the example. Cell G36 is then copied to column EH.

Cell F37 is equity investment, Cash Flow at Period 0 (CF_0). Therefore it references the equity requirement from Sources/Uses on the Summary_Inputs page. The formula for cell

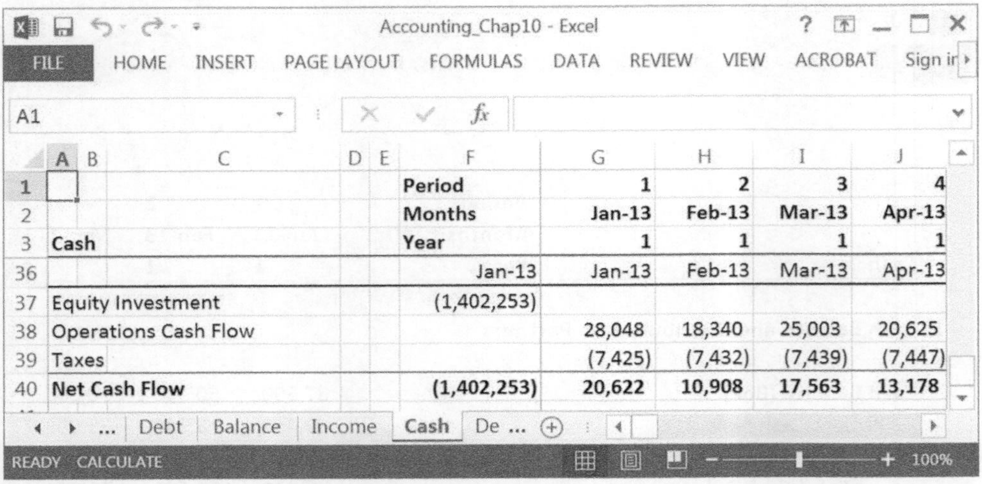

Accounting_Chap10 - Excel

FILE HOME INSERT PAGE LAYOUT FORMULAS DATA REVIEW VIEW ACROBAT Sign in

A1

	A B C	D E F	G	H	I	J
1		Period	1	2	3	4
2		Months	Jan-13	Feb-13	Mar-13	Apr-13
3	Cash	Year	1	1	1	1
36		Jan-13	Jan-13	Feb-13	Mar-13	Apr-13
37	Equity Investment	(1,402,253)				
38	Operations Cash Flow		28,048	18,340	25,003	20,625
39	Taxes		(7,425)	(7,432)	(7,439)	(7,447)
40	**Net Cash Flow**	**(1,402,253)**	**20,622**	**10,908**	**17,563**	**13,178**

... Debt Balance Income Cash De ...

READY CALCULATE 100%

Figure 10.29

F37 is "=-Summary_Inputs!D2". As this is an outflow, a negative is placed at the beginning of the formula to ensure it is shown (and valued) as an outflow.

Row 38 is Operations Cash Flow and references the cash available for distribution. As the final sales proceeds are included in this value, row 39 can be deleted. Cell G38 is therefore "=If(G1>Sale_Term_Yr*12,0,G28)". Finally the new row 39, Taxes, references the total income taxes from row 34. Cell G39 is "=If(G1>Sale_Term_Yr*12,0,-G34)". Cells G38 and G39 are then copied to column EH. The If/Then statements negate any cash flows occurring beyond the sale period. Row 40, Net Cash Flow, is the summary of equity investment, operations cash flow, and taxes, so cell F40 is "=Sum(G37:F39)". Cell G40 follows the same logic and is "=Sum(G37:G39)". Cell G40 is then copied to column EH to complete the Cash Flow Sheet for Net Cash Flow.

Balance sheet

The final step prior to the new valuation section on the Cash Flow page is to correct the balance sheet at time of sale. Note that in period 120 (year 10), the balance sheet is not in balance (Figure 10.30). The formulas must be adjusted to zero out the project at sale. Therefore, If/Then statements are added to reference the sale period.

Rows 5 and 6 in Current Assets are Cash and Accounts Receivable, respectively. Both of these accounts are zero at time of sale. The cash, row 5, is distributed to the shareholders; and Accounts Receivable, row 6, is considered paid. Therefore, cell DV5 is adjusted from "=Cash!DV29" to "=If(DV1>=Sale_Term_Yr*12,0,Cash!DV29)". The equivalent logic is applied to cell DV6, Accounts Receivable: "=If(DV1>=Sale_Term_Yr*12,0,Income!DV16/Day(DV2)*Day_Rec". Both cells DV5 and DV6 are copied to column G and to column EH for complete consistency.

Long-Term Assets, rows 11–17, follow the same logic as Current Assets. Cells DV11, DV12, DV13, and DV16 must have the If/Then logic added. These cells are then copied to column G and column EH, as in the previous paragraph. Assets in the period of sale and beyond are then zero.

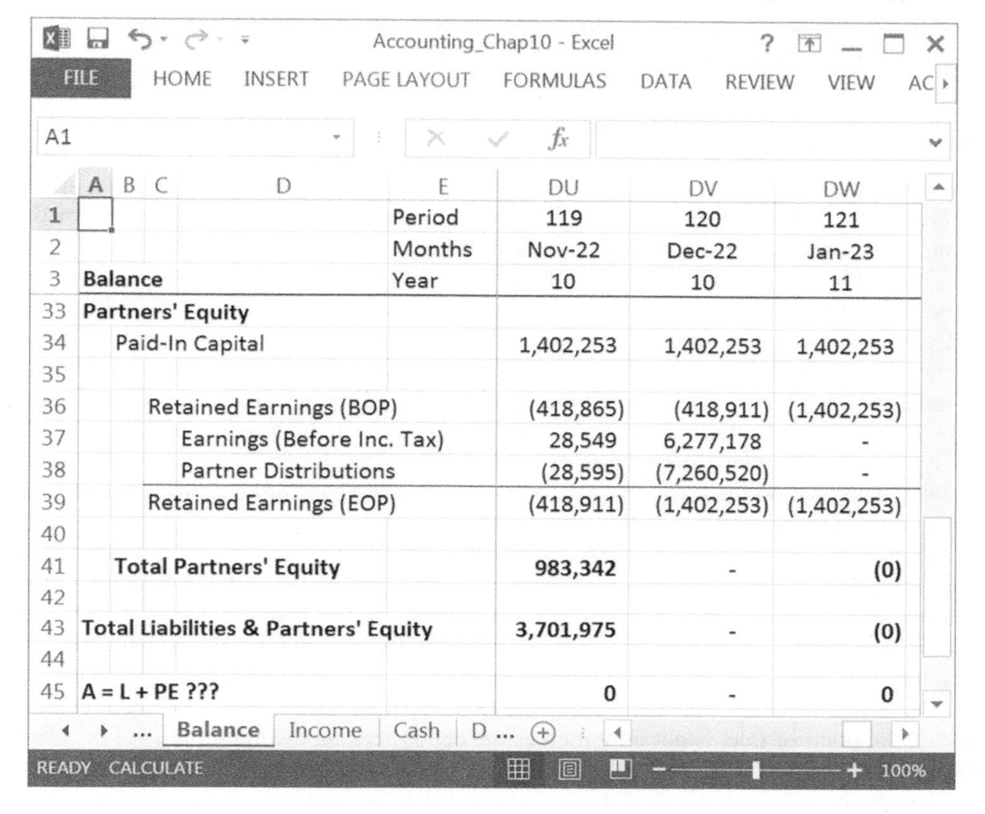

	A B C D	E	DU	DV	DW
1		Period	119	120	121
2		Months	Nov-22	Dec-22	Jan-23
3	Balance	Year	10	10	11
33	Partners' Equity				
34	Paid-In Capital		1,402,253	1,402,253	1,402,253
35					
36	Retained Earnings (BOP)		(418,865)	(418,911)	(1,402,253)
37	Earnings (Before Inc. Tax)		28,549	6,277,178	-
38	Partner Distributions		(28,595)	(7,260,520)	-
39	Retained Earnings (EOP)		(418,911)	(1,402,253)	(1,402,253)
40					
41	Total Partners' Equity		983,342	-	(0)
42					
43	Total Liabilities & Partners' Equity		3,701,975	-	(0)
44					
45	A = L + PE ???		0	-	0

Figure 10.30

As with Assets, Long-Term Liabilities are also nonexistent at time of sale and beyond. Cells DV22, DV23, and DV27 must have the If/Then logic added. The cells are then copied to the respective cells in column G and column EH. This will result in all liabilities being zero at time of sale and beyond.

Row 37 and Row 38 – Earnings (Before Inc. Tax) and Partner Distributions, respectively – must also be adjusted using the If/Then function. However, these are only zero after the point of sale and *not* at the point of sale. Therefore, the addition in column DV which must be copied to column G and column EH. The formulas for DV37 and DV38 are "=IF(DV1>Sale_Term_Yr*12,0,Income!DV49)" and "=IF(DV1>Sale_Term_Yr*12,0,-Cash!DV28)", respectively.

Statement of cash flow

The final logic adjustments for time of sale occurs in row 25, Minimum Cash Balance. This value is zero at time of sale and later points. The If/Then statement sets the minimum balance to zero at the point of sale and beyond.

Note: There are cash flows beyond period 120 (i.e. final sale) in the Statement of Cash Flow and other sheets. This is fine as the valuation will conclude in period 120. To eliminate these cash flows, If/Then logic must be added to cells to accommodate the zero values. While this could be done, it is recommended that the sheets be formatted to only the period of sale, thus simplifying the cell formulas.

Sheet adjustments

The accounting statements have been successfully added to the real estate pro forma model. This does not end the adjustments that must be completed, for example valuation. Now that cash is represented on the Statement of Cash Flow, the valuation must be completed (i.e. moved to this page). See Figure 10.31. To separate the valuation, row 41 is coloured grey. Cell A42 is titled 'Valuation:'.

Cells A43:A45 are 'Cash-on-Cash Minimum', 'Net Present Value', and 'Internal Rate of Return', respectively. Cell G43, Cash-on-Cash in respective period, is Free Cash Flow divided by invested equity. However, the cell formula is slightly different from the usual logic. Cell G43 is "=If(G1>Sale_Term_Yr*12,"",G40/-$F$40)". Prior to addressing the If/Then logic, the cash-on-cash is 'G40', period free cash flow, and '-$F$40' is the initial equity invested – i.e. the ratio is monthly cash-on-cash. Notice in the If/Then logic, rather than zero (0) being utilized for the TRUE result, double quotations were used. This was a deliberate action as for the periods exceeding the sales period the cell will be blank rather than zero. This allows for the minimum function for Cash-on-Cash to capture the entire row and the blank cells to not be utilized in the calculations because these are non-numeric (Figure 10.32). Cell G43 is then copied to column EH. Notice that cell DW43 is blank and does not have the characteristic dash associated with comma formatting. This indicates the cell is blank and not a numeric value, as stated earlier. Cell F43 is "=Min(G43:EH43)". The larger value in DV43, the point of sale, is not important as the cell is seeking the minimum, not the maximum, value.

The remaining valuation calculations, NPV and IRR, are then calculated based on row 40 (Net Cash Flow). Cell F44 is "=XNPV(Discount_Rate,F40:EH40,F36:EH36)". Cell F45 is

Accounting_Chap10 - Excel

FILE HOME INSERT PAGE LAYOUT FORMULAS DATA REVIEW VIEW ACRO

H43 | fx =IF(H1>Sale_Term_Yr*12,"",

	A B	C	D E	F	DV	DW	DX
1			Period		120	121	
2			Months		Dec-22	Jan-23	Feb
3	Cash		Year		10	11	
41							
42	Valuation:						
43	Cash-on-Cash Minimum			0.78%	342.11%		
44	Net Present Value						
45	Internal Rate of Return						

... Unit_Characteristics Summa ...

READY CALCULATE 100%

Figure 10.31

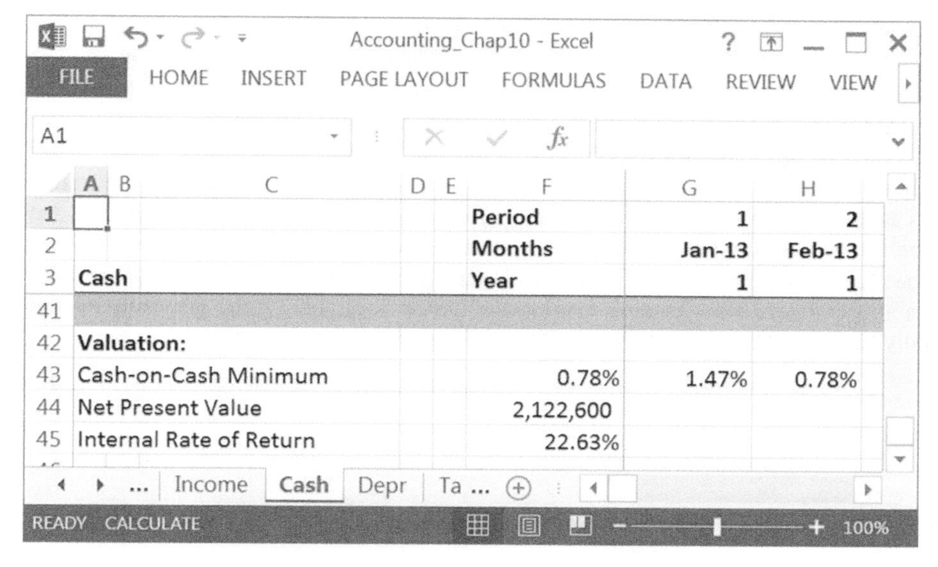

Figure 10.32

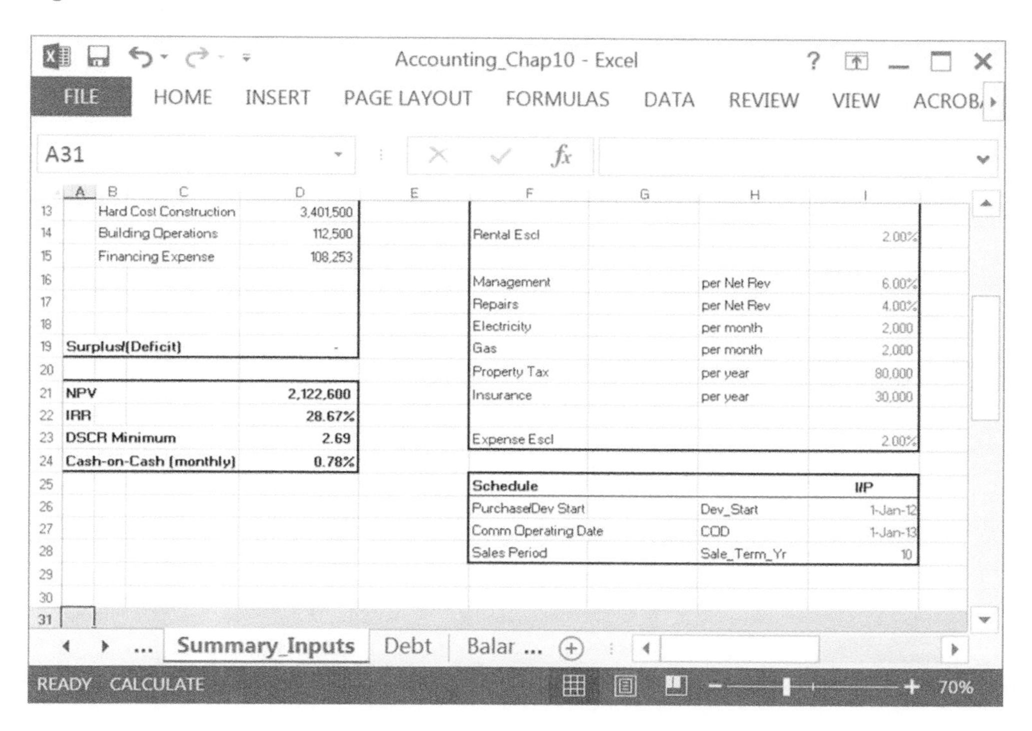

Figure 10.33

"=XIRR(F40:EH40,F36:EH36)". The completed valuation section on the cash flow statement is then demonstrated.

Returning to the Summary_Inputs page (Figure 10.33), the valuation section (cells A21:D29) must be adjusted to reference the Statement of Cash Flow page. Cell C25, NPV, is therefore "=Cash!F44". Cell C26, IRR, is therefore "=Cash!F45". Cell D28, Debt Service Coverage Ratio (DSCR), remains as this is calculated on the income statement and has not changed. However, cell D29, Cash-on-Cash (monthly), is adjusted to "=Cash!F43". Cells A21:D24, as these are no longer utilized, are deleted. Cells C25 and C26 are moved to column D and the section reformatted as there are fewer values.

The final adjustments are to be made to the income statement. On the income statement, rows 54:57 are deleted; these are Operating, Reversion, and Free Cash Flow. As this section added the accounting statements, which include the Statement of Cash Flow, these rows on the income statement are no longer necessary. The new rows 55:57, which include Valuation and FCF, are also deleted, leaving the DSCR logic below the grey line in row 54. Finally rows 57–66 are deleted, completing the Accounting section.

11 Hotel pro forma modelling

While this text has focused on model development and its scalability, the examples for the text are largely multifamily, single-use projects. As such, a very different real estate asset, hotels, will be modelled as a demonstration that the general structure proposed within this text is transferrable to all other real estate projects, albeit with some modifications.

As stated, this text is about pro forma construction, but a brief overview of hotels is important for the modeller. Hotels are at another spectrum of the real estate asset class because they are as much an operating business as they are about the real estate. One unique aspect of hotels is the tenor of the leases for the rooms available – i.e. generally a single day. Therefore, hotels are the real estate assets with the greatest sensitivity to economic cycles; they tend to be a leading indicator for economic performance, either entering recession early or recovering sooner. Hotels can re-price their leases on a daily basis as opposed to a multifamily product with year-long leases or commercial/industrial, which can have leases that extend decades, such as ground leases. As such, this makes hotel assets the best protection against inflationary concerns but also the most sensitive to changes in inflation.

For this text a hotel will be modelled using the same structure as earlier in the text, which is modular and scalable. The components will include the three basic sheets: Summary/Input, Debt (Amortization), and Pro Forma. However, a Development page will also be added to provide greater insight for a newly developed/renovated hotel. The hotel summary specifics are as follows:

- 125-key hotel
- 75,000 square feet; 10,000 banquet square feet
- Full service
- Develop/construct/stabilize

The inputs for the hotel will be similar to multifamily in category but with adjustments for the uniqueness of a hotel pro forma. These will include Occupancy Rate, Average Daily Rate with escalation, Food and Beverage (F&B), and other departments to include telephone/internet, garage, gift shop, spa, etc. The specifics, while modelled in this text, will largely be left to texts covering the hospitality industry and should be referenced there.

As demonstrated in Figure 11.1, the structure of the hotel pro forma is similar to the general structure for a real estate asset. The differences will be in Operational inputs – i.e. revenue and expense – as well as in some of the terms, for example RevPar (Revenue per available room). A review of all the inputs will follow during the pro forma construction, but the highlights of construction inputs are presented here. The Uses section will follow the same general summary sections as for Multifamily. This is to retain the consistent structure of all pro

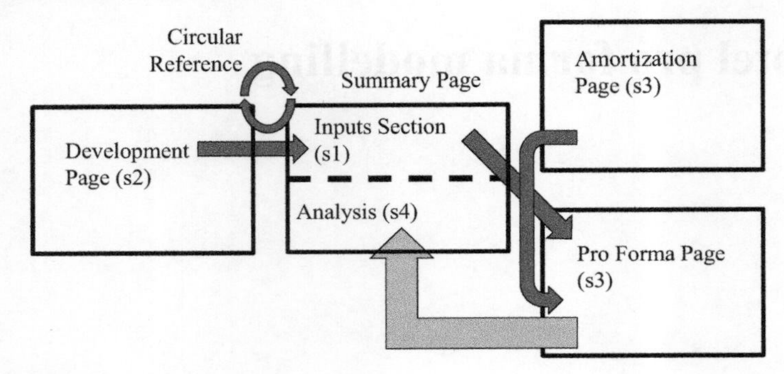

Figure 11.1 Hotel Diagram

forma models. The Uses will therefore include Land & Related, General & Administrative, Development Charges & Municipal Costs, Hard Cost Construction, Building Operations, and Financing Expense. Under Operations, the inputs for revenue are separated by sections: (1) Rooms, (2) Food & Beverage, and (3) Other Departments. A summary of each follows:

Rooms

- Occupancy
- Average Daily Rate

Food & beverage

- Restaurant
- Lounge
- Banquet
- Other F&B

Other departments

- Telephone
- Garage
- Gift Shop
- Spa
- Other Rents & Commissions
- Other

Expenses are similarly separated into three categories: (1) Department, (2) Unallocated Departments, and (3) Other Deductions. The subcategories of each follow:

Department

- Rooms
- Restaurant
- Lounge
- Banquet
- Other F&B
- Telephone/Internet

- Garage
- Spa
- Rents & Commissions

Unallocated departments

- General & Administrative (G&A)
- Repairs & Maintenance
- Sales & Marketing
- Utilities

Other deductions

- Management Fee
- FF&E Rep and Capital Reserves
- Property Taxes
- Insurance
- Permits & Licenses
- Equipment Rental

The remainder of the inputs are similar to the previous models presented. The Input section, while following the same logic, is completed slightly differently than the Multifamily (Figure 11.2).

The Sources/Uses section is the same structure only to the right, in column E, $/key is summarized for the major project costs. Please note that this could also be added to the multifamily and other pro formas, only $/sf is a better metric for that type of real estate asset. Generally with hotel projects the $/key metric can be used to determine cost of available rooms in a given area and is the general metric used in hospitality.

Notice that columns G–N include 'Revenue' in the top section and 'Characteristics' beneath. Further note that Revenue assumptions are separated into 3 years (i.e. ramp-up/start-up). This is to allow for a 3-year ramp-up for a hotel. Generally a new hotel in an area, even a flagged hotel, will require several years to become seasoned. As such there will need to be step-up provisions. For Rooms, Occupancy and Average Daily Rate (ADR) are estimated per year. As these are inputs they ought to be in blue; however, the individual cells have not been named. This is, of course, an analyst's choice but it allows for maximum flexibility. The names can later be adjusted/created, as per the analyst's intentions. The escalation for Average Daily Rate is named 'ADR_Escl' but, again, analyst's choice. The Metric (unit) of the values is POR or per occupied room (Figure 11.3).

Food & Beverage, the next section, discusses the amount of capital spent per occupied room using the equivalent 3-year step-up. The Banquet is the amount spent PBSF, that is per banquet square foot. The 'Other Food & Beverage' is a placeholder for additional revenue which may be applicable depending upon the type and level of hotel being modelled.

The final section of revenue is Other Departments. These include additional sources of revenue for the hotel from in-room (e.g. telephone/internet) services as well as additional services such as Garage, Gift Shop, Spa, and other Rents & Commissions. The Spa, which is open to others besides guests, is a stated capital amount that escalates with the seasoning of the hotel. The others are modelled as POR (per occupied room).

The section beneath Revenue is Characteristics (Figure 11.4). This allows the inputs of the number of rooms, total area (sf), and banquet area (sf). While each hotel and each real estate

	$	$/Key
Sources	70,974,294	567,794
Equity	17,743,574	141,949
Investor 1	17,743,574	141,949
Investor 2		
Debt	53,230,721	425,846
Tranche 1	53,230,721	425,846
Tranche 2		
Uses	70,974,294	567,794
Land & Related	6,500,000	52,000
General & Administrative	1,750,000	14,000
Dev Chrgs & Mun Costs	1,080,000	8,640
Hard Cost Construction	56,475,000	451,800
Building Operations	2,175,000	17,400
Financing Expense	2,994,294	23,954
Surplus/(Deficit)	-	

PV Reversion	4,908,155	28.46%
PV Operating	12,336,435	71.54%
Total	17,244,590	
CF_0	(18,943,049)	
New Present Value	(1,698,459)	
IRR	4.56%	

Revenue	Year 1	2	3	Escalation	Metric
Rooms					
Occupancy	68%	72%	75%		
Average Daily Rate	280.00	290.00	300.00	3%	POR
Food & Beverage					
Restaurant	50.00	65.00	70.00	3%	POR
Lounge	20.00	25.00	30.00	3%	POR
Banquet	525.00	575.00	650.00	3%	PBSF
Other Food & Beverage					
Other Departments					
Telephone/Internet	1.00	1.00	1.00	1%	POR
Garage	5.00	7.00	8.00	3%	POR
Gift Shop	1.00	2.00	2.00	3%	POR
Spa	1,000,000	1,200,000	1,500,000	5%	Total $
Other Rents & Commissions	8.00	9.00	10.00	3%	POR
Other					

Characteristics	
Number of Rooms	125
Total Area (sf)	75,000
Banquet Area (sf)	10,000

DSCR	
Average	1.66
Minimum	0.78

Cash-on-Cash/mo	
Average	0.77%
Minimum	-0.26%

Expense	Year 1	2	3	Escalation	Metric
Department					
Rooms	32.0%	31.0%	30.0%		% Rev
Restaurant	85.0%	80.0%	80.0%		% Rev
Lounge	65.0%	60.0%	60.0%		% Rev
Banquet	80.0%	72.0%	70.0%		% Rev
Other F&B					% Rev
Telephone/Internet	100.0%	100.0%	100.0%		% Rev
Garage	50.0%	50.0%	50.0%		% Rev
Gift Shop	90.0%	90.0%	90.0%		% Rev
Spa	85.0%	80.0%	75.0%		% Rev
Other Rents & Commissions	5.0%	5.0%	5.0%		% Rev
Other					% Rev
Unallocated Departments					
General & Administrative	30.00	31.00	32.00	3%	PAR
Repairs & Maintenance	22.40	22.40	22.40	3%	PAR
Sales & Marketing	22.40	22.40	22.40	3%	PAR
Utilities	0.50	0.50	0.50	3%	PSF
Other					
Other Deductions					
Management Fee	3.0%	3.0%	3.0%		% Ttl Rev
FF&E Rep and Capital Reserves	3.0%	4.0%	5.0%		% Ttl Rev
Property Taxes	600,000	625,000	650,000	3%	Total $
Insurance	100,000	100,000	100,000	4%	Total $
Permits & Licenses	20,000	20,000	20,000	2%	Total $
Equipment Rental	40,000	40,000	40,000	3%	Total $
Other					

Finance		I/P
Construction		
LtV	Const_LtV	75%
Principal		
Term (Years)		
Interest Rate		6.00%
Interest Only		Y
Origination Fee		1.00%
Permanent		
LtV	Perm_LtV	75%
Principal	Perm_Prir	53,230,721
Month Take-Out		
Term (Years)		30
Interest Rate		5%
Interest Only		Y
Origination Fee		0.50%

Valuation	I/P
Capitalization Rate	8%
Sales Expense	10%
Discount Rate	6%

Sales Period	10

Figure 11.2 Inputs

Revenue	Year	1	2	3	Escalation	Metric
Rooms						
Occupancy		68%	72%	75%		
Average Daily Rate		280.00	290.00	300.00	3%	POR
Food & Beverage						
Restaurant		50.00	65.00	70.00	3%	POR
Lounge		20.00	25.00	30.00	3%	POR
Banquet		525.00	575.00	650.00	3%	PBSF
Other Food & Beverage						
Other Departments						
Telephone/Internet		1.00	1.00	1.00	1%	POR
Garage		5.00	7.00	8.00	3%	POR
Gift Shop		1.00	2.00	2.00	3%	POR
Spa		1,000,000	1,200,000	1,500,000	5%	Total $
Other Rents & Commissions		8.00	9.00	10.00	3%	POR
Other						

Figure 11.3 Revenue

Characteristics	
Number of Rooms	125
Total Area (sf)	75,000
Banquet Area (sf)	10,000

Figure 11.4 Hotel Characteristics

asset is unique, the main characteristics for the property are summarized and can be adjusted. Given the fundamental nature of these inputs, they are named as follows:

Number of Rooms: Room_Number
Total Area (sf): Ttl_SF
Banquet Area (sf): BQ_SF

On the standard hotel pro forma, provided, the Characteristics section occupies G20:N23 in the Summary section.

Expenses are the next section, which occupies P1:W29 (Figure 11.5). As discussed, these are separated into major sections: Department, Unallocated Departments, and Other Deductions. Departments are ramped by 3 years of seasoning and expenses are allocated as a percentage of revenue (% Rev). None of the values in Departments are named. The next section, Unallocated Departments, is also ramped up using the 3-year model and allocated by PAR (per available room) and PSF (per square foot). Although the values are not in named cells, the escalation for each line item is named as follows:

General & Administrative: G_A_Escl
Repairs & Maintenance: RM_Escl
Sales & Marketing: SM_Escl
Utilities: Utilities_Escl

Finally, Other Deductions, which include project expenses, are listed in the third section. As with Unallocated Departments, only the escalation values for Property Taxes, Insurance,

Expense	Year	1	2	3	Escl	Metric
Department						
Rooms		32.0%	31.0%	30.0%		% Rev
Restaurant		85.0%	80.0%	80.0%		% Rev
Lounge		65.0%	60.0%	60.0%		% Rev
Banquet		80.0%	72.0%	70.0%		% Rev
Other F&B						% Rev
Telephone/Internet		100.0%	100.0%	100.0%		% Rev
Garage		50.0%	50.0%	50.0%		% Rev
Gift Shop		90.0%	90.0%	90.0%		% Rev
Spa		85.0%	80.0%	75.0%		% Rev
Other Rents & Commissions		5.0%	5.0%	5.0%		% Rev
Other						% Rev
Unallocated Departments						
General & Administrative		30.00	31.00	32.00	3%	PAR
Repairs & Maintenance		22.40	22.40	22.40	3%	PAR
Sales & Marketing		22.40	22.40	22.40	3%	PAR
Utilities		0.50	0.50	0.50	3%	PSF
Other						
Other Deductions						
Management Fee		3.0%	3.0%	3.0%		% Ttl Rev
FF&E Rep and Capital Reserves		3.0%	4.0%	5.0%		% Ttl Rev
Property Taxes		600,000	625,000	650,000	3%	Total $
Insurance		100,000	100,000	100,000	4%	Total $
Permits & Licenses		20,000	20,000	20,000	2%	Total $
Equipment Rental		40,000	40,000	40,000	3%	Total $
Other						

Figure 11.5 Hotel Expenses

Permits & Licenses, and Equipment Rental, are named cells. The escalation value for each line item is named as follows:

Property Taxes: Prop_Tax_Escl
Insurance: Insurance_Escl
Permits & Licenses: Permit_Lic_Escl
Equipment Rental: Equip_Rental_Escl

The final section of the Inputs are Finance/Valuation/Sales Period (Figure 11.6). This section occupies Y1:AB24. Notice that this is an abridged version of the amortization structure from earlier texts. This was done to simplify the hotel pro forma for modelling purposes. It is suggested that the analyst adjust and customize per project and as appropriate. Unlike the previous sections, the Finance section, while still including two debt tranches, separates them into Construction and Permanent Facility Loan-to-Value for each tranche and makes slight adjustments to the structure of each.

Valuation and Sales Period, while in this section of the Inputs, have remained largely unchanged. Note that when changing the sales period, the valuation structure will need to be adjusted.

The Development page is not going to be explained in detail in this chapter. The structure follows the equivalent structure for multifamily as modelled previously.

Finance		I/P
Construction		
LtV	Const_LtV	75%
Principal		
Term (Years)		
Interest Rate		6.00%
Interest Only		Y
Origination Fee		1.00%
Permanent		
LtV	Perm_LtV	75%
Principal	Perm_Prir	53,230,721
Month Take-Out		
Term (Years)		30
Interest Rate		5%
Interest Only		Y
Origination Fee		0.50%

Valuation		I/P
Capitalization Rate		8%
Sales Expense		10%
Discount Rate		6%

Sales Period		10

Figure 11.6 Hotel Finance

Hotel Pro Forma 17Mar13 (current) - Excel

FILE HOME INSERT PAGE LAYOUT FORMULAS DATA REVIEW VIEW ACROBAT Sign

F5 =EOMONTH(Dev_Start,0)

	A	B	C	D	E	F	G	H
1	**Development Start**	1-Jan-14						
2	**Construction Months**	24						
3	**Opening Date**	1-Jan-16						
4						1	2	3
5	**Cost Item**					**Jan-14**	**Feb-14**	**Mar-14**
6								
7	**Land & Related**		**Land_Rel**	**6,500,000**				
8	Land & Related			6,500,000		100.00%		
9								
10	**General & Administrative**		**G_A**	**1,750,000**				
11	Legal Fees			200,000		4.17%	4.17%	4.17%
12	Development Management Fees			600,000		4.17%	4.17%	4.17%

Development Summary Pro_Form ...

READY CALCULATE 80%

Figure 11.7

As a brief overview to complete the hotel model, the Development page, as discussed in previous chapters, is separated into three sections: Cost Allocation (%), Cost per Period ($), and Development Cash Flow (inflow/outflow). The first section appears in Figure 11.7.

The second section is shown in Figure 11.8.

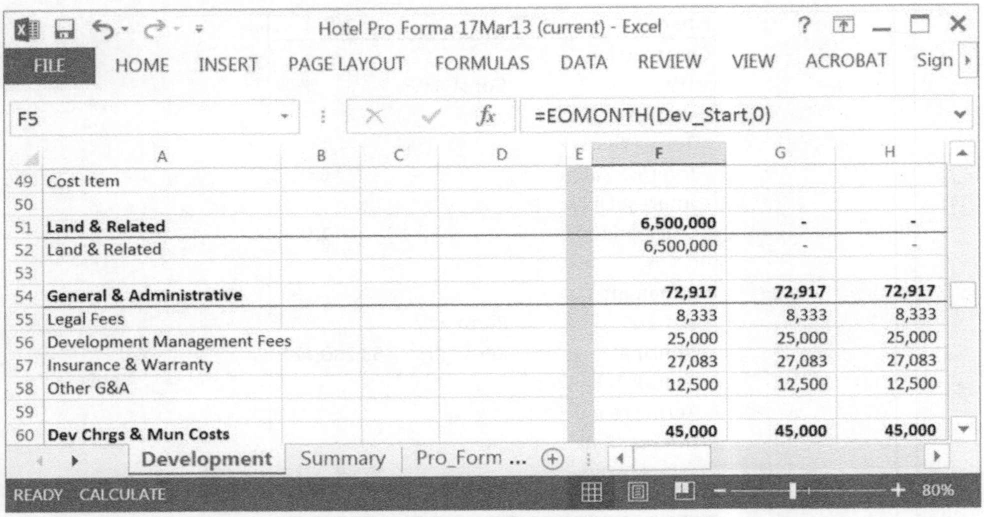

Hotel Pro Forma 17Mar13 (current) - Excel

FILE HOME INSERT PAGE LAYOUT FORMULAS DATA REVIEW VIEW ACROBAT Sign

F5 =EOMONTH(Dev_Start,0)

	A	B	C	D	E	F	G	H
49	Cost Item							
50								
51	**Land & Related**					**6,500,000**	**-**	**-**
52	Land & Related					6,500,000	-	-
53								
54	**General & Administrative**					**72,917**	**72,917**	**72,917**
55	Legal Fees					8,333	8,333	8,333
56	Development Management Fees					25,000	25,000	25,000
57	Insurance & Warranty					27,083	27,083	27,083
58	Other G&A					12,500	12,500	12,500
59								
60	**Dev Chrgs & Mun Costs**					**45,000**	**45,000**	**45,000**

Development Summary Pro_Form ...

READY CALCULATE 80%

Figure 11.8

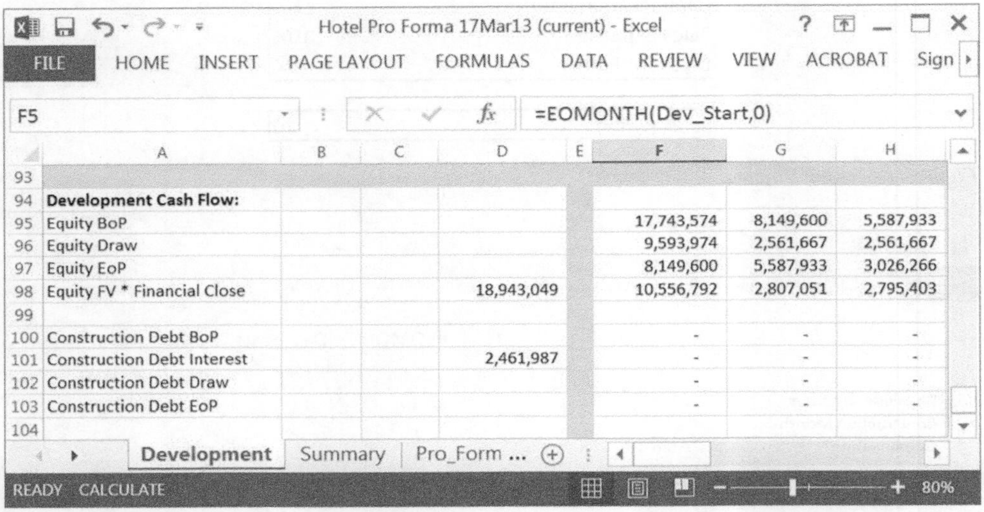

Hotel Pro Forma 17Mar13 (current) - Excel

FILE HOME INSERT PAGE LAYOUT FORMULAS DATA REVIEW VIEW ACROBAT Sign

F5 =EOMONTH(Dev_Start,0)

	A	B	C	D	E	F	G	H
93								
94	**Development Cash Flow:**							
95	Equity BoP					17,743,574	8,149,600	5,587,933
96	Equity Draw					9,593,974	2,561,667	2,561,667
97	Equity EoP					8,149,600	5,587,933	3,026,266
98	Equity FV * Financial Close			18,943,049		10,556,792	2,807,051	2,795,403
99								
100	Construction Debt BoP					-	-	-
101	Construction Debt Interest			2,461,987		-	-	-
102	Construction Debt Draw					-	-	-
103	Construction Debt EoP					-	-	-
104								

Development Summary Pro_Form ...

READY CALCULATE 80%

Figure 11.9

The third section is the Development Cash Flow (Figure 11.9), which denotes the inflow/outflow of capital during the development/construction periods.

Again, the development section is virtually identical to the multifamily and therefore the construction is not described here. Instead, the pro forma page will be discussed in detail because there are subtle yet significant changes to the structure and methodology required for a hotel pro forma asset.

As mentioned, the length of lease for a hotel real estate asset is unique at one day. As such, revenues (and expenses) for hotels vary month-to-month depending upon the days per month. Therefore, the schedule part of a hotel pro forma must accommodate a day count. While maintaining consistency of schedule (i.e. starting all schedules in column G) to the greatest possible extent, days will be added in the pro forma tab. The timing captured will therefore include Periods, Days, Month, and Year (Figure 11.10).

Periods, as in cell G1, will be the numerical count starting at 1, and in cell H1 will be "=G1+1". Months will be keyed off of the opening month, which can be from the Development sheet or from the summary Sheet. This should be a named cell. Days, in row 2, is therefore referenced using the "=Day()" function. This provides the days per given month. Finally, row 4, Year, is completed as with previous models using the "=Roundup()" function.

The header section, rows 6–11, is the hotel's operating characteristics: Occupancy, Average Daily Rate (ADR), RevPar (Revenue per available room), Occupied Rooms (per month), and Available Rooms (per month). From these characteristics the hotel revenues and expenses are derived (Figure 11.11).

Occupancy and Average Daily Rate are referenced from the Summary page using a HLOOKUP. On the Summary page the revenue section is named 'Revenue' (i.e. G1:L19).

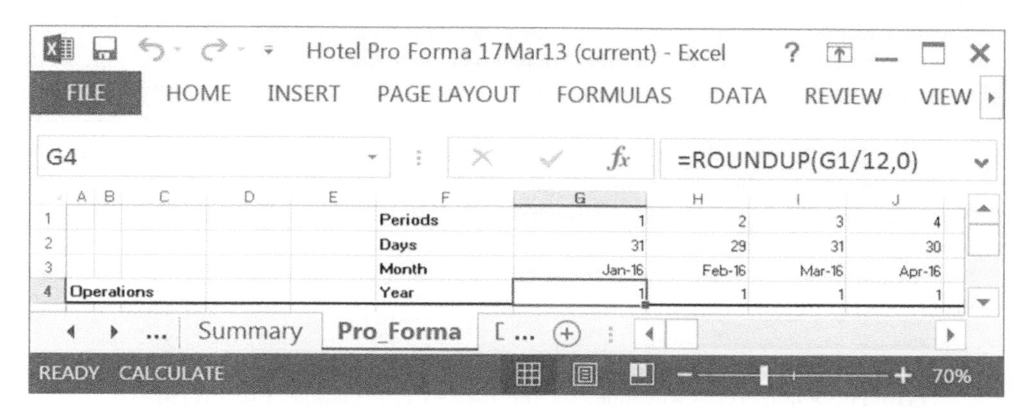

Row	A–E	F	G	H	I	J
1		Periods	1	2	3	4
2		Days	31	29	31	30
3		Month	Jan-16	Feb-16	Mar-16	Apr-16
4	Operations	Year	1	1	1	1

Figure 11.10

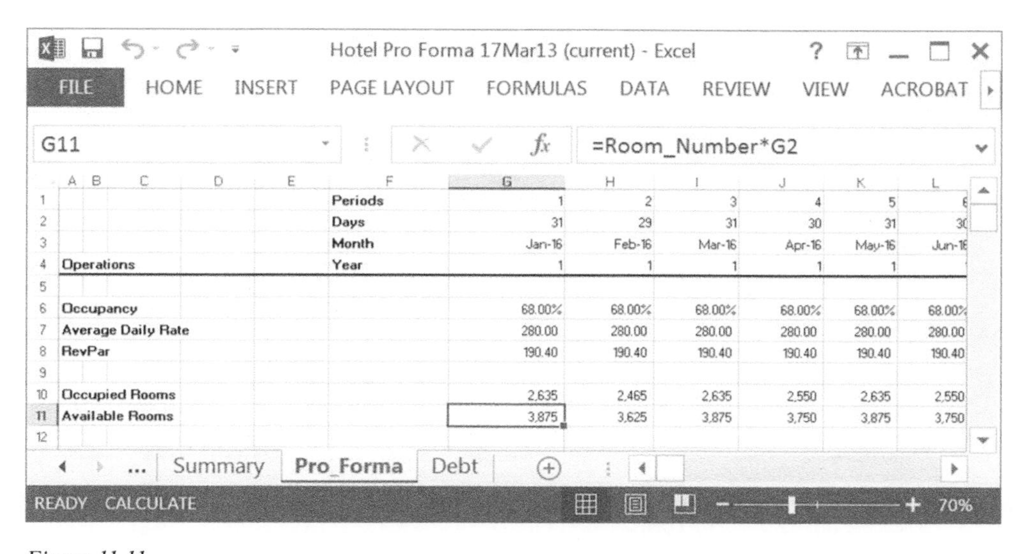

Row	A–E	F	G	H	I	J	K	L
1		Periods	1	2	3	4	5	6
2		Days	31	29	31	30	31	30
3		Month	Jan-16	Feb-16	Mar-16	Apr-16	May-16	Jun-16
4	Operations	Year	1	1	1	1	1	1
5								
6	Occupancy		68.00%	68.00%	68.00%	68.00%	68.00%	68.00%
7	Average Daily Rate		280.00	280.00	280.00	280.00	280.00	280.00
8	RevPar		190.40	190.40	190.40	190.40	190.40	190.40
9								
10	Occupied Rooms		2,635	2,465	2,635	2,550	2,635	2,550
11	Available Rooms		3,875	3,625	3,875	3,750	3,875	3,750
12								

Figure 11.11

This captures the step-up provisions for the Hotel for the first 3 years of operations. This array is also referenced for the revenue section.

For Occupancy in cell G6 the formula is "=HLOOKUP(G$4,Revenue,3)". The cell references the current year and pulls the value for this year. For years after the third year of step-up, the year 3 values remain and this is used as a stabilized value. The formula for Average Daily Rate is similar but uses an inflation adjustment to increase for later years. The formula for cell G7 is "=HLOOKUP(G$4,Revenue,4)*IF(G4<=3,1,(1+ADR_Escl)^(G4–3)). The If/Then applies no escalation for the first 3 years of step-up and then escalates annually thereafter.

RevPar, Revenue per available room, is the product of Occupancy and ADR. Cell G8 is then "=Product(G6:G7)".

Occupied Rooms and Available Rooms, cells G10 and G11 respectively, reference stated values. Occupied Rooms is the product of Available Rooms and Occupancy. Available Rooms is a static value referenced from the Summary page and multiplied by the number of days in the month. Cell G11 is therefore "=Room_Number*G2".

The next section of the pro forma page is Department Revenue (Figure 11.12). The first section of this includes Rooms, Restaurant, Lounge, Banquet, and Other Food & Beverage (left blank as a placeholder). Cell G11, Rooms, is ADR multiplied by Occupied Rooms: "=G7*G10". Restaurant is a function of Occupied Rooms with an escalator. Cell G15 is therefore "=G$10*HLOOKUP(G$4,Revenue,7)*IF(G$4<=3,1,(1+Restaurant_Escl)^(G$4–3))". This follows the same logic as stated earlier for a similar logic cell, Average Daily Rate. Lounge and Banquet follow equivalent logic only lounge references cell G10 (Occupied Rooms) while Banquet references banquet square feet (BQ_SF) from the Summary page.

The remaining items: Telephone, Garage, Gift Ship, Spa, and Other Rents & Commissions (cells G20:G24) follow in order:

=G$10*HLOOKUP(G$4,Revenue,13)*IF(G$4<=3,1,(1+Telephone_Escl)^(G$4–3))
=G$10*HLOOKUP(G$4,Revenue,14)*IF(G$4<=3,1,(1+Garage_Escl)^(G$4–3))
=G$10*HLOOKUP(G$4,Revenue,15)*IF(G$4<=3,1,(1+Gift_Escl)^(G$4–3))
=HLOOKUP(G$4,Revenue,16)/365*G$2*IF(G$4<=3,1,(1+Spa_Escl)^(G$4–3))
=G$10*HLOOKUP(G$4,Revenue,17)*IF(G$4<=3,1,(1+Other_Rent_Escl)^(G$4–3))

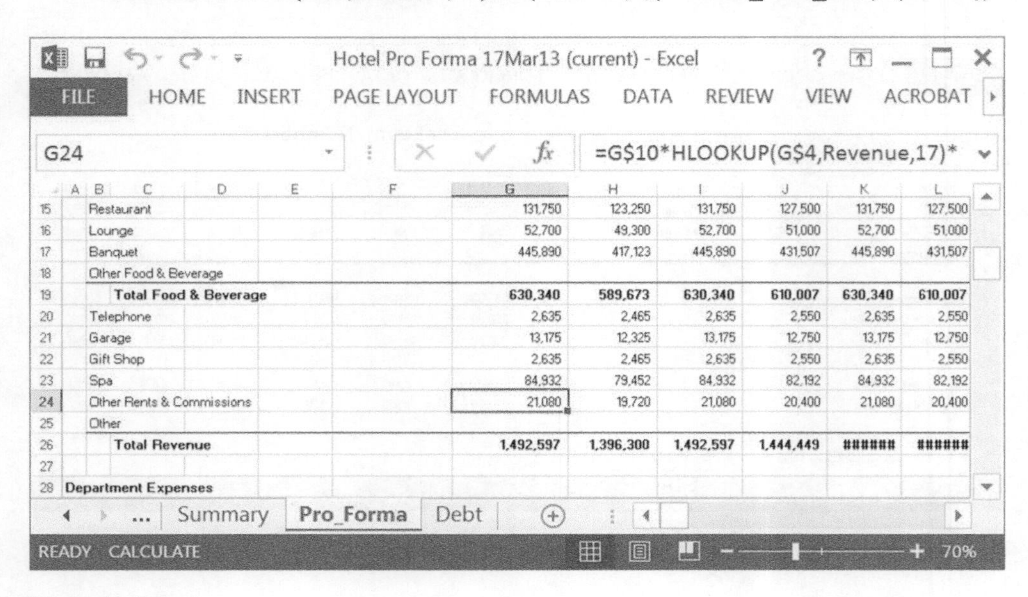

	B/C	G	H	I	J	K	L
15	Restaurant	131,750	123,250	131,750	127,500	131,750	127,500
16	Lounge	52,700	49,300	52,700	51,000	52,700	51,000
17	Banquet	445,890	417,123	445,890	431,507	445,890	431,507
18	Other Food & Beverage						
19	**Total Food & Beverage**	**630,340**	**589,673**	**630,340**	**610,007**	**630,340**	**610,007**
20	Telephone	2,635	2,465	2,635	2,550	2,635	2,550
21	Garage	13,175	12,325	13,175	12,750	13,175	12,750
22	Gift Shop	2,635	2,465	2,635	2,550	2,635	2,550
23	Spa	84,932	79,452	84,932	82,192	84,932	82,192
24	Other Rents & Commissions	21,080	19,720	21,080	20,400	21,080	20,400
25	Other						
26	**Total Revenue**	**1,492,597**	**1,396,300**	**1,492,597**	**1,444,449**	**######**	**######**
27							
28	**Department Expenses**						

Figure 11.12

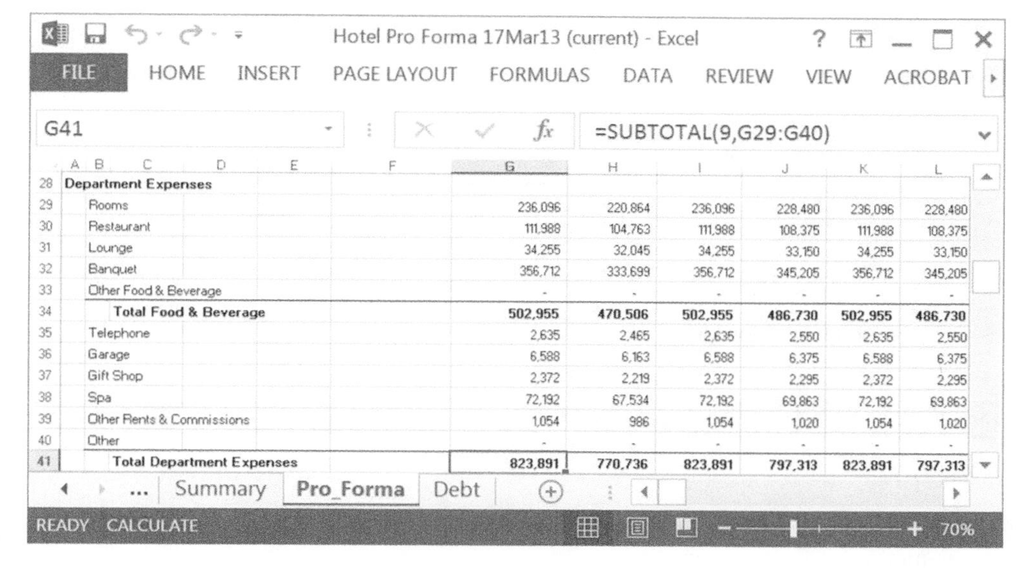

Hotel Pro Forma 17Mar13 (current) - Excel

G41 =SUBTOTAL(9,G29:G40)

Row	A–F	G	H	I	J	K	L
28	**Department Expenses**						
29	Rooms	236,096	220,864	236,096	228,480	236,096	228,480
30	Restaurant	111,988	104,763	111,988	108,375	111,988	108,375
31	Lounge	34,255	32,045	34,255	33,150	34,255	33,150
32	Banquet	356,712	333,699	356,712	345,205	356,712	345,205
33	Other Food & Beverage	-	-	-	-	-	-
34	**Total Food & Beverage**	**502,955**	**470,506**	**502,955**	**486,730**	**502,955**	**486,730**
35	Telephone	2,635	2,465	2,635	2,550	2,635	2,550
36	Garage	6,588	6,163	6,588	6,375	6,588	6,375
37	Gift Shop	2,372	2,219	2,372	2,295	2,372	2,295
38	Spa	72,192	67,534	72,192	69,863	72,192	69,863
39	Other Rents & Commissions	1,054	986	1,054	1,020	1,054	1,020
40	Other	-	-	-	-	-	-
41	**Total Department Expenses**	**823,891**	**770,736**	**823,891**	**797,313**	**823,891**	**797,313**

Summary | Pro_Forma | Debt

Figure 11.13

Each uses an escalator but the Spa uses the HLOOKUP logic reference as mentioned earlier.

For expenses, rows 29–41, similar logic follows (Figure 11.13). A listing for the logic for column G follows:

Rooms: =G14*HLOOKUP(G$4,Expense,3)
Restaurant: =G15*HLOOKUP(G$4,Expense,4)
Lounge: =G16*HLOOKUP(G$4,Expense,5)
Banquet: =G17*HLOOKUP(G$4,Expense,6)
Other Food & Beverage: =G18*HLOOKUP(G$4,Expense,7)
Total Food & Beverage: =SUBTOTAL(9,G30:G33)
Telephone: =G20*HLOOKUP(G$4,Expense,8)
Garage: =G21*HLOOKUP(G$4,Expense,9)
Gift Shop: =G22*HLOOKUP(G$4,Expense,10)
Spa: =G23*HLOOKUP(G$4,Expense,11)
Other Rents & Commissions: =G24*HLOOKUP(G$4,Expense,12)
Other: =G25*HLOOKUP(G$4,Expense,13)
Total Department Expenses: =SUBTOTAL(9,G29:G40)

To remain consistent, the logic follows the equivalent of revenue. Following Expenses is row 43, which is Gross Operating Profit. Cell G43 is "=G26-G41".

Rows 45–51 are Unallocated Departments expenses (Figure 11.14). The expenses largely consist of direct expenses that are difficult to allocate to a particular room or do not vary depending upon occupancy. Column G for these expenses is as follows:

General & Administrative:
=G$11*HLOOKUP(G$4,Expense,16)*IF(G$4<=3,1,(1+G_A_Escl)^(G$4–3))
Repairs & Maintenance:
=G$11*HLOOKUP(G$4,Expense,17)*IF(G$4<=3,1,(1+RM_Escl)^(G$4–3))
Sales & Marketing:

Hotel Pro Forma 17Mar13 (current) - Excel

FILE HOME INSERT PAGE LAYOUT FORMULAS DATA REVIEW VIEW ACRO

G51 =SUBTOTAL(9,G46:G50)

Row	Label	G	H	I	J	K
45	**Unallocated Departments**					
46	General & Administrative	116,250	108,750	116,250	112,500	116,250
47	Repairs & Maintenance	86,800	81,200	86,800	84,000	86,800
48	Sales & Marketing	86,800	81,200	86,800	84,000	86,800
49	Utilities	37,500	37,500	37,500	37,500	37,500
50	Other					
51	**Total Unallocated Departments**	**327,350**	**308,650**	**327,350**	**318,000**	**327,350**
52						

Summary | Pro_Forma | Debt

READY CALCULATE 70%

Figure 11.14

=G$11*HLOOKUP(G$4,Expense,18)*IF(G$4<=3,1,(1+SM_Escl)^(G$4–3))
Utilities:
=Ttl_SF*HLOOKUP(G$4,Expense,19)*IF(G$4<=3,1,(1+Utilities_Escl)^(G$4–3))
Other:
Total Unallocated Departments: =SUBTOTAL(9,G46:G50)

The final and unique section to the Hotel pro forma is 'Other Deductions', rows 53–61 (Figure 11.15). These consist of the expenses listed as well as the formulas for each expense from column G, the first period. They are as follows:

Management Fee:
=G$26*HLOOKUP(G$4,Expense,23)
FF&E Rep and Capital Rsvs:
=G$26*HLOOKUP(G$4,Expense,24)

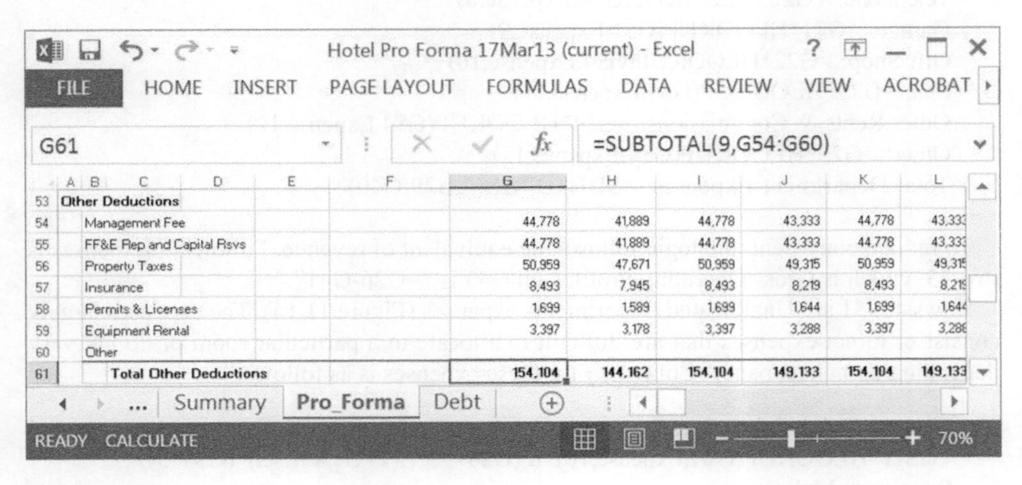

Hotel Pro Forma 17Mar13 (current) - Excel

FILE HOME INSERT PAGE LAYOUT FORMULAS DATA REVIEW VIEW ACROBAT

G61 =SUBTOTAL(9,G54:G60)

Row	Label	G	H	I	J	K	L
53	**Other Deductions**						
54	Management Fee	44,778	41,889	44,778	43,333	44,778	43,333
55	FF&E Rep and Capital Rsvs	44,778	41,889	44,778	43,333	44,778	43,333
56	Property Taxes	50,959	47,671	50,959	49,315	50,959	49,315
57	Insurance	8,493	7,945	8,493	8,219	8,493	8,219
58	Permits & Licenses	1,699	1,589	1,699	1,644	1,699	1,644
59	Equipment Rental	3,397	3,178	3,397	3,288	3,397	3,288
60	Other						
61	**Total Other Deductions**	**154,104**	**144,162**	**154,104**	**149,133**	**154,104**	**149,133**

Summary | Pro_Forma | Debt

READY CALCULATE 70%

Figure 11.15

Property Taxes:
=HLOOKUP(G$4,Expense,25)/365*G$2*IF(G$4<=3,1,(1+Prop_Tax_Escl)^(G$4–3))
Insurance:
=HLOOKUP(G$4,Expense,26)/365*G$2*IF(G$4<=3,1,(1+Insurance_Escl)^(G$4–3))
Permits & Licenses:
=HLOOKUP(G$4,Expense,27)/365*G$2*IF(G$4<=3,1,(1+Permit_Lic_Escl)^(G$4–3))
Equipment Rental:
=HLOOKUP(G$4,Expense,28)/365*G$2*IF(G$4<=3,1,(1+Equip_Rental_Escl)^(G$4–3))
Other:
Total Other Deductions: =SUBTOTAL(9,G54:G60)

The remainder of the Pro Forma section is similar to the other pro formas already developed (Figure 11.16).

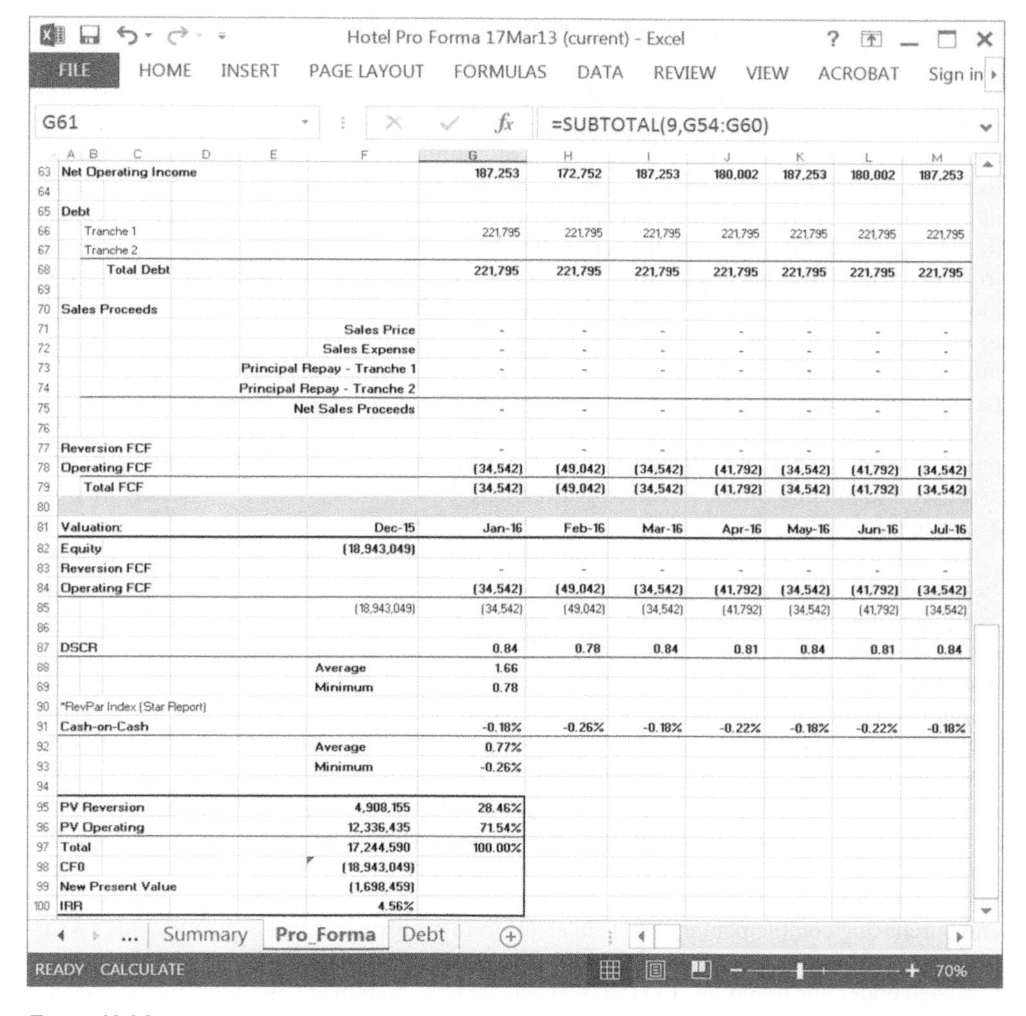

Row	Label	F	G	H	I	J	K	L	M
63	Net Operating Income		187,253	172,752	187,253	180,002	187,253	180,002	187,253
64									
65	Debt								
66	Tranche 1		221,795	221,795	221,795	221,795	221,795	221,795	221,795
67	Tranche 2								
68	Total Debt		221,795	221,795	221,795	221,795	221,795	221,795	221,795
69									
70	Sales Proceeds								
71	Sales Price		-	-	-	-	-	-	-
72	Sales Expense		-	-	-	-	-	-	-
73	Principal Repay - Tranche 1		-	-	-	-	-	-	-
74	Principal Repay - Tranche 2								
75	Net Sales Proceeds		-	-	-	-	-	-	-
76									
77	Reversion FCF		-	-	-	-	-	-	-
78	Operating FCF		(34,542)	(49,042)	(34,542)	(41,792)	(34,542)	(41,792)	(34,542)
79	Total FCF		(34,542)	(49,042)	(34,542)	(41,792)	(34,542)	(41,792)	(34,542)
80									
81	Valuation:	Dec-15	Jan-16	Feb-16	Mar-16	Apr-16	May-16	Jun-16	Jul-16
82	Equity	(18,943,049)							
83	Reversion FCF		-	-	-	-	-	-	-
84	Operating FCF		(34,542)	(49,042)	(34,542)	(41,792)	(34,542)	(41,792)	(34,542)
85		(18,943,049)	(34,542)	(49,042)	(34,542)	(41,792)	(34,542)	(41,792)	(34,542)
86									
87	DSCR		0.84	0.78	0.84	0.81	0.84	0.81	0.84
88	Average		1.66						
89	Minimum		0.78						
90	*RevPar Index (Star Report)								
91	Cash-on-Cash		-0.18%	-0.26%	-0.18%	-0.22%	-0.18%	-0.22%	-0.18%
92	Average		0.77%						
93	Minimum		-0.26%						
94									
95	PV Reversion	4,908,155	28.46%						
96	PV Operating	12,336,435	71.54%						
97	Total	17,244,590	100.00%						
98	CF0	(18,943,049)							
99	New Present Value	(1,698,459)							
100	IRR	4.56%							

Figure 11.16

12 Pro forma portfolio modelling

As demonstrated by the previous chapters of this book, the single-asset real estate pro forma is a complex and detailed financial tool that takes both time and expertise to successfully construct. Aside from the application of high-level financial formulas, pro forma financial evaluations require the input of various detailed assumptions. Throughout the development of a pro forma, these assumptions transform from the analyst's initial instincts into refined estimates that are grounded in concrete market research and real-world data. These assumptions are evaluated continuously and subsequently polished, each time making the model more precise and providing a higher degree of confidence in the resulting critical metrics (net present value, internal rate of return, etc.). The proper construction and continued maintenance of a project pro forma is an essential exercise to perform in order to evaluate the profitability, and ultimately the feasibility, of investing in the project. For a manager to truly understand the intricacies of a project and anticipate the potential shortcomings, he must take the time to lay out the minute components of a pro forma. It is, however, important to understand that no matter how much time and effort is spent building and refining the model, there are always factors that over (or under) influence the project as initially forecasted. The inadequate assumptions made (or missed) by the analyst have the ability to greatly influence the overall project and can quickly result in extreme outcomes, either positive or negative. Knowing this, the asset manager must be diligent about analyzing each asset with a balance of conservative yet realistic input assumptions to generate the most accurate results.

For one to understand the input assumptions and the potential problem areas, it is first essential to highlight and understand some of the primary differences between a real estate asset and other asset classes. Unlike the behavior of equities (e.g. stocks, mutual funds, exchange-traded funds, etc.) and bonds (which are continuously traded on open, efficient markets), real estate transactions and valuations are non-liquid[1] and significantly influenced by human factors. It is important not to overlook the fact that real estate assets have an actual presence in the physical world and are therefore subject to location and natural forces, whereas many other equities are electronic blips that exist in cyberspace (1). At a minimum, real estate can be classified by property type, geographic location, and its stage in the property life cycle (e.g. construction, lease-up, and full occupancy); equities, however, are usually classified by growth type or size of market capitalization (2). Given the multitude of factors that affect a project's outcome, a manager will attempt to shield his single asset by pursuing and purchasing complementary assets in an effort to mitigate risk and exposure. This assemblage of multiple single-asset projects, in effect, creates a portfolio of projects that are now managed under one umbrella. To effectively create this assemblage, we must first understand what a portfolio is, discuss the importance of identifying complementary assets to create a balanced and diversified portfolio, and understand the different metrics used to evaluate portfolios.

Portfolios and portfolio theory

A portfolio, in its most basic form, is simply a grouping of financial assets (3). It can be as small as two items and as many as a few hundred – practically speaking, a portfolio would never grow to be that large, but in theory it could. Assets are typically purchased and sold at different times and in quantities throughout the life of the portfolio. Each asset is its own unique component and, as such, performs independently of every other asset in the portfolio. Although certain natural or global events will affect all assets simultaneously, each asset may be affected at different times and to varying degrees. As will be discussed later, the concept of independent performance is critical to the function of a successful portfolio. The notion of independent assets relies on each asset having its own fundamentals, associated risks, and returns. Although the asset manager, now more appropriately called the portfolio manager, cares very much about achieving positive individual returns on each asset, he should care more about the overall performance of the portfolio. A manager would likely neither be considered successful nor be employed very long if half his assets consistently made money and half the assets lost money. It is then extremely important that the manager understand the relationships between individual assets and their associated contribution to the portfolio as a whole. In general, the rationale for aggregating assets into a portfolio – versus holding them independently – relies heavily on the concept of risk-adjusted return, diversification, and optimization (which will be discussed later).

As assets are assembled, the manager shifts his focus to the performance of the portfolio, thereby changing his analysis from single-asset analysis to the realm of Portfolio Theory. This theoretical foundation was established by the American economist Harry Markowitz in the early 1950s. His theory, then just called Portfolio Theory, relates the importance of individual asset allocations within a portfolio to the overall performance of the portfolio.[2] This concept later became known as Markowitz Portfolio Theory or Modern Portfolio Theory (MPT). The theory establishes that by adjusting the allocation (i.e. weighting[3]) of the assets within the portfolio, the resulting returns of the portfolio itself should change. It is important to note that the manager is not changing the *return* of that asset, but rather adjusting the weight of an asset within the portfolio, thereby altering the return of the portfolio. Applying this weight-adjusted technique, the manager would likely seek an allocation that would result in the (1) maximum return for a given level of risk, (2) minimum risk for a given level of return, or (3) optimum efficiency of the portfolio.[4] There are multiple measures of efficiency; however, the primary three are the Sharpe Ratio, the Treynor Ratio, and the Coefficient of Variation (4). This chapter will focus solely on efficiency as measured by the Coefficient of Variation, or CV. Mathematically, CV is defined as the risk (represented by the standard deviation of the asset) divided by the expected return of the asset. The ratio is any number between zero and positive infinity. The smaller the value, the more efficient the portfolio.

In its most fundamental form, Modern Portfolio Theory identifies sets or groups of these efficient portfolios. It is important to understand that, by definition, an efficient portfolio is one in which no other allocation combination of portfolio assets could offer a higher expected return at a precise level of risk (5). That is to say, in order to increase (or decrease) a return, the manager would have to subsequently increase (or decrease) the tolerance for risk. The manager can chart the varying levels of return against the associated increasing levels of risk to generate what is known as the Efficient Frontier (see Figure 12.1). The theory underlying

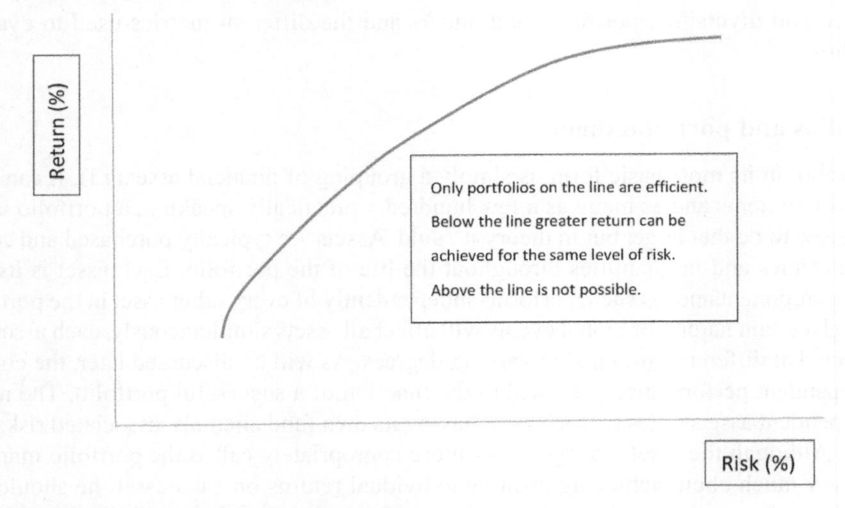

Figure 12.1 Theoretical Graphical Representation of the Efficient Frontier (6)

the Efficient Frontier states that no greater return is possible at a given level of risk without the use of positive or negative leverage, or in the absence of financial instruments.[5] Later this chapter will focus on achieving an asset allocation, which seeks to exist on the frontier rather than above it.

As illustrated in Figure 12.1, the lower the risk (x-axis), the lower the expected return (y-axis) and vice versa. As one metric increases (or decreases), so too does the other in order to maintain the position on the frontier. If, for example, risk were to increase without a proportional increase in the return, the resulting portfolio would fall below the curve. Although the vast majority of real-world portfolios exist below the curve, they are not considered efficient. Given that a goal of a manager is often to provide the optimum risk-adjusted return, he must evaluate his current allocations and determine if the portfolio falls along the curve and, if it does not, then consider a reallocation.

Finally, it is extremely important to consider the size of a real estate portfolio because, unlike other asset classes, real estate is considered indivisible[6] and takes substantial effort to obtain and divest each singular asset. Because of this, portfolios often take considerable time to grow and even longer to satisfactorily contract. Therefore, a manager must be extremely mindful of his ability to anticipate changing market conditions – whether the conditions are quick swings up or down or more long-term trends – with reasonably illiquid assets.

Portfolio diversification

If an investor knew the future return of an asset with absolute certainty, thereby eliminating all possible risk of losing money, then the investor would invest heavily in only that one asset. With the exception of engaging in insider trading, there is no certainty in the future return of an asset, so investors must instead reduce the uncertainty as much as possible (7). This uncertainty can be categorized into two broad concepts: systematic and unsystematic risk. Systematic risk, also known as market risk, cannot be physically reduced or diversified away as it is equally common to all assets in the market. Unsystematic risk, also called unique risk, however, is specific to each asset and varies widely from one to another. In contrast to market risk,

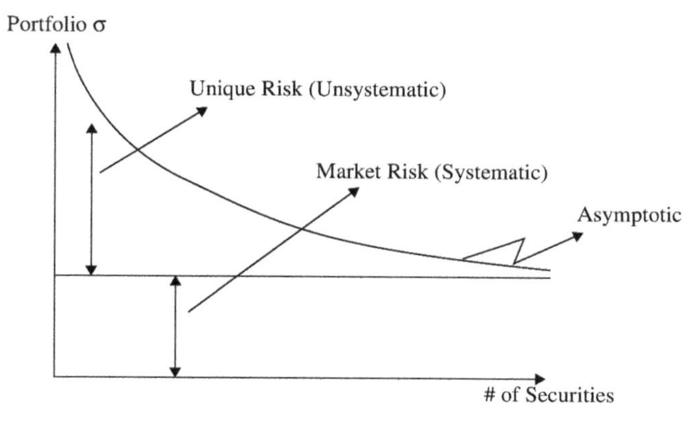

Figure 12.2 Theoretical Graphical Representation of Diversification (8)

unsystematic risk can be dramatically reduced through the incremental addition of supplementary assets that exhibit a less-than-perfect correlation to one another (1). This reduction of risk through the addition of assets is, in essence, the concept of diversification.

By definition, diversification is the combination of multiple assets to benefit from the less-than-perfect correlation of each in order to minimize unique/unsystematic risk. When sufficient assets have been aggregated, this results in a portfolio with only market/systematic risk (8). That is to say, with the strategic addition of select, complementary assets, the unique risk of any one asset is dramatically minimized when compared to the entirety of the portfolio's risk. This concept is best depicted in Figure 12.2, which illustrates declining portfolio risk in proportion to the incremental increase in assets. Note that the risk is asymptotic to the total market risk. That is, if a portfolio contained every known asset, the portfolio would still be subject to the risk of the market. The only way this market risk can be reduced is through the implementation of financial instruments (as defined in note 5), derivatives, or options (note that these concepts are beyond the scope of this chapter).

The essential element and subsequent advantage in real estate diversification is rooted in the notion that both the return and the risk of the investment will vary according to the geographic location of the asset, even if all other aspects (e.g. type, size, etc.) remain constant. Geographic location is, without question, inherently more localized to a real estate asset than any other asset type and thus presents a greater unsystematic risk if it is not properly managed and anticipated. Thus, real estate portfolios can substantially benefit from efficient diversification as it reduces the effects of unsystematic risk (1). This type of physical diversification is something that is unobtainable with more traditional investments (a stock is not subject to change no matter if it is being held in Phoenix, Anchorage, or London). This type of physical diversification also provides isolation from sudden shocks to local market factors. That is, a downswing in one city or region does not necessarily result in immediate downswings in other national markets. Even if the downward trend persists and spreads from city to city, certain scenarios are likely to have some time delay across regions, allowing for asset managers to act (or counteract) accordingly. This is in stark contrast to stocks and bonds, in which their associated price reflects market news within minutes of public release.

Finally, it is also important to understand the concept of diversification across property types (single and multifamily residential, commercial, industrial, etc.) (9). As shown in Figure 12.3, real estate diversification can be thought of in three dimensions. First, it can be

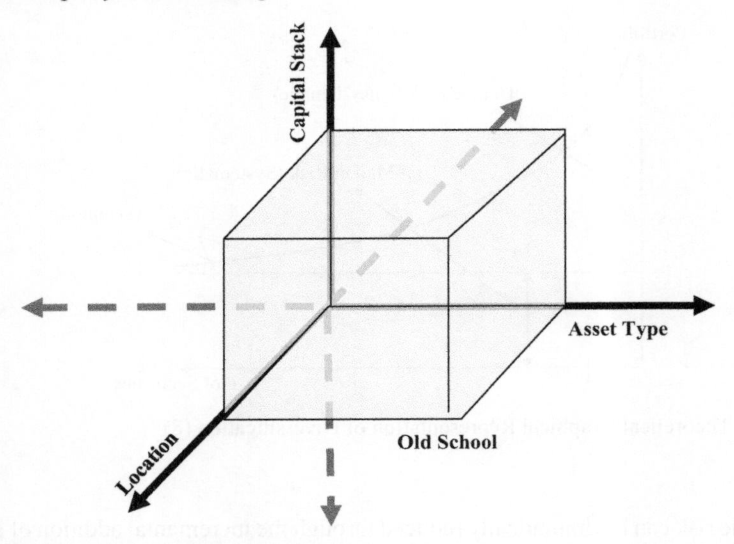

Figure 12.3 Full-Spectrum Real Estate Diversification (10)

divided into type (e.g. office or residential), then by its life cycle stage (e.g. preconstruction), and finally by location. Each of these sectors (and subsectors) provides for an extremely high degree of customization. Even if the manager owned all stabilized Class A properties, each asset could be located across any one of the geographic regions or sub-regions in the country. By diversifying geographically, the portfolio manager would minimize the total accumulated risk.

Calculation of multi-asset return and risk

The final step in understanding the basics of portfolios and Portfolio Theory is to learn and identify the differences between single- and multi-asset risk and return. The following sections will explain these differences and provide an example.

Single-asset return, variance, and standard deviation

The expected return of a single-asset project is the result of its individual scenario returns multiplied by that scenario's probability of occurrence. For example, consider the information shown in Table 12.1.

Table 12.1 Sample Single-Asset Return and Probability Scenarios

Scenario	$E(R_i)$	P_i
Best Case	40%	10%
Most Likely	10%	60%
Worst Case	(20%)	30%

The expected return, $E(R_i)$, can be calculated as follows:

$$E(r) = \sum_{i=1}^{N} E(R_i)p_i$$

$$E(r) = (0.4 \times 0.1) + (0.1 \times 0.6) + (-0.2 \times 0.3) = 0.04 = 4\%$$

The variance, σ^2, is then defined as:

$$\sigma^2 = \sum_{i=1}^{N} [E(r_i) - E(r)]^2 \, p_i$$

$$\sigma^2 = (0.4 - 0.04)^2 \times (0.1) + (0.1 - 0.04)^2 \times (0.6) + (-0.2 - 0.04)^2 \times (0.3) = 0.03240$$

Therefore, standard deviation, σ, is defined as:

$$\sigma = \sqrt{\sigma^2}$$

$$\sigma = \sqrt{0.03240} = 0.18 = 18\%$$

The project parameters (return, risk) are therefore (4%, 18%)
where

$E(R_i)$: Expected return of the scenario$_i$

P_i: Probability of scenario$_i$ occurring

$E(r)$: Expected return

σ^2: Variance of returns

σ : Standard deviation of returns

Multi-asset (portfolio) return, variance, and standard deviation

The expected return of a multi-asset portfolio is the result of each individual asset's return multiplied by its weight, or percentage makeup, in relation to the entire portfolio. In the previous single-asset example, the weight would be equal to 100% as it is the only asset in the portfolio. The following example, based on the information shown in Table 12.2 and Table 12.3, will demonstrate the effects of asset weighting on the portfolio's expected return.

Table 12.2 Sample Portfolio Asset Returns, Weights, and Standard Deviations

Asset	$E(R_i)$	x_i	σ
1	11%	30%	15%
2	8%	25%	10%
3	13%	45%	18%

(Note that the sum of x_i must total 100%)

Table 12.3 Sample Correlation Coefficients

Asset	*1*	*2*	*3*
1	1.00	0.70	0.85
2	0.70	1.00	0.40
3	0.85	0.40	1.00

The portfolio's expected return can be calculated by the summation of an asset's expected return multiplied by its weight in the portfolio, shown as follows:

$$\text{Portfolio}_{\text{Return}} = \sum_{i=1}^{N} x_i E(R_i)$$

$$\text{Portfolio}_{\text{Return}} = (0.3 \times 0.11) + (0.25 \times 0.08) + (0.45 \times 0.13) = 0.1115 = 11.15\%$$

The variance, however, is far more complex at the portfolio level than at the single-asset level. One way the variance can be calculated is by the double summation of an asset's return multiplied by the product of each other asset's return multiplied by the covariance. It can be calculated as follows:

$$\text{Portfolio}_{\text{Variance}} = \sum_{i=1}^{N} \sum_{j=1}^{N} x_i x_j \sigma_{ij}$$

$$\text{Portfolio}_{\text{Variance}} = x_1 x_1 \sigma_{11} + x_1 x_2 \sigma_{12} + x_1 x_3 \sigma_{13} + x_2 x_1 \sigma_{21} + x_2 x_2 \sigma_{22} + x_2 x_3 \sigma_{23} + x_3 x_1 \sigma_{31} + x_3 x_2 \sigma_{32} + x_3 x_3 \sigma_{33}$$

This can be reduced to the following (note that this is just a reengagement of the previous):

$$\text{Portfolio}_{\text{Variance}} = x_1 x_1 \sigma_{11} + x_2 x_2 \sigma_{22} + x_3 x_3 \sigma_{33} + 2(x_1 x_2 \sigma_{12}) + 2(x_1 x_3 \sigma_{13}) + 2(x_2 x_3 \sigma_{23})$$

$$\begin{aligned}\text{Portfolio}_{\text{Variance}} = {} & (0.30)(0.30)(0.15)^2 + (0.30)(0.25)(0.70)(0.15)(0.10) \\ & + (0.30)(0.45)(0.85)(0.15)(0.18) + (0.25)(0.30)(0.70)(0.15)(0.10) \\ & + (0.25)(0.25)(0.1)^2 + (0.25)(0.45)(0.40)(0.10)(0.18) \\ & + (0.45)(0.30)(0.85)(0.15)(0.18) \\ & + (0.45)(0.25)(0.40)(0.10)(0.18) + (0.45)(0.45)(0.18)^2\end{aligned}$$

$$\text{Portfolio}_{\text{Variance}} = 0.00203 + 0.00079 + 0.00310 + 0.00079 + 0.00063 + 0.00081 + 0.00310 + 0.00081 + 0.00656$$

$$\text{Portfolio}_{\text{Variance}} = 0.01862$$

Therefore, standard deviation is then defined as:

$$\sigma = \sqrt{\sigma^2}$$

$$\sigma = \sqrt{0.01862} = 0.13646 = 13.64\%$$

The portfolio parameters (return, risk) are therefore (11.2%, 13.6%)
where

x_i : Weight of Asset i in Portfolio

$E(R_i)$: Expected Return of Asset i

σ_{ij} = Covariance of Assets i and j

$\sigma_{ij} = \rho_{ij}\sigma_i\sigma_j$

ρ_{ij} : Correlation Coefficient of Assets i and j

σ_i : Standard Deviation of Asset i

As can be seen from the previous example of the three-asset portfolio, the calculations can prove to be tedious and become exceedingly more so as the number of assets increases. For this reason, we can apply higher-level matrix algebra to represent each component from Table 12.2 and Table 12.3. However, before doing so we first need a foundational understanding of matrix algebra.

Matrix algebra

Matrices are simply a block of data represented by a defined number of rows and columns. The size of the block is always described using M rows by N columns. For example, Figure 12.4 is a 2 (row) by 3 (column) matrix – traditionally just called a '2x3 matrix' or a '2x3'. Note that the subscript represents the position of the element within the matrix (e.g. element "$a_{2,1}$" is found in the second row, first column, of matrix A).

Matrices can be modified using all four of the basic algebraic operations – addition, subtraction, division, and multiplication. This chapter, however, primarily discusses the process of non-scalar matrix multiplication and its application to portfolio variance. As an aside, scalar multiplication is when a numeric constant is multiplied into a matrix. See Figure 12.5.

Two (or more) matrices can be multiplied together using a process known as the 'dot product'. This, essentially, can be thought of as the order of operations – PEMDAS – of matrix multiplication. As the addition or subtraction of two fractions requires that the denominators be identical, matrix multiplication requires that the matrices being multiplied be of a certain size pairing. The fundamental requirement of matrix multiplication is that the number of *columns* in matrix 1 be equal to the *rows* in matrix 2. From this, we can extrapolate that matrix 1, of p x w size, can be multiplied by matrix 2, of w x t size, and result in a product matrix of p x t. However, it is important to notice that a p x w matrix cannot be multiplied by a t x w matrix, as the column-to-row requirement is not satisfied. When the requirement is satisfied, the multiplication is carried out through a series of smaller multiplications and sums

$$A = \begin{bmatrix} a_{1,1} & a_{1,2} & a_{1,3} \\ a_{2,1} & a_{2,2} & a_{2,3} \end{bmatrix}$$

Figure 12.4 Example 2x3 Matrix

$$2A = 2\times\begin{bmatrix} a_{1,1} & a_{1,2} & a_{1,3} \\ a_{2,1} & a_{2,2} & a_{2,3} \end{bmatrix} = \begin{bmatrix} 2a_{1,1} & 2a_{1,2} & 2a_{1,3} \\ 2a_{2,1} & 2a_{2,2} & 2a_{2,3} \end{bmatrix}$$

Figure 12.5 Example of Scalar Multiplication

$$A = \begin{bmatrix} a_{1,1} & a_{1,2} & a_{1,3} \\ a_{2,1} & a_{2,2} & a_{2,3} \end{bmatrix}$$

$$B = \begin{bmatrix} b_{1,1} & b_{1,2} \\ b_{2,1} & b_{2,2} \\ b_{3,1} & b_{2,3} \end{bmatrix}$$

$$A \times B = AB = \begin{bmatrix} \overbrace{(a_{1,1} \times b_{1,1}) + (a_{1,2} \times b_{2,1}) + (a_{1,3} \times b_{3,1})}^{AB_{1,1}} & \overbrace{(a_{1,1} \times b_{1,2}) + (a_{1,2} \times b_{2,2}) + (a_{1,3} \times b_{2,3})}^{AB_{1,2}} \\ \underbrace{(a_{2,1} \times b_{1,1}) + (a_{2,2} \times b_{2,1}) + (a_{2,3} \times b_{3,1})}_{AB_{2,1}} & \underbrace{(a_{2,1} \times b_{1,2}) + (a_{2,2} \times b_{2,2}) + (a_{2,3} \times b_{2,3})}_{AB_{2,2}} \end{bmatrix}$$

Figure 12.6 Example of Multiplication of Two Matrices

to create the solution matrix. See the example in Figure 12.6 outlining the process of the multiplication.

It is important to notice the following:

- A is a 2x3 matrix
- B is a 3x2 matrix
- The product, AB, is a 2x2 matrix

Three-asset portfolio numeric example

The following is reworking the previous three-asset example using matrices and matrix algebra. It is important to note that even though the problem will be solved using matrices, the underlying formulas will not change – just the operational math. The example information from this book's companion website (www.routledge.com/cw/staiger) can be broken down into the matrices in Figure 12.7.

$$W = \begin{bmatrix} 30\% \\ 25\% \\ 45\% \end{bmatrix}$$

$$R = \begin{bmatrix} 11\% \\ 8\% \\ 13\% \end{bmatrix}$$

$$S = \begin{bmatrix} 15\% \\ 10\% \\ 18\% \end{bmatrix}$$

$$W \times S = \begin{bmatrix} 5.5\% \\ 2.5\% \\ 8.1\% \end{bmatrix}$$

$$C = \begin{bmatrix} 1 & 0.7 & 0.85 \\ 0.7 & 1 & 0.4 \\ 0.85 & 0.4 & 1 \end{bmatrix}$$

Figure 12.7 Matrix Breakdown

where

W: Weight Asset Martix

R: Return Asset Martix

S: Standard Deviation Asset Martix

C: Correlation Coefficient Martix

Step 1: compute the expected return

To compute the expected return, you multiply the transposed Weight matrix (1x3) by the Return matrix (3x1).

$$Portfolio\ Return = \begin{bmatrix} 30\% & 25\% & 45\% \end{bmatrix} \times \begin{bmatrix} 11\% \\ 8\% \\ 13\% \end{bmatrix}$$

This yields a 1x1 Portfolio Return of 11.15%.

Step 2: compute the variance

This step is a combination of finding the double summation of the asset weight times its covariance. To simplify the calculations, a third matrix was created (which is the result of multiplying the asset's weight by its standard deviation). This takes the place of the correlation coefficient and streamlines the calculations.

$$Portfolio\ Variance = \left(\begin{bmatrix} 4.5\% & 2.5\% & 8.1\% \end{bmatrix} \times \begin{bmatrix} 1 & 0.7 & 0.85 \\ 0.7 & 1 & 0.4 \\ 0.85 & 0.4 & 1 \end{bmatrix} \right) \times \begin{bmatrix} 4.5\% \\ 2.5\% \\ 8.1\% \end{bmatrix}$$

This yields a 1x1 portfolio variance of 1.86%.
This takes the place of the following:

$$\text{Portfolio}_{\text{Variance}} = \sum_{i=1}^{3} \sum_{j=1}^{3} x_i x_j \sigma_{ij}$$

$$\text{Portfolio}_{\text{Variance}} = x_1 x_1 \sigma_{11} + x_1 x_2 \sigma_{12} + x_1 x_3 \sigma_{13} + x_2 x_1 \sigma_{21} + x_2 x_2 \sigma_{22} + x_2 x_3 \sigma_{23} + x_3 x_1 \sigma_{31} + x_3 x_2 \sigma_{32} + x_3 x_3 \sigma_{33}$$

Step 3: compute the standard deviation

This formula is simply the square root of the variance. Essentially,

$$\text{Portfolio}_{\text{Standard Deviation}} = \sqrt{\text{Portfolio}_{\text{Variance}}}$$

Example of real estate portfolio construction

Now, with a solid foundation and understanding of portfolios, diversification, and the calculations used to measure performance, we can examine specifically how this applies to real

Table 12.4 MSA Summary

Property	*Assumed Purchase Price*	*Debt Ratio*	*Regional Capitalization Rate*	*URental Rate*	*Calculated Risk, Return*	*Approximate P(Loss)*
New York, New York	±$9,500,000	20%	±3.5%	±$4,000/mo.	(8%, 10%)	8%
Los Angeles, California	±$1,200,000	20%	±7.0%	±$1,500/mo.	(4%, 8%)	2%
Washington, DC	±$5,200,000	20%	±5.0%	±$3,000/mo.	(5%, 9%)	3%
Chicago, Illinois	±$200,000	20%	±13.0%	±$700/mo.	(8%, 15%)	3%

estate assets through a fictional portfolio construction. Often, projects (or assets) are evaluated independently from one another and there is little consideration of a project's impact on a group of already-owned assets. It is essential for asset managers to understand how asset types relate and correlate so that projects can be purchased and paired in an effort to minimize the unique risk of any one asset and thus reduce the portfolio risk through diversification. The following fictional portfolio example (referred to as Example Portfolio or Example Real Estate Portfolio throughout this section) considers four multifamily residential investment properties for purchase and operation. The key metrics for each of the properties are detailed in Table 12.4. The four properties are based on actual assets that were randomly chosen from available major metropolitan market listings from June 2014. These properties are located in New York, Los Angeles, Washington, DC, and Chicago, representing four of the major metropolitan regions in the United States. The order in which the properties were acquired by the portfolio was randomized in an effort to (1) simplify the analysis, (2) simulate the order in which an actual manager would purchase real-world properties, and (3) highlight the insignificance of asset permutations.

Analysis

Each property was evaluated using a discounted pro forma cash flow model and Oracle's Crystal Ball modelling software over an assumed hold period of 5 years. This software evaluated the effects of 33 changing variables over a series of one million simulated trials. Each trial had a resulting internal rate of return (IRR) distribution, which was used as the primary metric of success for the investment. For each of the input variables (e.g. potential rental income, vacancy rate, interest rate, capitalization rate) certain logical assumptions were made in order to transform the information from static inputs into dynamic ranges (giving the range an actual shape within the software). These assumptions varied between properties and were based on regional market conditions and available actual market data.

The benefits of using the Crystal Ball simulation software are evident in the creation of the underlying input assumptions and confidence of the final results. When using Crystal Ball, any input assumption becomes a dynamic range and thus has an infinite number of values within the bounds of the range. Although all possible numbers within the range could be chosen, most often the analyst will restrict the choices to integers. For example, rather than an analyst assuming that the monthly rent is a static value of $1,000/month, he can create a range of probable values for the rent. This means that he can set the assumed rent to be any integer between $900/month and $1,100/month, allowing every possible whole dollar

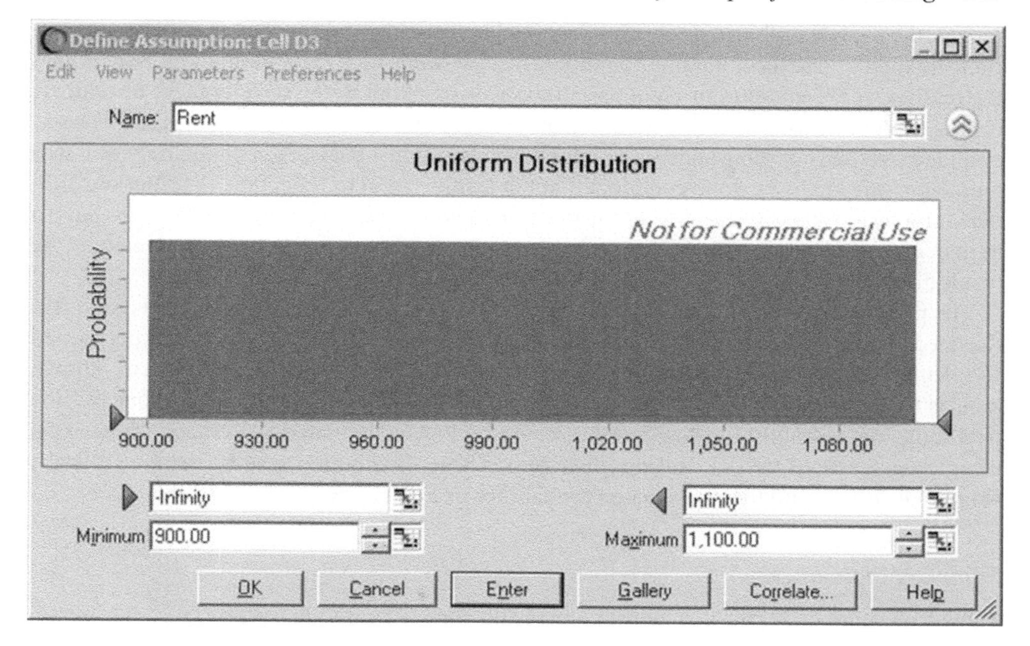

Figure 12.8 Uniform Input Distribution

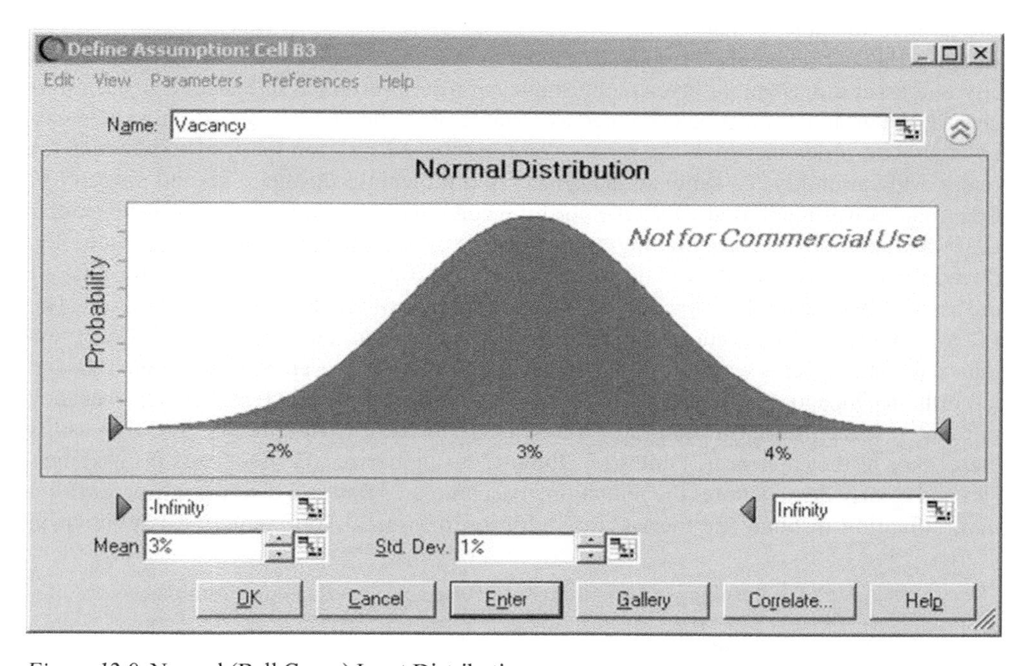

Figure 12.9 Normal (Bell Curve) Input Distribution

between the bounds to have an equal probability of occurrence, as shown in Figure 12.8. Likewise, the analyst could assume that the vacancy rate is normally distributed with a mean of 3% and a standard deviation of 1%, as shown in Figure 12.9, rather than a static 3%. When models are created using Crystal Ball and harnessing the power of inputs with dynamic ranges, the analyst is able to establish likely (and realistic) bounds on the range of an assumption and then utilize the computer to randomly draw values for each simulation. Remember, the realistic representation of input assumptions is essential to the development of realistic results (output data). When this randomization occurs for each of the variables and is compounded over tens of thousands – or in this case, a million – trials, then by the Law of Large Numbers,[7] the analyst can be confident that the results begin to reflect the distributions fundamental to those assumed in Modern Portfolio Theory. Therefore, the analyst can have a high degree of confidence in that all probable (and possible) outcomes are represented by the result distribution.

The primary reason for detailed valuations (such as the ones performed here) is rooted in the need to quantify the risk of the investment. More specifically, the evaluation is performed to calculate the probability that an investor will lose all of the initial capital invested – known here as the probability of loss or P(Loss). Given that the software simulation ran one million trial runs, the probability of loss is represented by the accumulated percentage chance of a return, as measured by IRR, of less than zero. These distribution results are shown in the following section and their associated probability of loss is summarized in Table 12.4.

Summary of findings

Table 12.4 summarizes the key metrics and results for each individual real estate asset, evaluated independently. Each asset has desirable levels of return in relation to the modest levels of risk. This balance results in a relatively low estimated probability of loss on the initial capital investment (less than 10%).

The estimated probabilities of losses shown in this table are low, making the purchase of any one asset a reasonable investment; however, after more than one asset has been purchased, the grouping must be evaluated as a single portfolio and the metrics must be recalculated. The order in which the assets were purchased was randomized to account for real-world variability. To simulate the effects of a growing portfolio, a second property in Los Angeles was purchased. Once the portfolio contained two assets, a Crystal Ball simulation was then performed, including both properties. Like the single-asset simulation, this dual-asset simulation evaluated the changing effects of the same 33 variables for each of the assets and measured the resulting outcomes for the combined internal rate of return. The P(Loss) metric was the accumulated percent chance of a return, as measured by IRR, of less than zero. This process was repeated for the addition of a third asset, in Washington, DC, and a fourth, in Chicago. No underlying input assumptions (e.g. potential rental income, vacancy rate, interest rate, capitalization rate) were changed when assembling the portfolio; assets were taken as they appear in Table 12.4. Table 12.5 summarizes the associated P(Loss) metrics and clearly demonstrates the reduction in accumulated risk as a direct result of portfolio diversification through the incremental addition of assets. This reduction is evident in

Table 12.5 Resulting P(Loss) from Asset Assemblage in Example Portfolio

Portfolio Composition	*P(Loss)*
New York	5.04%
New York and Los Angeles	2.91%
New York, Los Angeles, and Washington, DC	1.90%
New York, Los Angeles, Washington, DC, and Chicago	1.86%

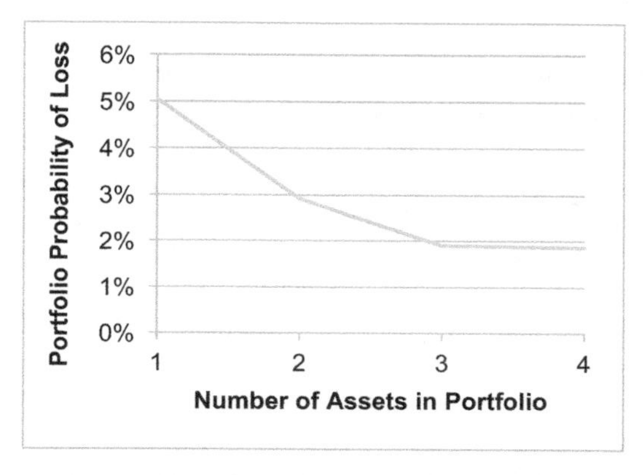

Figure 12.10 Diminishing Portfolio Risk through Diversification in the Example Portfolio

Table 12.6 Reduction in IRR between Iterations of the Example Portfolio

Number of Assets in Portfolio	*P(Loss)*	*Reduction in P(Loss) as Compared To:*		
1	5.04%	v 1 asset	v 2 assets	v 3 assets
2	2.91%	42.2%		
3	1.90%	62.3%	34.7%	
4	1.86%	63.1%	36.1%	2.3%

Figure 12.10. As shown in the figure, the total portfolio probability of loss (along the y-axis) is reduced with each additional asset to the portfolio.

It is important to note that this fictitious example portfolio – comprising real-world properties and assumptions – follows the purely theoretical behavior shown in Figure 12.2, on page 207. As telling as Figure 12.10, the information shown in Table 12.6 summarizes the diminishing effect of additional assets. That is, when the portfolio contained only one asset, the total probability of loss was 5.04%. However, with the addition of the second asset, the probability was reduced to 2.91% – a 42% reduction compared to the single asset. This reduction continues at each incremental asset addition; however, it is important to notice that the resulting percent reduction versus the previous portfolio combination becomes increasingly smaller. If this asset addition were to continue, the resulting net reduction compared to the previous portfolio combination would approach (but never reach) a zero percent change. This asymptotic relationship begins to form in Figure 12.10 and is highlighted in Figure 12.2, on page 207.

Seeking the efficient frontier

Although the addition of strategically selected assets can result in the reduction of the probability of losing all of the initially invested capital, as demonstrated in the previous section, this practice is only one of many portfolio management techniques. Oftentimes, the specific objective of a portfolio is to maximize the expected returns over a certain time horizon. It is worth noting that in certain circumstances – university endowments, pension funds, or insurance funds – the portfolio objective may be to minimize risk (or to manage risk within prescribed thresholds) and thus accept lower levels of return. This balance between risk and return, known as the risk-adjusted return, is essential in the application of the Efficient Frontier analysis. As discussed in the Portfolios and Portfolio Theory section in this chapter, and depicted in Figure 12.1, on page 206, there exists a specific allocation of assets that, when combined, results in the highest possible return at a given level of risk.

In order to determine the best possible combination of assets, we must first understand how the assets relate to or interact with each other. Since each asset is unique and thus is considered independent, an analysis must be performed to determine some justifiable mathematical relationship that can be used to model asset interactions. This relationship is best quantified through statistical correlation, which is a measure of linear strength between independent variables. Specifically, this quantification is represented by correlation coefficients which have an inclusive bounded numerical range between negative one (−1) and positive one (+1). At a correlation coefficient value of positive one, the two independent variables tend to move synchronously with each other; that is, if one variable increases, the other is likely to increase as well. In contrast, at a correlation coefficient value of negative one, the independent variables tend to move opposite each other; when one variable increases, the other is likely to decrease. Although these coefficients are a mathematically justifiable quantification of the relationship of independent variable behavior, it is extremely important to note that *correlation* in no way represents *causation*. The second variable does not move *because* the first moved; rather, each variable moved because they each independently reacted to a common outside influence.

To determine the relationship between the independent properties included in the example portfolio, a 5-year retrospective analysis of the relevant Standard & Poor's Case-Shiller Home Price Indices[8] was performed. Each city's Case-Shiller Home Price Index (e.g. Los Angeles) served as a representative of the value of home prices in that city. These representative values were analyzed to determine the associated correlation coefficients that would represent the four real estate assets included in the example portfolio. The resulting correlation coefficients for the properties are summarized in Table 12.7 and charted in Figure 12.11.

As can be seen in Table 12.7, and more clearly depicted in Figure 12.11, all four cities generally display a strong positive linear relationship between properties. As values increase in one region, they also increase (although at a different rate) in another region. Again, it is important to emphasize that this strong correlation between regions does not amount to one region causing another to increase, merely that there is a mathematical relationship between regions.

The correlation coefficient matrix, shown on page 210, and the individual risk and return profiles calculated in the Summary of Findings section provide a more thorough representation of the potential performance of the portfolio. This data becomes the input criteria for

Table 12.7 Example Portfolio Correlation Coefficients

5 Year	*Washington, DC*	*New York, New York*	*Chicago, Illinois*	*Los Angeles, California*
Washington, DC	1.00	0.88	0.91	0.97
New York, NY	0.88	1.00	0.91	0.90
Chicago, IL	0.91	0.91	1.00	0.91
Los Angeles, CA	0.97	0.90	0.91	1.00

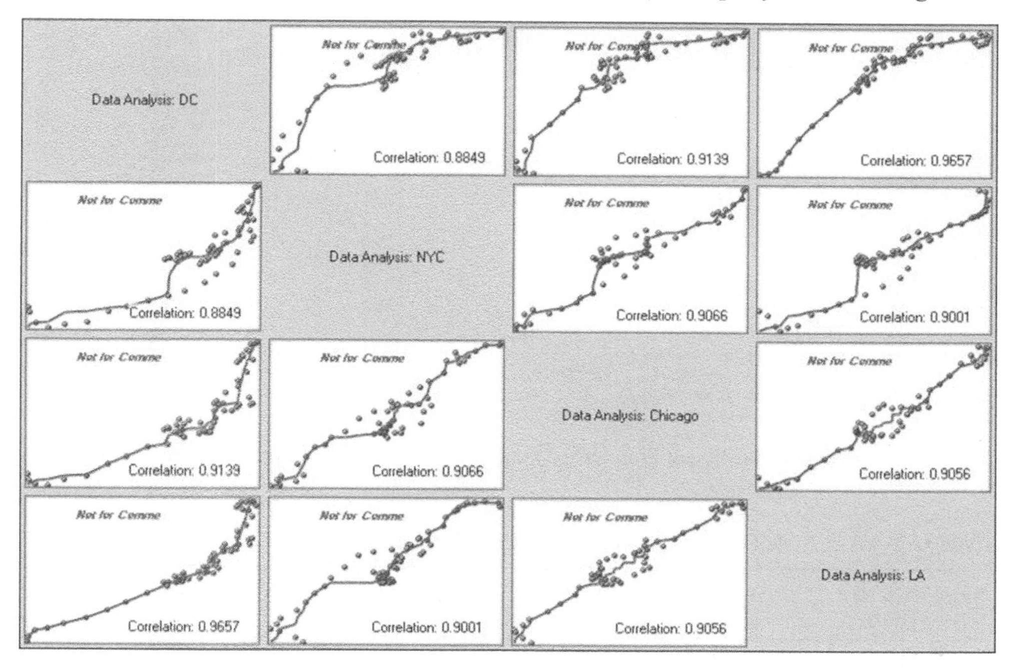

Figure 12.11 Five-Year Historic Scatter Chart Analysis of Case-Shiller Home Price Indices for Washington, DC, New York, Chicago, and Los Angeles (11)

Crystal Ball's OptQuest optimization engine and allows for the construction of the Efficient Frontier. The primary advantage of OptQuest's optimization engine is that, rather than representing the range of possible outcomes for one asset (as was done to establish the P(Loss) metrics in the Analysis portion of the Example Real Estate Portfolio section), the engine acts as a decision analyzer. This means that the optimization learns from its previous simulations and automatically adjusts asset combinations as it searches for optimal solutions to the simulation model (12). The optimization evaluates the risk and return profiles for each asset in combination with that asset's associated correlation coefficients to determine the asset combination resulting in the highest possible return at a specific level of risk. Once the highest return is established, the risk tolerance is increased and a new maximum return is sought. This process continues for advancing levels of risk until all possible combinations of asset allocations under consideration have been evaluated. The resulting Efficient Frontier is shown in Figure 12.12.

Although the resulting calculated frontier in Figure 12.12 is jagged, it mimics the general shape of the purely theoretical curve shown in Figure 12.1, on page 206. The calculated frontier clearly depicts that as the tolerance for acceptable risk increases, so too does the resulting maximum portfolio return. Put simply, if a manager is willing to accept a high degree of risk, theoretically, he could generate a higher degree of return.

Throughout the optimization, the software evaluates each of the possible asset combinations. Given that the example portfolio has four assets, there are a total of 11 feasible allocation combinations containing two or more assets.[9] These combinations, ordered by increasing levels of return, are summarized in Table 12.8.

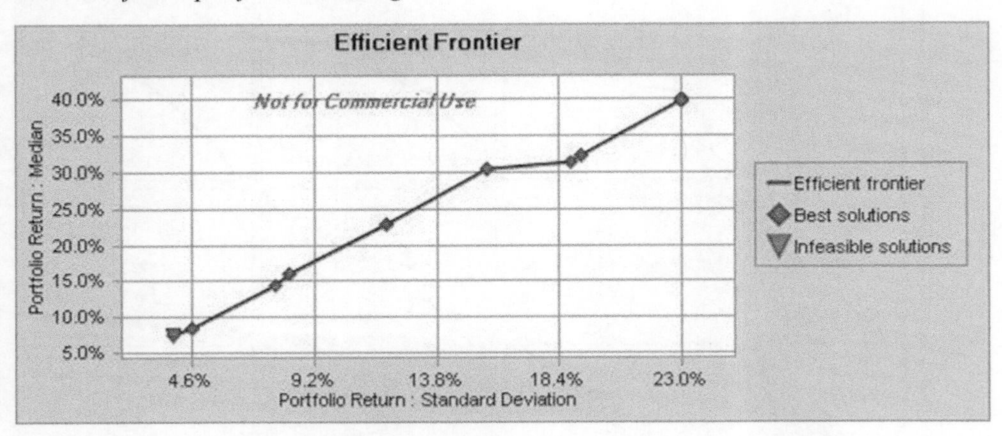

Figure 12.12 Calculated Efficient Frontier for Properties within the Example Portfolio

Table 12.8 All Possible Example Portfolio Asset Combinations

Portfolio Return	*Portfolio Risk*	*Chicago*	*Washington, DC*	*Los Angeles*	*New York*
—	—	*N*	*N*	*N*	*N*
7.6%	*3.9%*	*N*	*N*	*Y*	*N*
8.4%	*4.5%*	*N*	*Y*	*N*	*N*
9.4%	*7.6%*	*N*	*N*	*N*	*Y*
14.3%	*7.7%*	*Y*	*N*	*N*	*N*
16.0%	8.2%	N	Y	Y	N
17.0%	11.3%	N	N	Y	Y
17.8%	11.8%	N	Y	N	Y
22.0%	11.5%	Y	N	Y	N
22.8%	11.9%	Y	Y	N	N
23.8%	15.1%	Y	N	N	Y
25.4%	15.5%	N	Y	Y	Y
30.4%	15.7%	Y	Y	Y	N
31.4%	18.8%	Y	N	Y	Y
32.2%	19.2%	Y	Y	N	Y
39.8%	**23.0%**	**Y**	**Y**	**Y**	**Y**

Y = Included in Portfolio | N = Excluded from Portfolio

It should come as no surprise that the highest level of return and risk, respectively, is possible only with the inclusion of all four properties. However, the purpose of this table is to show that each possible combination of properties was considered, as well as to indicate the resulting portfolio risk and return associated with that combination. It is important to note that each one of the feasible combinations that lies along the Efficient Frontier is considered efficient. A portfolio manager can then justifiably select any one of the combinations and be assured that he is obtaining the highest return at the selected risk level. In some cases, the portfolio manager may choose a lower risk-adjusted return if he feels that the return is not worth the potential associated risk at the highest extreme. The frontier does not identify one combination as 'better' than another one, but rather it charts the possible returns that exist along the defined spectrum of risk.

The Efficient Frontier could be constructed for any number of assets within the portfolio. As the number of assets increased, the potential maximum returns and associated risk continued to increase. Although it is not considered in this chapter, additional real estate assets could be analyzed and added to the example portfolio in the same nature as was described through the previous sections. Doing so would result in additional points along the Efficient Frontier that did not exist with the four-asset portfolio.

Limitations

As with any analysis or application of a broad-based theorem, such as Modern Portfolio Theory, there exist many potential shortcomings or uncertainties when applied to areas other than its initial foundation. This section attempts to identify without bias some of the issues associated with the application of Modern Portfolio Theory and the calculation of the Efficient Frontier on real estate investment portfolios. To begin, real estate valuation is often obtained through infrequent property appraisals that tend to overstate and therefore misrepresent actual market conditions.[10] This slight distortion of actual market conditions results from appraisals that are conducted exclusive or independent of an actual sale transaction. In other words, a value is theoretical and set by the appraiser based on assessed regional market conditions and like-kind properties, yet the value cannot be verified since the property was neither purchased nor sold at this assessed value. It is worth noting that at times properties are assessed as part of the underwriting requirements of a bank's refinancing process. As such, the assessed value is never exposed to, and subsequently never validated by, actual market conditions. Also, since the assets are 'traded' – that is, bought and sold in micro-economies – there is little possible impetus to achieve a stable market efficiency. In addition, unlike equities, real estate property can have inherently high emotional attachment, which is neither easily nor accurately numerically quantifiable yet greatly affects the 'value'.[11] In general, the following are primary factors that contribute to the inefficiency found in real estate markets: high transaction disposition and acquisition costs; lack of audited, publicly available information; high entry capital requirements; complexity and variety of state and local laws affecting the sale and operation of real estate; and other unique and unknown factors of each particular asset (1). It is also important to understand that because real estate assets have a physical presence in our cities, they are often very visible to the public and can become the target of political or civil actions in favor of or against the asset (13).

Academics Ping Cheng, Zhenguo Lin, and Yingchun Liu have conducted extensive research in the area of Modern Portfolio Theory as it pertains to real estate asset portfolios. In the findings of this research, they discuss some of the primary pitfalls that have been uncovered when Modern Portfolio Theory is applied in its purest form to real estate asset portfolios. These authors' research seeks to develop an alternative model that expands Modern Portfolio Theory to incorporate three unique characteristics of real estate assets. According to the authors' hypothesis, these characteristic are (1) that real estate returns are not typically independent and identically distributed over the asset's life; (2) that real estate bears a major liquidity risk; and (3) that real estate involves very high "barrier to entry costs."[12] Modern Portfolio Theory assumes that there is a centralized and efficient market that trades continuously (i.e. that it performs like the equities market); however, the authors contend that the reality is that real estate is traded fairly infrequently and has neither stable nor normally distributed capital returns. This inherently conflicts with one of the fundamental Modern Portfolio Theory assumptions – that returns are normally distributed.

One of the most essential variables with any asset/investment is the optimal holding period. When accounting for the systematic ex ante risk[13] of illiquidity and high transaction costs of real estate, the resulting return is more vulnerable the longer the asset is held. The authors argue that real estate becomes increasingly riskier the longer the asset is owned. The issue, however, is the exact circumstance that makes an asset highly illiquid, and considering the high disposition/acquisition costs, real estate properties cannot viably be held as short-term investments. According to research conducted by Cheng, Lin, and Liu, the paradoxical relationship between entry and disposition cost and ideal holding periods tends to roughly balance in the range of the 4- to 6-year holding periods (14).[14]

One thing to consider about the illiquidity risk (as defined on page X) is that unlike a stock, which can be sold within seconds, real estate must be marketed, survive a tedious and cumbersome offer and negation process, and then must successfully close. At each stage of this process, the seller has little control over expediting the timeline. It is quite conceivable that months if not years could lapse after an owner makes the decision to sell an asset. Therefore, the optimal holding period is greatly affected by both the systematic risk of the market conditions and the unique risks of the specific property. Finally, it is worth noting that the authors also identify empirical studies that show that real estate has historically low correlation with financial assets. Unlike the Case-Shiller correlation matrix calculated in the Example of Real Estate Portfolio Construction section, which showed positive relationships between the cities, the correlation between real estate assets and equities tends to have a weaker linear relationship. This means that real estate values may not inflate from the increase in value of stock or bonds. This should be a great consideration when a manager is blending real estate assets into a portfolio of equities and bonds.

Conclusion

As was detailed in the preceding chapters of this book, and reiterated in the introduction of this chapter, the diligent construction of an accurate pro forma is essential to successful asset management. Throughout the creation of the pro forma, it is critical to research and establish realistic values (or ranges of values) for input assumptions. Doing so will provide for more precise forecast results, ultimately leading to more confident investment decisions and lower probability of capital loss. Once a sound pro forma has been established for a single asset, the next logical progression is to consider the effects and associated performance of aggregating multiple assets managed within one portfolio.

Although each asset within the portfolio is considered independent – meaning that the asset has its own unique input assumptions, risks, and returns – the portfolio must be evaluated in a way that considers the complementary effects of the assets. This advanced multi-asset analysis can be best described under the principles vested in Modern Portfolio Theory, which outlines the importance of asset allocation within the portfolio and this allocation's effect on overall performance. Again, it is important to note that at no point in the portfolio analysis are the underlying fundamentals or assumptions of an individual asset altered; only the asset's proportional weight within the portfolio is changed. A primary intent of Portfolio Theory analysis is to achieve one of the following objectives: (1) maximize the return for a given level of risk, (2) minimize the risk for a given level of return, or (3) optimize the efficiency of the portfolio. These goals are achieved primarily through the concepts of diversification and optimization.

The first step in analyzing a group of assets using the concepts found in Modern Portfolio Theory is to evaluate each individual asset thoroughly. With regard to real estate assets, this evaluation, more correctly termed underwriting, is often done using a discounted pro forma cash flow analysis. As discussed, this type of detailed analysis intends to flesh out and mitigate (through conservative input assumptions) any potential problem areas of an asset and evaluate the feasibility of investment in the asset. At this level, each individual asset will undergo its own thorough pro forma evaluation and provide the analyst with the basic picture of the property's performance. This performance is typically measured by the calculated risk, the expected return, and the approximate probability of losing all the initially invested capital (P(Loss)). Once each asset passes this level of screening, the analyst must transition from evaluating individual performances to assessing the success of assets in combination with each other.

After each asset has been underwritten and has its own calculated risk, expected return, and P(Loss), the assets are then combined and evaluated as a multi-asset portfolio. The first advantage of portfolio analysis, versus single-asset management, is the reduction of portfolio-level risk through diversification. Remember that diversification is the concept of strategic asset combinations that attempt to minimize the unique (unsystematic) risk of any one asset, thereby decreasing the overall risk of the portfolio. The minimization of unique risk, and by association, the portfolio risk, was highlighted in the first section of the four-asset example of real estate portfolio construction in this chapter. The resulting compression of unique risk continues with the incremental addition of each new asset; however, the reduction is ultimately asymptotic to the upper threshold of market risk (as shown in Figure 12.2 on page 207). In reality, the market risk can also be diminished, but only through the implementation of financial instruments, derivatives, or options strategies.

The primary principle of Modern Portfolio Theory is the optimization of the portfolio. Here optimization refers to obtaining the highest possible return at a precise level of risk, known as the risk-adjusted return. To obtain the optimum allocation, the asset combination must balance the risk and return. This balance is known as the efficiency of the portfolio. Therefore, the definition of an efficient portfolio is one in which no other allocation combination of portfolio assets could offer a higher expected return at a precise level of risk. Once the maximum return has been generated for a given level of risk, the risk tolerance is increased and the new maximum return is calculated. This process is repeated until all feasible levels of risk have been evaluated or all possible asset combinations have been exhausted. Each optimum coordinate (risk, return) is plotted to generate what is known as the Efficient Frontier (as shown in Figure 12.1 on page 206). The frontier for the example real estate portfolio was generated using Oracle's Crystal Ball OptQuest optimization software, which evaluated all possible asset allocation combinations. The software evaluated the risk and return profiles for each individual asset in combination with that asset's associated correlation coefficient to determine the asset grouping that resulted in the highest possible return at specific levels of risk. Once the highest return was established, the risk tolerance was increased and a new maximum return was calculated. This process continued for advancing levels of risk until all possible combinations of asset allocations under consideration had been evaluated. The Efficient Frontier does not identify one combination as 'better' than another, but rather it charts the possible returns that exist along the defined spectrum of risk. Although portfolio combinations that exist below the frontier are valid portfolios, they are not considered efficient since there is an alternate asset combination which would result in a higher return for the same level of risk. Therefore, it is essential for successful portfolio managers to continually assess the efficiency of the portfolio and, where necessary, adjust the allocation to increase returns without increasing risk.

As demonstrated through the theoretical construct and the creation of the practical example portfolio, the principles of diversification and Modern Portfolio Theory can be successfully applied to portfolios composed entirely of real estate assets. Although the application requires the acceptance of some foundational assumptions, the overall analysis and subsequent results clearly align with the expected theoretical outcomes. It behooves any asset manager aspiring to responsible portfolio management to learn, understand, and master the concepts outlined herein to successfully obtain optimum portfolio returns.

Notes

1 Throughout this chapter, the liquidity of an asset refers to its ability to be quickly (within a few days) sold on the open market. Likewise, illiquidity is the inability of an asset to be disposed of quickly.
2 Mr Markowitz would receive the Nobel Memorial Prize in Economic Sciences in 1990 for this theory.
3 Throughout this chapter, the concept of weighting refers to an asset's specific contribution compared to the overall portfolio. For example, one asset in a portfolio of five would have a weighting value of 20% (1/5).
4 The investment strategies of the portfolio will vary widely and will depend on, among other things, the objectives and risk tolerances of those involved in the management of the portfolio. The examples given throughout this chapter are academic in nature and are not intended to provide investment direction or guidance.
5 Here financial instruments are paper contracts that subdivide or otherwise manipulate portions of the real estate asset, often splitting the debt and equity interests. These split interests are then repackaged and resold in a secondary market.
6 This excludes raw, undeveloped land that could be subdivided pursuant to local zoning regulations. This also excludes partial ownership in portions of real assets.
7 The Law of Large Numbers refers to the statistical theorem which states that as the sample size increases (i.e. more simulations are run), the sample's calculated average will tend to converge with the actual (true) expected value for the entire population (of which the sample is a subset). In this context, the more trial simulations that are performed, the closer the simulation's expected value should reflect the true expected value of all possible outcomes.
8 All data was accessed via publicly available Federal Reserve Economic Research databases found at http://research.stlouisfed.org/fred2/release?rid=199
9 Note that although 16 combinations exist mathematically, the first trial (shown in Table 12.8), which is devoid of all assets, as well as the next four trials, which contained only one asset, have been excluded from the analysis because a portfolio, by definition, must contain two or more assets.
10 For this reason, the analysis performed in this chapter uses the Case-Shiller Home Price Index as a proxy of asset value to reduce the impact of subjective appraisals.
11 The emotional attachment can be so influencing that the industry term "Trophy Asset" is often given to those properties that will never be sold because they represent the culture of the fund or business that they are owned by.
12 A "barrier to entry cost" refers to the high capital initially required to purchase and control an asset.
13 Ex ante analysis considers events that are forecasted into the future.
14 For this reason, the analysis performed in this chapter uses an investment horizon of 5 years.

References

(1) Sanders, Anthony B., Pagliari, Joseph L., Jr., and Webb, James R. Portfolio Management Concepts and Their Application to Real Estate. In Joseph L. Pagliari Jr. [ed.], *The Handbook of Real Estate Portfolio Management*. N.p.: Irwin, 1995, chapter 2, pp. 117–172.

(2) Webb, James R., and Pagliari, Joseph L., Jr. The Characteristics of Real Estate Returns and Their Estimation. In Joseph L. Pagliari Jr. [ed.], *The Handbook of Real Estate Portfolio Management.* N.p.: Irwin, 1995, chapter 3, p. 173.

(3) Investopedia US, A Division of IAC. Portfolio. *Investopedia.* [Online] http://www.investopedia.com/terms/p/portfolio.asp.

(4) Staiger, Roger, III. Love Is Efficient! *LiveValuation Magazine.* February 2011, pp. 18–19.

(5) Lieblich, Frederich. The Real Estate Portfolio Management Process. In Joseph L. Pagliari Jr. [ed.], *The Handbook of Real Estate Portfolio Management.* N.p.: Irwin, 1995, chapter 25, pp. 998–1058.

(6) Investopedia US, A Division of IAC. Explaining the Efficient Frontier. *Investopedia.* [Online] 2003. http://www.investopedia.com/video/play/explaining-efficient-frontier/.

(7) Markowitz, Harry M. Foundations of Portfolio Theory. *Journal of Finance*, Vol. 46, No. 2 (June 1991), pp. 469–477.

(8) Staiger, Roger, III. Diversification (cont'd). *Risk and Return (Portfolio).* Presentation made for Stage Capital, LLC, June 2013, p. 38.

(9) Del Casino, Joseph J. Portfolio Diversification Considerations. Joseph L. Pagliari Jr. [ed.], *The Handbook of Real Estate Portfolio Management.* N.p.: Irwin, 1995, chapter 23, pp. 912–966.

(10) Pagliari, Joseph L., Jr. Real Estate in 3-D: See It Now! *Real Estate Issues*, Vol. 15, No. 2 (Fall 1990/Winter 1991), pp. 16–19.

(11) Federal Reserve Bank of St. Louis. S&P/Case-Shiller Home Price Indices. *Federal Reserve Economic Data (FRED).* Federal Reserve Bank of St. Louis, April 2014. http://research.stlouisfed.org/fred2/release?rid=199.

(12) Oracle. Introduction. *Oracle Crystal Ball Decision Optimizer OptQuest User's Guide.* 2013. http://docs.oracle.com/cd/E17236_01/epm.1112/cb_user.pdf

(13) Mueller, Glenn R., and Louargand, Marc A. Developing a Portfolio Strategy. In Joseph L. Pagliari Jr. [ed.], *The Handbook of Real Estate Portfolio Management.* N.p.: Irwin, 1995, chapter 24, pp. 967–997.

(14) Cheng, Ping, Lin, Zhenguo, and Liu, Yingchun. Is There a Real Estate Allocation Puzzle? *Journal of Portfolio Management*, Vol. 39, No. 5 (2013), pp. 61–74.

13 Structured products

Structured products are various investment strategies that are based on derivatives. They are highly customized, and usually the payoffs are derived not from the issuer's own cash flows but from the values of one or more underlying assets. Many structured products offer a principal guarantee, which guarantees the return of principal if held to maturity. Investors consider structured products an alternative to direct investment or a means of portfolio diversification. Structured products are attractive to investors because they may offer higher returns, principal protection, or tax benefits. Disadvantages include lack of liquidity, credit risk, and the highly complex structure, which few investors completely understand.

Mortgage-backed securities

A mortgage-backed security (MBS), as the name suggests, is a type of asset-backed security that is secured by a collection of mortgages. The collection, or 'pool', of mortgages, which may be made up of over a thousand mortgages, is sold to a government-sponsored entity (GSE), such as Fannie Mae or Freddie Mac, or to a private entity, such as a bank or financial institution. The new owner then groups the loans together with other mortgage pools into a security, which is then sold to investors. The monthly payments made on the mortgages, residential or commercial, serve as the revenue stream used to pay investors.

The shares, which are sold to investors through various structures, are not identical. These shares, referred to as tranches, are issued with different criteria for repayment and interest; each tranche ultimately has a different level of risk. One tranche may have a lower yield, for example, but it may have an earlier principal due date than one with a higher yield. This tranche may have a lower interest rate, but it is also less risky because the payments are being made ahead of other tranches. Another tranche with a higher yield may have other characteristics such that it may accrue interest, while payments are not made until a higher-priority tranche is retired.

There are several risks associated with mortgage-backed securities. One major risk taken by investors is prepayment, which occurs when a mortgage is repaid early. Typically prepayment occurs when homeowners sell or refinance, usually as a result of lower interest rates. This may also occur in the event of a tragedy – a fire, for instance, in which case insurance proceeds would be used to repay a mortgage. A rate of prepayment, calculated using historical data, is factored into the price and yield of an MBS.

Collateralized mortgage obligations

A collateralized mortgage obligation (CMO) is a structured product that uses mortgage-backed securities as collateral. Mortgages are pooled and interests in these pools are sold to investors in classes, or 'tranches'. Bondholders buy into these mortgage pools and receive cash flows. The payments are prioritized according to their class. Some bondholders receive cash flows automatically while others choose to defer cash flows based upon a future higher return or some other greater return (i.e. tax purposes). Deferring the payments may yield a greater return but this is in exchange for a greater risk taken by the bondholders. The following case study will help to further define a collateralized mortgage obligation.

Case study: Anderson Insurance Corporation

In this case the Anderson Insurance Corporation has purchased – from several regional banks, Fannie Mae, and Ginnie Mae – mortgages totaling $959,000,000 (see Figure 13.1).

The CMO will be offered in three classes of securities – a senior, a junior, and a sub-junior tranche; and bonds will be issued in minimum denominations of $1,000 (Figure 13.2). Interest on Class C bonds is to be paid semiannually. Class B is a compound interest bond: interest will accrue and be added to the principal amount until the Class A bonds are fully retired. Class C bonds will not be paid until Class A and B bonds are fully paid. The portion not offered to the public will be retained by Anderson Insurance Corporation, and any residual income will be held by the corporation (see Figures 13.3, 13.4, and 13.5).

	Underwriter	Outstanding Principal	Coupon Rate	Average Remaining Term	Approximate Bond Value	Bondholders' Percentage
1	GNMA	1,250,000	6.50%	30	1,245,897	99.67%
2	FNMA	345,000,000	5.25%	29.7	319,800,000	92.70%
3	GNMA	315,000,000	7.50%	28.6	315,000,000	100%
4	Private Issuers	2,750,000	8%	30	2,715,000	98.73%
5	FNMA	295,000,000	7.25%	27.1	295,000,000	100%

Figure 13.1 Mortgages Purchased by Anderson Insurance

Class	Original Principal Amount	Interest Rate	Price to Public (%)	Price to Public ($)
A	455,000,000	4%	87.65%	398,807,500
B	295,000,000	5.50%	79.25%	233,787,500
C	183,760,897	6%	96.50%	177,329,266

Figure 13.2 CMO Class Offerings

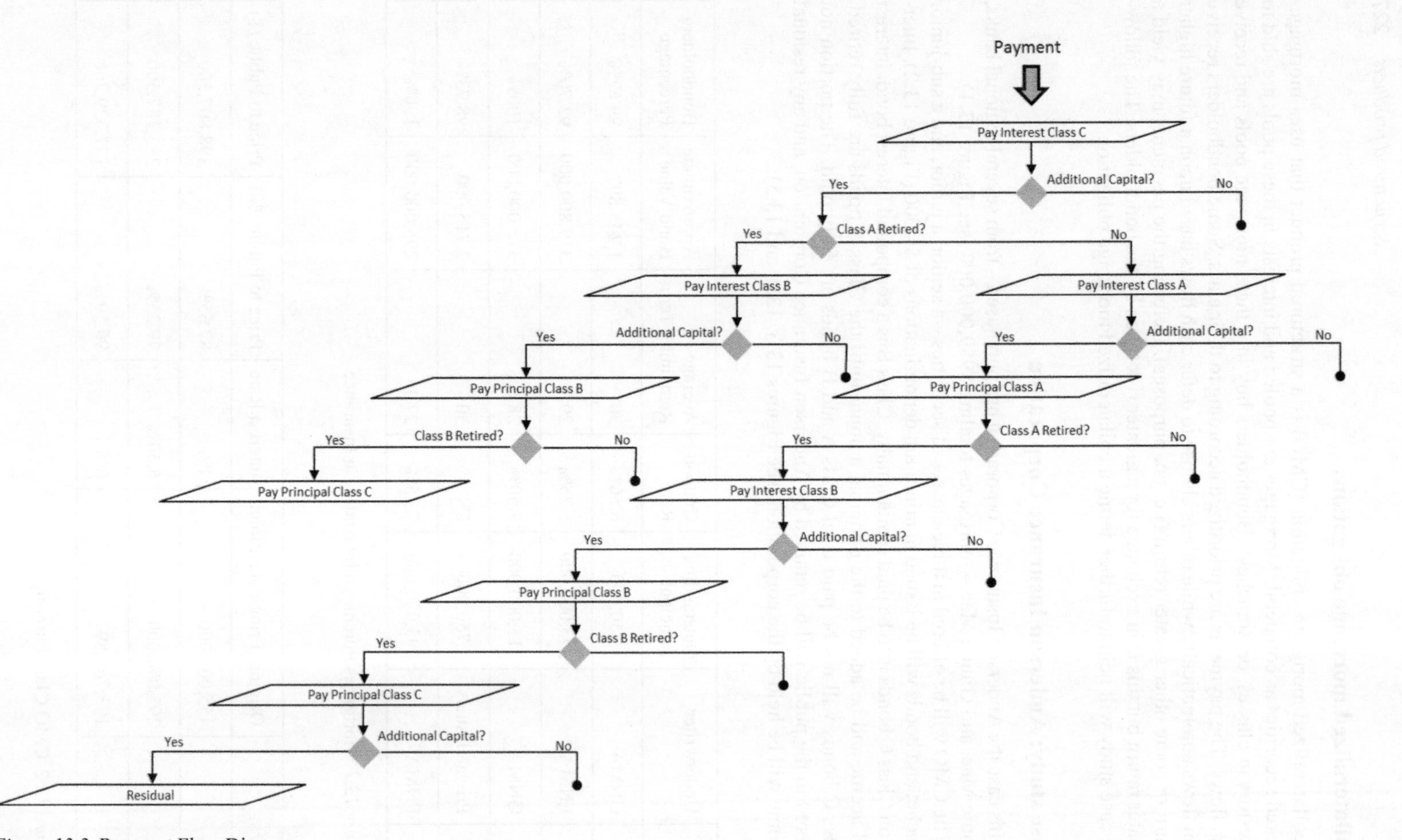

Figure 13.3 Payment Flow Diagram

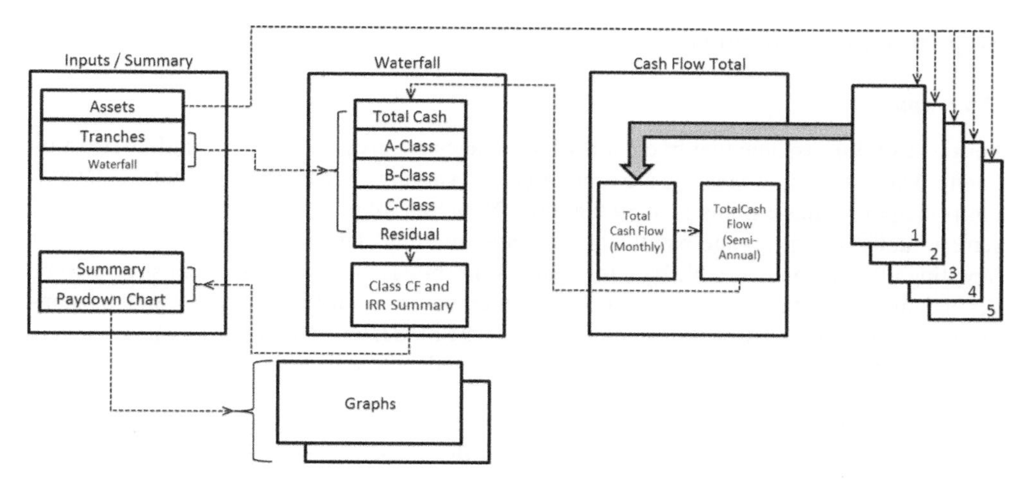

Figure 13.4 Model Flow Diagram

This flow diagram should aid in creating the model in Excel.

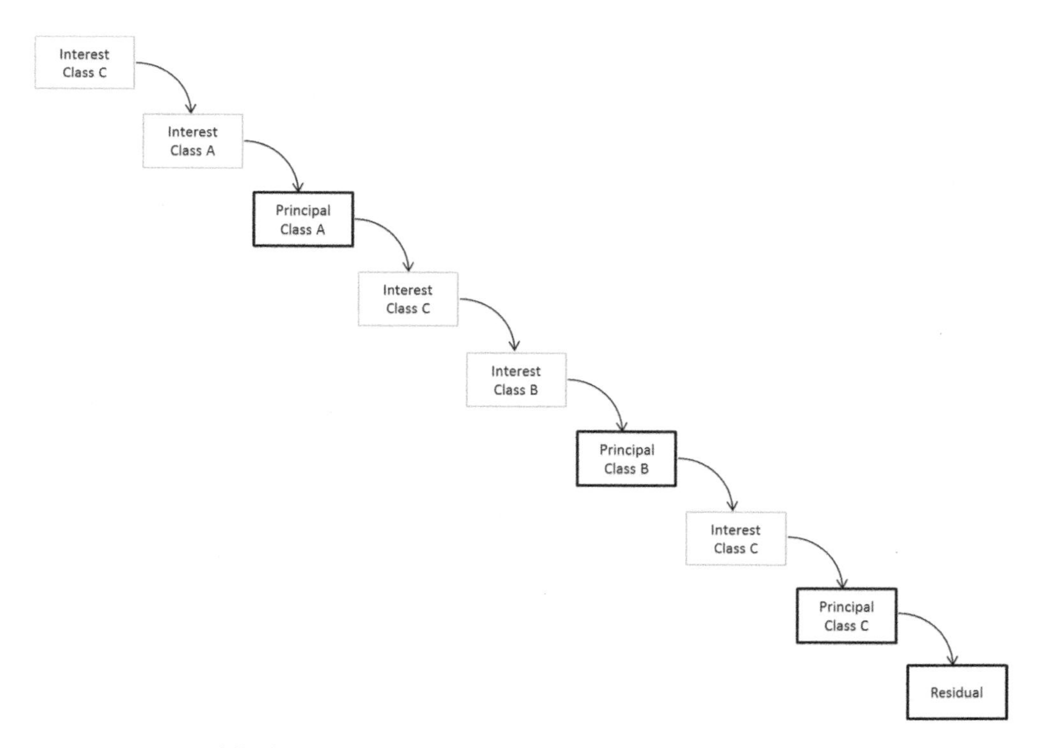

Figure 13.5 Waterfall Diagram

Step-by-step instructions

Data input

For this example assume that five pools of mortgages are created. They are underwritten by Fannie Mae and Ginnie Mae and are each offered at varying coupon rates. The total outstanding principal on these pools is $959,000,000. Note that although fixed income is paid out semiannually, the mortgage payments are made monthly. The first sheet will be used for inputting all static and variable data for analysis. Formulas on each subsequent sheet will be dependent upon data entered here. The data generated on the following sheets will be summarized on this page below the input data. Begin the model by changing the name of Sheet1 to 'Summary'. Do this by right-clicking the mouse where the page title says "Sheet1", then select 'Rename' (Figure 13.6).

The title will be highlighted and the analyst may rename the page as desired (Figure 13.7).

The first step on the Summary page is to enter the analysis date in cell C1; then recreate the chart in Figure 13.8.

Notice the numbers 1 through 5 in cells A5 through A9. This will be useful when creating the cash flow tables later on. The sum of all outstanding principal can be entered by selecting cell C10 and typing Alt + H + U + S, or by using the SUM function (Figure 13.9).

The AVERAGE function may be used for determining the average coupon rate and average remaining term; however, since the pools are not equally weighted, this would not return an accurate value. The Excel function SUMPRODUCT shall be used instead. This takes the coupon rate of each component of the CMO and weights them proportionally. The SUMPRODUCT will multiply arrays of numbers; in this case the coupon rate is multiplied by its corresponding outstanding principal (i.e. $1,250,000*6.50%). That number is then divided by the total outstanding principal. Thus, the formula for cell D10 is "=SUMPRODUCT(D5:D9,$C$5:$C$9)/$C$10" (see Figure 13.10).

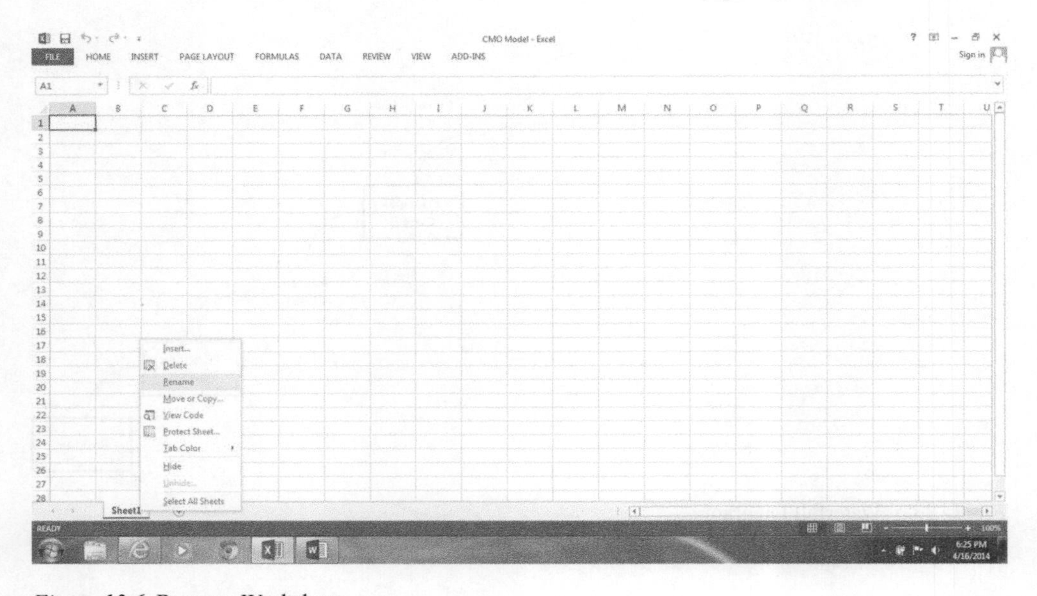

Figure 13.6 Rename Worksheet

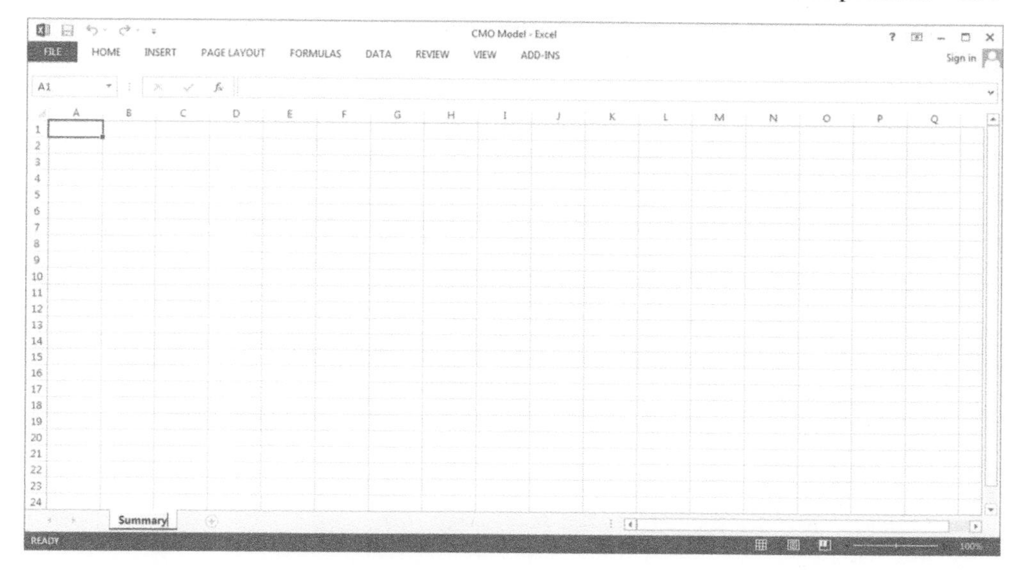

Figure 13.7 Rename Worksheet as Summary

	A	B	C	D	E	F	G
1	Analysis Date		4/20/2014				
2							
3	Assets						
4		Underwriter	Outstanding Principal	Coupon Rate	Average Remaining Term	Approx. Bond Value	Bondholders Percentage
5	1	GNMA	1,250,000	6.50%	30	1,245,897	99.67%
6	2	FNMA	345,000,000	5.25%	29.7	319,800,000	92.70%
7	3	GNMA	315,000,000	7.50%	28.6	315,000,000	100.00%
8	4	Private Issuers	2,750,000	8%	30	2,715,000	98.73%
9	5	FNMA	295,000,000	7.25%	27.1	295,000,000	100.00%

Figure 13.8 Data Entry

If the '$' is used for direct references, the formula may be copied and pasted to the right in order to determine the average remaining term for the entire CMO. The formula in cell E10 should read "=SUMPRODUCT(E5:E9,$C$5:$C$9)/$C$10". Note that using keyboard shortcuts Ctrl + C and Ctrl + V to cut and paste will also copy the percentage format from cell C10 to D10. Therefore, copy cell C10, select cell D10, and right click. Then select 'Paste Options' and choose the icon f_x, which will copy the formula only (Figure 13.11).

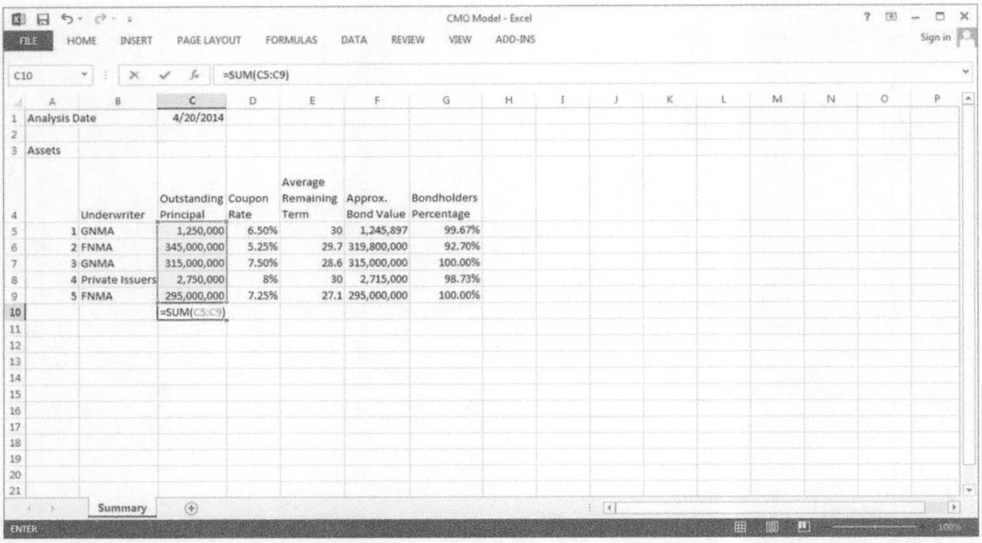

CMO Model - Excel

C10 =SUM(C5:C9)

	A	B	C	D	E	F	G
1	Analysis Date		4/20/2014				
2							
3	Assets						
4		Underwriter	Outstanding Principal	Coupon Rate	Average Remaining Term	Approx. Bond Value	Bondholders Percentage
5	1	GNMA	1,250,000	6.50%	30	1,245,897	99.67%
6	2	FNMA	345,000,000	5.25%	29.7	319,800,000	92.70%
7	3	GNMA	315,000,000	7.50%	28.6	315,000,000	100.00%
8	4	Private Issuers	2,750,000	8%	30	2,715,000	98.73%
9	5	FNMA	295,000,000	7.25%	27.1	295,000,000	100.00%
10			=SUM(C5:C9)				

Figure 13.9 SUM Function

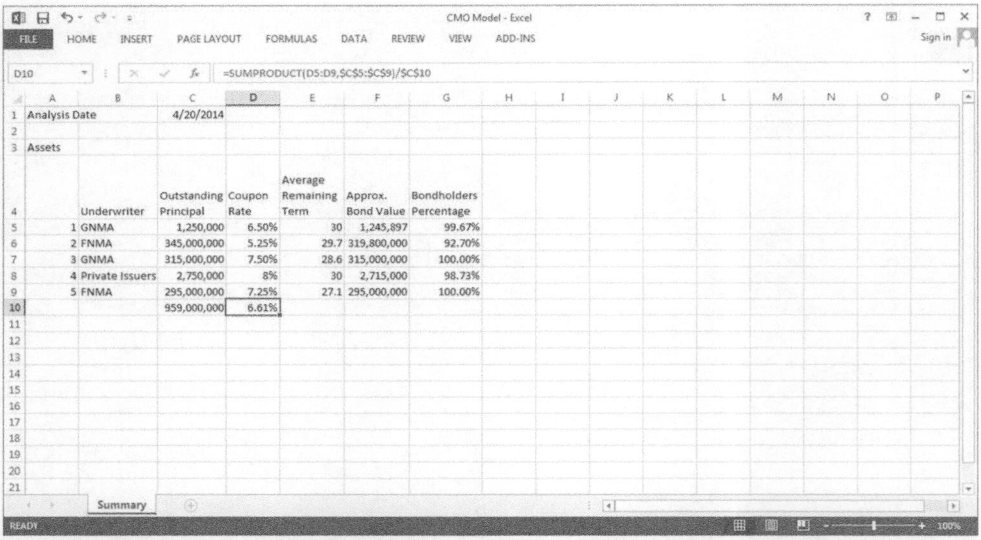

CMO Model - Excel

D10 =SUMPRODUCT(D5:D9,$C$5:$C$9)/$C$10

	A	B	C	D	E	F	G
1	Analysis Date		4/20/2014				
2							
3	Assets						
4		Underwriter	Outstanding Principal	Coupon Rate	Average Remaining Term	Approx. Bond Value	Bondholders Percentage
5	1	GNMA	1,250,000	6.50%	30	1,245,897	99.67%
6	2	FNMA	345,000,000	5.25%	29.7	319,800,000	92.70%
7	3	GNMA	315,000,000	7.50%	28.6	315,000,000	100.00%
8	4	Private Issuers	2,750,000	8%	30	2,715,000	98.73%
9	5	FNMA	295,000,000	7.25%	27.1	295,000,000	100.00%
10			959,000,000	6.61%			

Figure 13.10 SUMPRODUCT Function

The keyboard shortcut for pasting the formula would be Shift + F10, which will produce the same menu as right-clicking. Then use the arrow keys to navigate to 'Paste Options' and then formulas.

Cells may be formatted as desired; however, consistency is important. Thus, in this model the dollar amounts will be formatted with a comma and no currency symbol; and the percentages will be formatted with two decimal places. Finally, select cells A4:G10 and name the entire array by typing "Assets" in the reference bar above column A (Figure 13.12).

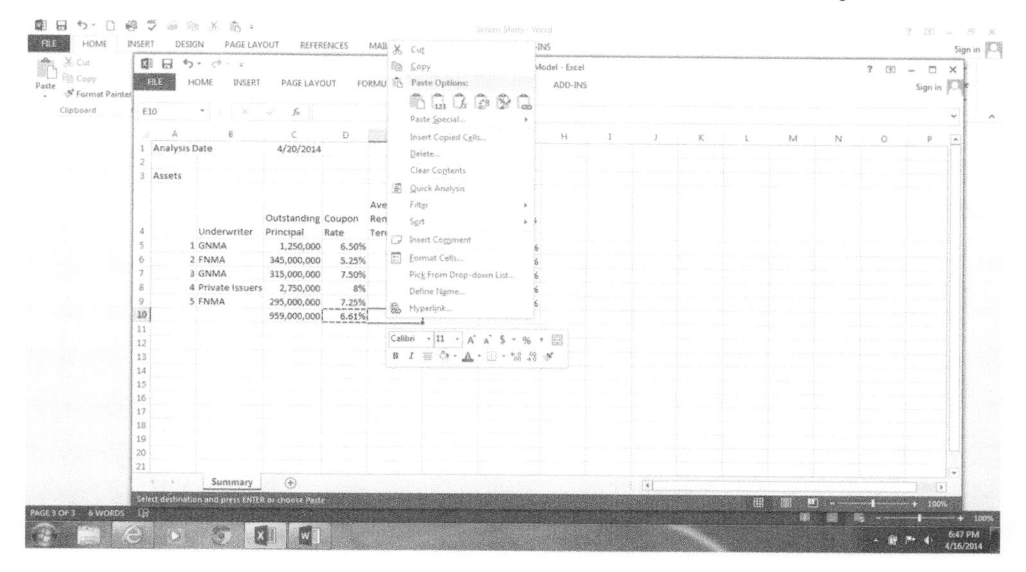

Figure 13.11 Direct References in Copying Formulas

	Underwriter	Outstanding Principal	Coupon Rate	Average Remaining Term	Approx. Bond Value	Bondholders Percentage
1	GNMA	1,250,000	6.50%	30	1,245,897	99.67%
2	FNMA	345,000,000	5.25%	29.7	319,800,000	92.70%
3	GNMA	315,000,000	7.50%	28.6	315,000,000	100.00%
4	Private Issuers	2,750,000	8%	30	2,715,000	98.73%
5	FNMA	295,000,000	7.25%	27.1	295,000,000	100.00%
		959,000,000	6.61%	28.540146	933,760,897	97.37%

Figure 13.12 Name the Array

The next information used when creating the waterfall is about the varying tranches. This information was given in the case study and should be entered into the model as seen in Figure 13.13; be sure to name the entire array "Tranches":

Note that the 'Price to Public ($)' values are entered as formulas: Price to Public (%) * Original Principal Amount. For future use, it is advisable to name the cells likely to be used in subsequent analysis. Cell B14, for example, which represents the total outstanding

CMO Model - Excel

E14 =D14*B14

	Underwriter	Outstanding Principal	Coupon Rate	Average Remaining Term	Approx. Bond Value	Bondholders Percentage
1	GNMA	1,250,000	6.50%	30	1,245,897	99.67%
2	FNMA	345,000,000	5.25%	29.7	319,800,000	92.70%
3	GNMA	315,000,000	7.50%	28.6	315,000,000	100.00%
4	Private Issuers	2,750,000	8%	30	2,715,000	98.73%
5	FNMA	295,000,000	7.25%	27.1	295,000,000	100.00%
		959,000,000	6.61%	28.54	933,760,897	97.37%

Analysis Date 4/20/2014

Assets

Tranches

Class	Original Principal Amount	Interest Rate	Price to Public (%)	Price to Public ($)
A	455,000,000	4%	87.65%	398,807,500
B	295,000,000	5.50%	79.25%	233,787,500
C	183,760,897	6%	96.50%	177,329,266
Total	933,760,897			809,924,266

Figure 13.13 Tranche Data

CMO Model - Excel

ClassA_Pr... 455000000

	Underwriter	Outstanding Principal	Coupon Rate	Average Remaining Term	Approx. Bond Value	Bondholders Percentage
1	GNMA	1,250,000	6.50%	30	1,245,897	99.67%
2	FNMA	345,000,000	5.25%	29.7	319,800,000	92.70%
3	GNMA	315,000,000	7.50%	28.6	315,000,000	100.00%
4	Private Issuers	2,750,000	8%	30	2,715,000	98.73%
5	FNMA	295,000,000	7.25%	27.1	295,000,000	100.00%
		959,000,000	6.61%	28.54	933,760,897	97.37%

Analysis Date 4/20/2014

Assets

Tranches

Class	Original Principal Amount	Interest Rate	Price to Public (%)	Price to Public ($)
A	455,000,000	4%	87.65%	398,807,500
B	295,000,000	5.50%	79.25%	233,787,500
C	183,760,897	6%	96.50%	177,329,266
Total	933,760,897			809,924,266

Figure 13.14 Cell Designation

principal for Class A, is named 'ClassA_Principal' (Figure 13.14). When naming cells it is important to remember that no spaces can be used; hence the use of the underscore between words. Cell C14 is named 'ClassA_Interest'.

The final data to be input are the assumptions which can be taken from the case study. The number of payments per year is important to note because it refers to payments made to the investors. Since this is a collateralized mortgage obligation, however, and mortgages are paid

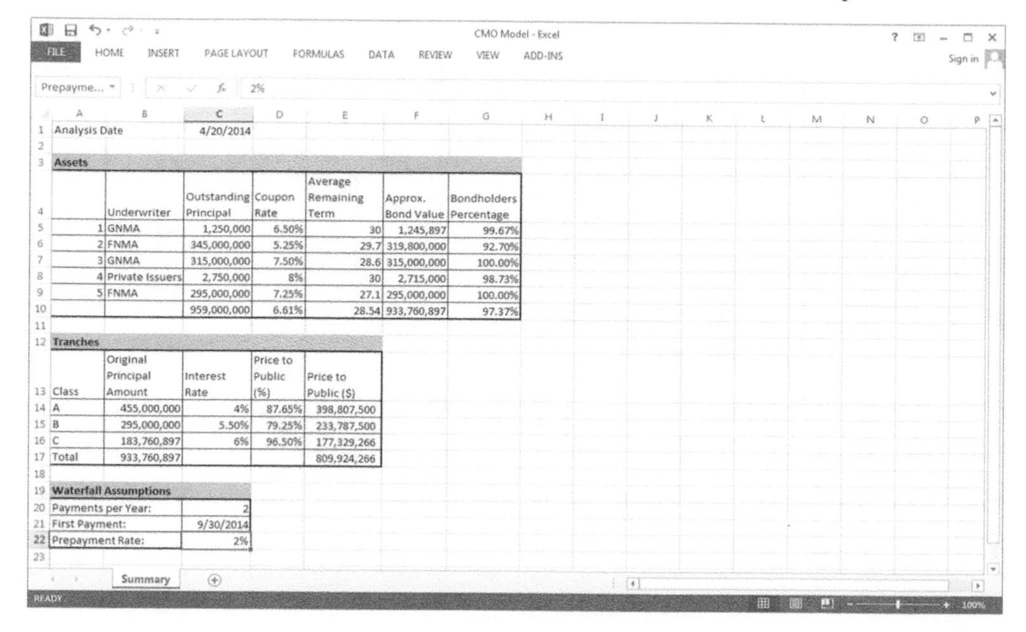

Analysis Date	4/20/2014

Assets

	Underwriter	Outstanding Principal	Coupon Rate	Average Remaining Term	Approx. Bond Value	Bondholders Percentage
1	GNMA	1,250,000	6.50%	30	1,245,897	99.67%
2	FNMA	345,000,000	5.25%	29.7	319,800,000	92.70%
3	GNMA	315,000,000	7.50%	28.6	315,000,000	100.00%
4	Private Issuers	2,750,000	8%	30	2,715,000	98.73%
5	FNMA	295,000,000	7.25%	27.1	295,000,000	100.00%
		959,000,000	6.61%	28.54	933,760,897	97.37%

Tranches

Class	Original Principal Amount	Interest Rate	Price to Public (%)	Price to Public ($)
A	455,000,000	4%	87.65%	398,807,500
B	295,000,000	5.50%	79.25%	233,787,500
C	183,760,897	6%	96.50%	177,329,266
Total	933,760,897			809,924,266

Waterfall Assumptions

Payments per Year:	2
First Payment:	9/30/2014
Prepayment Rate:	2%

Figure 13.15 Waterfall Assumptions

monthly, the cash flow will indicate monthly deposits. Payments are made on 31 March and 30 September of each year. The analysis date for this model is 20 April 2014; the first payment, therefore, will be 30 September of that year (Figure 13.15). The final assumption will be the prepayment rate. Begin by entering 2% in cell C22. Each of the assumptions entered here is a variable and may be altered as needed to completely analyze the model. A 2% prepayment rate will yield a different return than a 5% rate, for instance. This information will be summarized later in this chapter.

Cash flows

In a Collateralized Mortgage Obligation there will be several cash flows, one for each pool of mortgages. The cash flows will be combined to create one overall cash flow, but they must be modelled individually and then combined (Figure 13.16).

The analyst will begin by setting up a sheet for the first cash flow; this will be very similar to the amortization tables demonstrated in Chapter 4. Name the page '1'. This will correspond to the assets input on the Summary page (Figure 13.17).

Set up sheet 1 by entering the following labels in cells A1 through A5: 'Security', 'Underwriter', 'Principal', 'Interest Rate', and 'Term'; enter 'Payment' in cell A7. The number 1 will be entered into cell B1 as the Security for which this cash flow is being created.

Instead of reentering the data already on the Summary page, the VLOOKUP function, which was introduced in Chapter 5, will be used to automatically fill in cells B2 through B5. The first argument in each formula will be cell B1. By entering 'Assets' as the second argument, the formula determines that the value will come from the Assets array on the Summary page and will be found in the row where '1' is the corresponding reference point

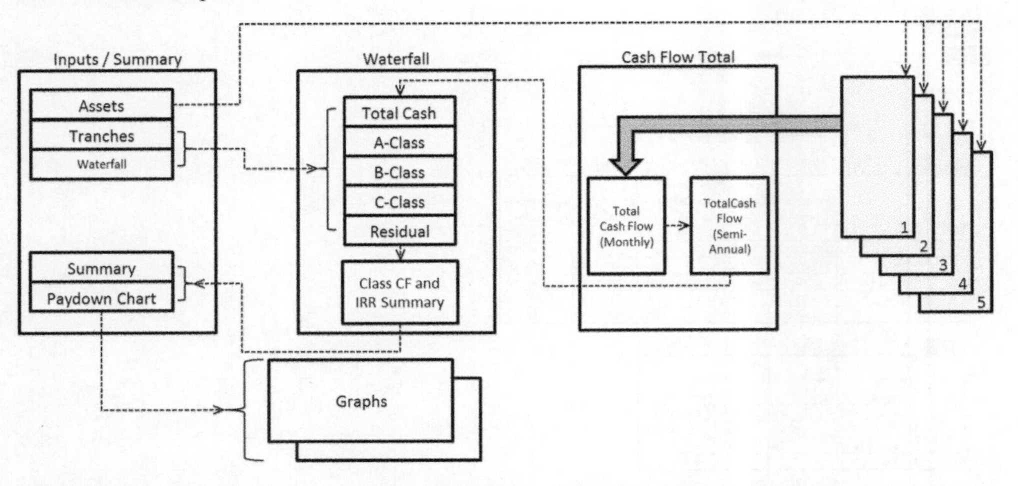

Figure 13.16 Model Flow Diagram (Mortgage Pools – Individual Worksheets)

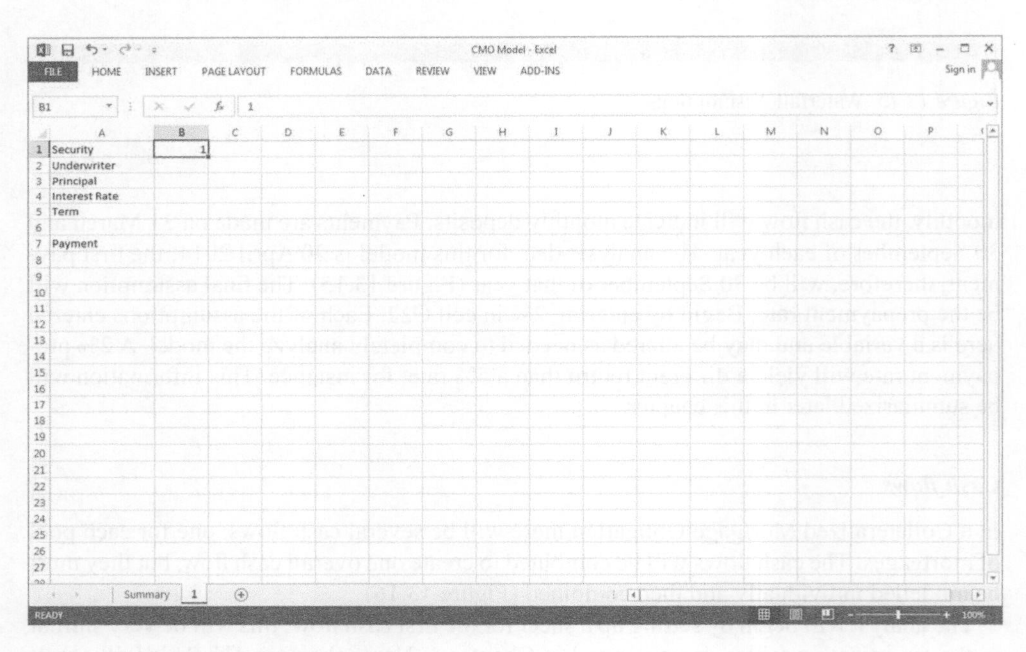

Figure 13.17 Cash Flow Sheet Setup

(Figure 13.18). The third argument is the column in which the value is found. The formula for cell B2, the underwriter of the first asset, is thus "=VLOOKUP(B$1,Assets,2,FALSE)". Formulas for cells B3, B4, and B5 are similar, with a different column designation as the third argument.

Calculate payment as demonstrated in Chapter 4. Since the mortgage payments are made monthly and the interest rates are quoted annually, be sure to adjust the interest rate and payment periods accordingly. Thus, the formula in cell B7 should read "=PMT(B4/12,B5*12,B3)"

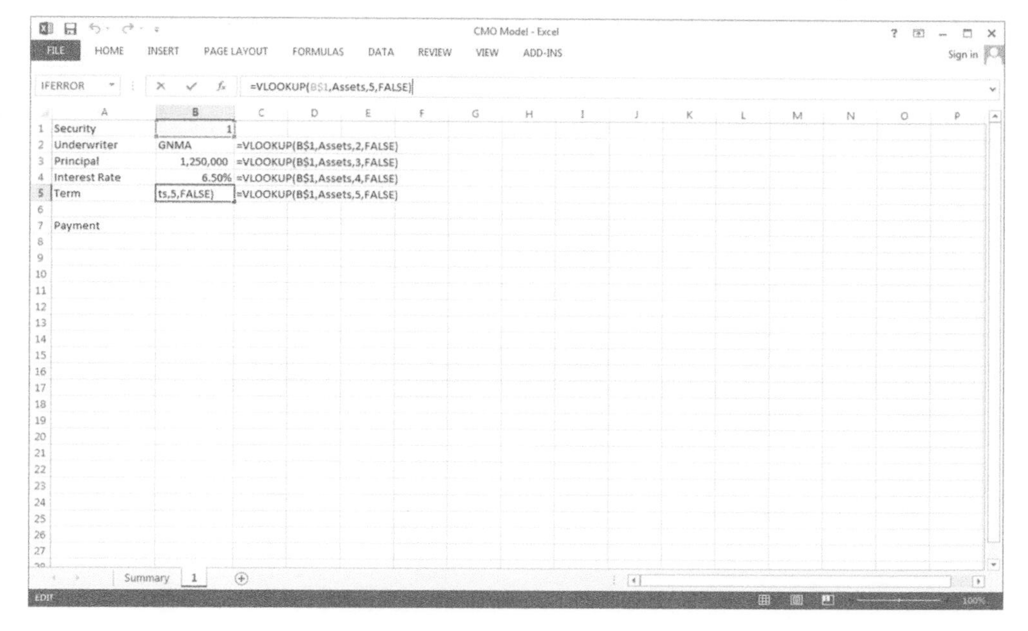

Figure 13.18 VLOOKUP Function

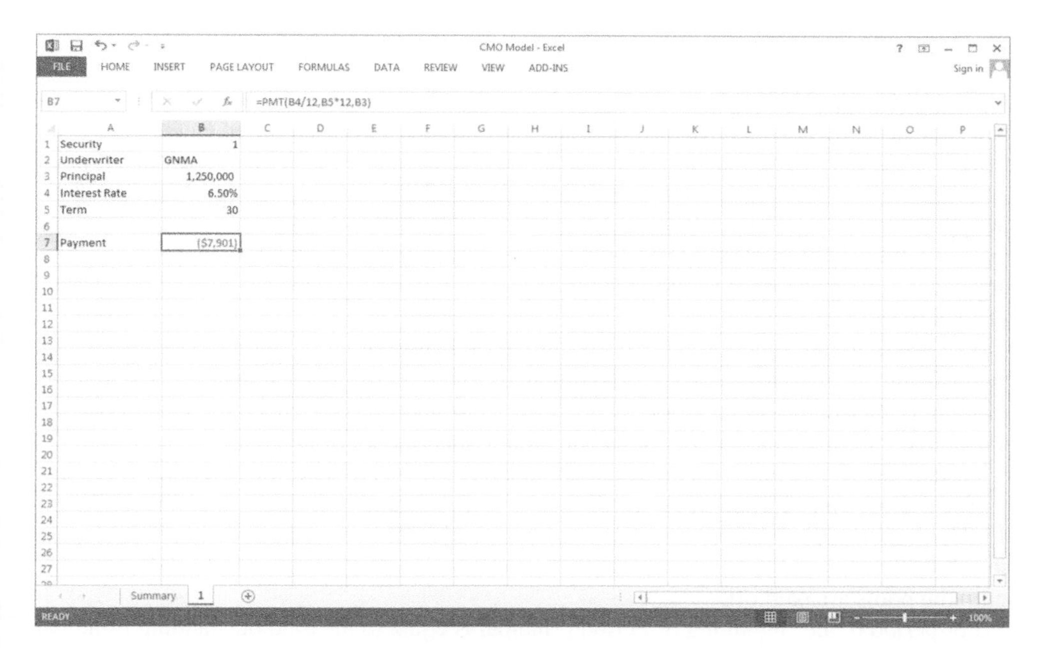

Figure 13.19 Payment Calculation Using PMT Function

(Figure 13.19). Notice the value returned is negative. Negative values represent payments, while positive values represent deposits.

Next label cells A9 through H9 with the following: 'Payment Date', 'Period', 'BoP', 'Int. PMT', 'Princ. PMT', 'Total PMT', 'Prepayment', and 'EoP' (Figure 13.20). BoP and EoP

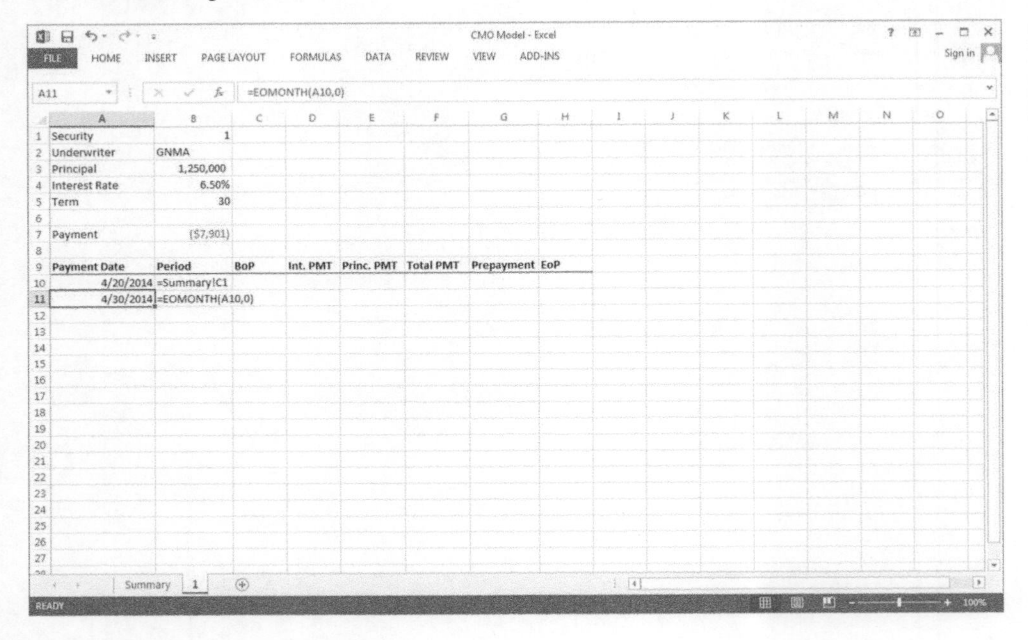

Figure 13.20 EOMONTH Function

represent principal balances at the beginning and end of the period, respectively. The first row represents period 0, which occurs at the analysis date. The information entered in period 0 will be carried through the model. It simply indicates that an initial amount was invested and the numbers will not be used until IRR is calculated during analysis. Cell A10, therefore, is "=Summary!C1". The first period will end on 30 April. This can be entered automatically by using the EOMONTH function, which was introduced in a previous chapter. The first argument will be cell A10; and the second argument will be '0' since the desired outcome is the end of the present month.

The end of month function can also be used to fill in the remaining months (Figure 13.21). When using this function there are two arguments to be entered, the start date and the number of months. Start date, in this case, is 30 April 2014, or cell A10. Since the cash flows are calculated monthly and each period is one month, the number of months is 1.

Next, fill in the remaining payment dates by copying cell A11 as far down column A as desired.

The periods can be filled in by entering '0' in cell B10 and "=B10+1" in cell B11 (Figure 13.22). Again, copy cell B11 down as far as required. In this case the cell should be copied to cell B370. Because the longest term is 30 years, there will, theoretically, be 360 payments until the pools are paid off.

For period 0, the Beginning of Period principal is equal to the total outstanding principal (Figure 13.23). This can be entered simply using an = sign, or in cell C10 enter "=B3". This value will remain positive since it represents a deposit. On the first day (i.e. period 0) there are no interest payments, so the EoP balance is simply equal to the total principal.

For period 1, and all subsequent periods, the beginning principal balance is equal to the ending principal balance of the prior period (Figure 13.24). Therefore, cell C11 is "=H10".

The equation for the interest payment is simply the BoP balance multiplied by the monthly interest rate. Note that the interest rate given is an annual rate and must be converted by dividing

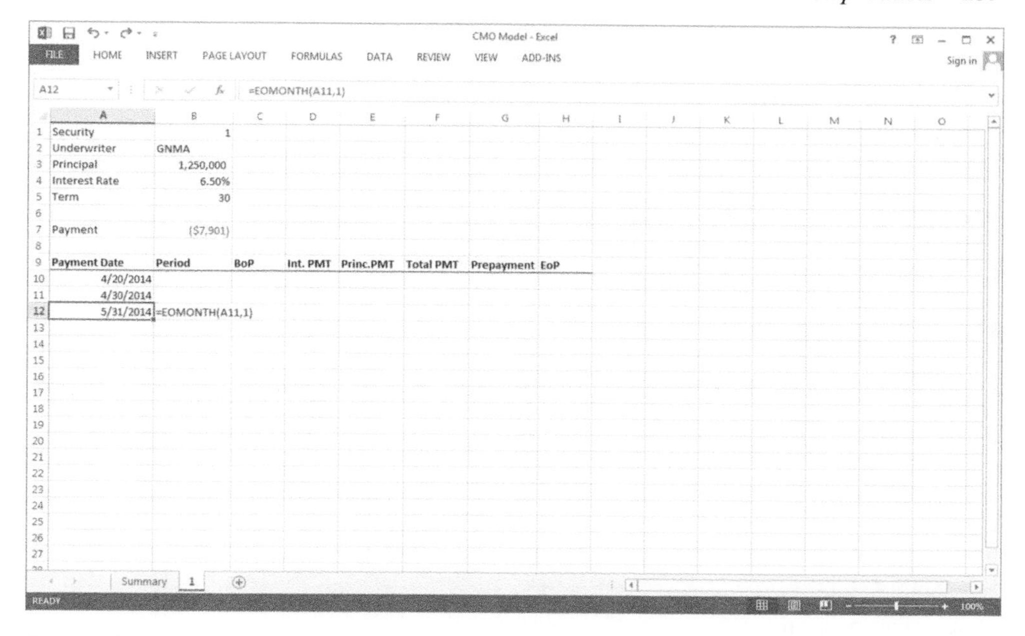

Figure 13.21 Using EOMONTH Function to Create Chart

	A	B	C	D	E	F	G	H
1	Security	1						
2	Underwriter	GNMA						
3	Principal	1,250,000						
4	Interest Rate	6.50%						
5	Term	30						
6								
7	Payment	($7,901)						
8								
9	Payment Date	Period	BoP	Int. PMT	Princ.PMT	Total PMT	Prepayment	EoP
10	4/20/2014	0						
11	4/30/2014	1	=B10+1					
12	5/31/2014	2	=B11+1					
13	6/30/2014	3	=B12+1					
14	7/31/2014	4	=B13+1					
15	8/31/2014	5	=B14+1					
16	9/30/2014							
17	10/31/2014							
18	11/30/2014							
19	12/31/2014							
20	1/31/2015							
21	2/28/2015							
22	3/31/2015							
23	4/30/2015							
24	5/31/2015							
25	6/30/2015							
26	7/31/2015							
27	8/31/2015							

Figure 13.22 Designating Payment Periods

by 12. The BoP value is positive, so it is important to remember to insert a negative sign in front of the equation. The cash flow for the first security should now look like Figure 13.25.

Note the use of direct references. The interest rate does not change and will always be as indicated in cell B4. The column, B, and row, 4, must therefore both be static.

CMO Model - Excel

H10 | =C10

	A	B	C	D	E	F	G	H
1	Security	1						
2	Underwriter	GNMA						
3	Principal	1,250,000						
4	Interest Rate	6.50%						
5	Term	30						
6								
7	Payment	($7,901)						
8								
9	Payment Date	Period	BoP	Int. PMT	Princ.PMT	Total PMT	Prepayment	EoP
10	4/20/2014	0	1,250,000	0	0	0	0	1,250,000
11	4/30/2014	1						
12	5/31/2014	2						
13	6/30/2014	3						
14	7/31/2014	4						
15	8/31/2014	5						
16	9/30/2014	6						
17	10/31/2014	7						
18	11/30/2014	8						
19	12/31/2014	9						
20	1/31/2015	10						
21	2/28/2015	11						
22	3/31/2015	12						
23	4/30/2015	13						
24	5/31/2015	14						
25	6/30/2015	15						
26	7/31/2015	16						

Figure 13.23 Period 0

CMO Model - Excel

H10 | =H10

	A	B	C	D	E	F	G	H
1	Security	1						
2	Underwriter	GNMA						
3	Principal	1,250,000						
4	Interest Rate	6.50%						
5	Term	30						
6								
7	Payment	($7,901)						
8								
9	Payment Date	Period	BoP	Int. PMT	Princ.PMT	Total PMT	Prepayment	EoP
10	4/20/2014	0	1,250,000	0	0	0	0	1,250,000
11	4/30/2014	1	=H10					
12	5/31/2014	2						
13	6/30/2014	3						
14	7/31/2014	4						
15	8/31/2014	5						
16	9/30/2014	6						
17	10/31/2014	7						
18	11/30/2014	8						
19	12/31/2014	9						
20	1/31/2015	10						
21	2/28/2015	11						
22	3/31/2015	12						
23	4/30/2015	13						
24	5/31/2015	14						
25	6/30/2015	15						
26	7/31/2015	16						

Figure 13.24 Beginning of Period Balance (BoP)

The principal payment is where the equations can be complicated. Typically the principal payment is equal to the total payment minus the interest payment. This is true unless that number is larger than the outstanding principal amount, in which case the principal would be overpaid. This consideration is important at later payment dates when an otherwise negative principal balance would be recognized. In this example a minimum formula could be used in Excel. This model, however, uses negative numbers to represent payments and thus, a maximum function will be used instead.

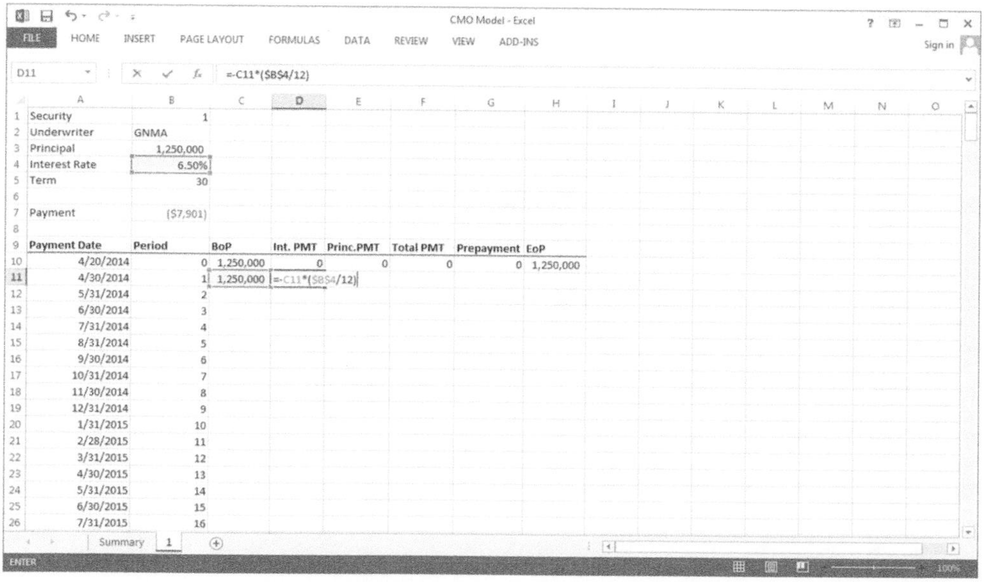

CMO Model - Excel

D11 =-C11*($B$4/12)

	A	B	C	D	E	F	G	H
1	Security	1						
2	Underwriter	GNMA						
3	Principal	1,250,000						
4	Interest Rate	6.50%						
5	Term	30						
6								
7	Payment	($7,901)						
8								
9	Payment Date	Period	BoP	Int. PMT	Princ.PMT	Total PMT	Prepayment	EoP
10	4/20/2014	0	1,250,000	0	0	0	0	1,250,000
11	4/30/2014	1	1,250,000	=-C11*($B$4/12)				
12	5/31/2014	2						
13	6/30/2014	3						
14	7/31/2014	4						
15	8/31/2014	5						
16	9/30/2014	6						
17	10/31/2014	7						
18	11/30/2014	8						
19	12/31/2014	9						
20	1/31/2015	10						
21	2/28/2015	11						
22	3/31/2015	12						
23	4/30/2015	13						
24	5/31/2015	14						
25	6/30/2015	15						
26	7/31/2015	16						

Figure 13.25 Interest Payment Calculation (Int. PMT)

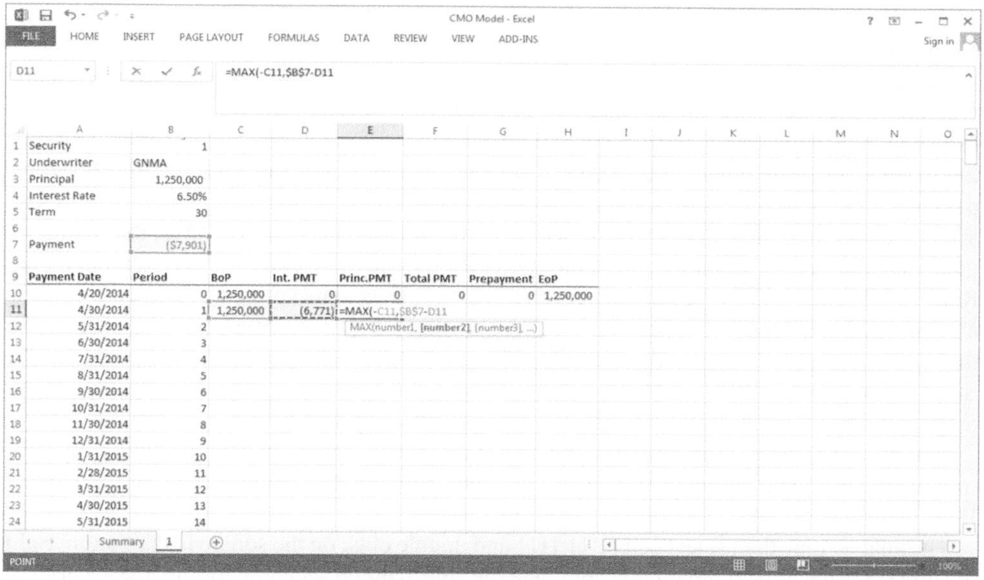

CMO Model - Excel

D11 =MAX(-C11,$B$7-D11

	A	B	C	D	E	F	G	H
1	Security	1						
2	Underwriter	GNMA						
3	Principal	1,250,000						
4	Interest Rate	6.50%						
5	Term	30						
6								
7	Payment	($7,901)						
8								
9	Payment Date	Period	BoP	Int. PMT	Princ.PMT	Total PMT	Prepayment	EoP
10	4/20/2014	0	1,250,000	0	0	0	0	1,250,000
11	4/30/2014	1	1,250,000	(6,771)	=MAX(-C11,$B$7-D11			
12	5/31/2014	2						
13	6/30/2014	3						
14	7/31/2014	4						
15	8/31/2014	5						
16	9/30/2014	6						
17	10/31/2014	7						
18	11/30/2014	8						
19	12/31/2014	9						
20	1/31/2015	10						
21	2/28/2015	11						
22	3/31/2015	12						
23	4/30/2015	13						
24	5/31/2015	14						

Figure 13.26 Principal Payment Calculation (Princ. PMT)

By entering "=MAX" into a cell, one must enter two values for which a maximum must be returned. As stated earlier, one of these will be the outstanding principal amount, or cell C11. The next argument is the difference between the total payment and the interest payment. Thus, the formula to be entered into cell E11 is "=MAX(-C11,$B$7-D11)", and the returned value is the principal amount paid (Figure 13.26).

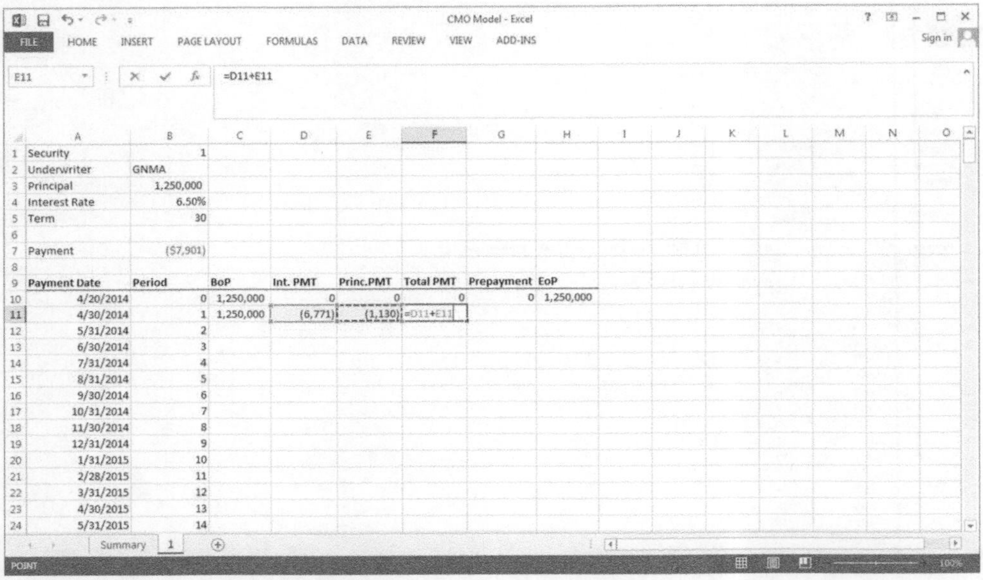

CMO Model - Excel

FILE HOME INSERT PAGE LAYOUT FORMULAS DATA REVIEW VIEW ADD-INS

E11 =D11+E11

	A	B	C	D	E	F	G	H
1	Security	1						
2	Underwriter	GNMA						
3	Principal	1,250,000						
4	Interest Rate	6.50%						
5	Term	30						
6								
7	Payment	($7,901)						
8								
9	Payment Date	Period	BoP	Int. PMT	Princ.PMT	Total PMT	Prepayment	EoP
10	4/20/2014	0	1,250,000	0	0	0	0	1,250,000
11	4/30/2014	1	1,250,000	(6,771)	(1,130)	=D11+E11		
12	5/31/2014	2						
13	6/30/2014	3						
14	7/31/2014	4						
15	8/31/2014	5						
16	9/30/2014	6						
17	10/31/2014	7						
18	11/30/2014	8						
19	12/31/2014	9						
20	1/31/2015	10						
21	2/28/2015	11						
22	3/31/2015	12						
23	4/30/2015	13						
24	5/31/2015	14						

Summary | 1

POINT

Figure 13.27 Total Payment

The total payment is simply the sum of the interest payment and the principal payment. The equation is therefore "=D11+E11" (Figure 13.27). This is a good check to be sure that the payment is equal to the number calculated in cell B7.

The prepayment rate will also be taken into account. A rate of 2% per year is assumed while creating the model. This means that 2% of outstanding mortgages are paid off each year. Prepayment usually occurs when a home is sold, an inheritance is used to pay off an outstanding balance, or some other form of bulk payment, such as a bonus or raise, is used. The prepayment rate input on the Summary page will be used here. Since the prepayment rate can affect the outcome, this model will be set up so that the rate may be altered on the Summary page and the results will calculate automatically. In cell G11 enter "=-C11*Prepayment_Rate/12" (Figure 13.28). Again, the negative symbol needs to be entered because the value is a payment, not a deposit. Notice that the prepayment rate is given as a yearly percentage, so the number must be converted to monthly. This is done by simply dividing the rate by 12. Since the prepayment rate, cell C24 on the Summary page, was named previously, the analyst may type 'Prepayment_Rate' instead of moving between spreadsheets.

Finally, the principal remaining at the end of the period can be calculated simply by subtracting the principal payment and the prepayment from the BoP balance. The formula in cell H11, therefore, is "=C11+E11+G11" (Figure 13.29).

The analyst may then select cells C11:H11 and double click on the square in the bottom right of cell H11. This will automatically complete the amortization schedule for the first security in the mortgage pool. In this instance, with a prepayment rate of 2%, all mortgages are paid off by 30 April 2034, the 241st period. In Figure 13.30 is a summary of the formulas used.

Once the table for the first asset is completed, the entire sheet may be copied. Create a copy of sheet '1' and rename it '2'. By simply changing the referenced asset, in cell B1, the entire table should update accordingly (Figure 13.31).

Repeat this three more times for assets 3, 4, and 5. At this point the analyst should have 5 cash flows, one for each asset.

CMO Model - Excel

FILE HOME INSERT PAGE LAYOUT FORMULAS DATA REVIEW VIEW ADD-INS

C11 =-C11*Prepayment_Rate/12

	A	B	C	D	E	F	G	H
1	Security	1						
2	Underwriter	GNMA						
3	Principal	1,250,000						
4	Interest Rate	6.50%						
5	Term	30						
6								
7	Payment	($7,901)						
8								
9	**Payment Date**	**Period**	**BoP**	**Int. PMT**	**Princ. PMT**	**Total PMT**	**Prepayment**	**EoP**
10	4/20/2014	0	1,250,000	0	0	0	0	1,250,000
11	4/30/2014	1	1,250,000	(6,771)	(1,130)	(7,901)	=-C11*Prepayment_Rate/12	
12	5/31/2014	2						
13	6/30/2014	3						
14	7/31/2014	4						
15	8/31/2014	5						
16	9/30/2014	6						
17	10/31/2014	7						
18	11/30/2014	8						
19	12/31/2014	9						
20	1/31/2015	10						
21	2/28/2015	11						
22	3/31/2015	12						
23	4/30/2015	13						
24	5/31/2015	14						

Summary | 1

ENTER

Figure 13.28 Prepayment Calculation

CMO Model - Excel

FILE HOME INSERT PAGE LAYOUT FORMULAS DATA REVIEW VIEW ADD-INS

G11 =C11+E11+G11

	A	B	C	D	E	F	G	H
1	Security	1						
2	Underwriter	GNMA						
3	Principal	1,250,000						
4	Interest Rate	6.50%						
5	Term	30						
6								
7	Payment	($7,901)						
8								
9	**Payment Date**	**Period**	**BoP**	**Int. PMT**	**Princ. PMT**	**Total PMT**	**Prepayment**	**EoP**
10	4/20/2014	0	1,250,000	0	0	0	0	1,250,000
11	4/30/2014	1	1,250,000	(6,771)	(1,130)	(7,901)	(2,083)	=C11+E11+G11
12	5/31/2014	2						
13	6/30/2014	3						
14	7/31/2014	4						
15	8/31/2014	5						
16	9/30/2014	6						
17	10/31/2014	7						
18	11/30/2014	8						
19	12/31/2014	9						
20	1/31/2015	10						
21	2/28/2015	11						
22	3/31/2015	12						
23	4/30/2015	13						
24	5/31/2015	14						

Summary | 1

POINT

Figure 13.29 End of Period Balance (EoP)

Payment Date	Period	BoP	Int. PMT	Princ. PMT	Total PMT	Prepayment	EoP
=Summary!C1	0	=B3	0	0	0	0	=C10
=EOMONTH(A10,0)	=B10+1	=H10	=-C11*($B$4/12)	=MAX(-C11,$B$7-D11)	=D11+E11	=-C11*Prepayment_Rate/12	=C11+E11+G11
=EOMONTH(A11,1)	=B11+1	=H11	=-C12*($B$4/12)	=MAX(-C12,$B$7-D12)	=D12+E12	=-C12*Prepayment_Rate/12	=C12+E12+G12
=EOMONTH(A12,1)	=B12+1	=H12	=-C13*($B$4/12)	=MAX(-C13,$B$7-D13)	=D13+E13	=-C13*Prepayment_Rate/12	=C13+E13+G13
=EOMONTH(A13,1)	=B13+1	=H13	=-C14*($B$4/12)	=MAX(-C14,$B$7-D14)	=D14+E14	=-C14*Prepayment_Rate/12	=C14+E14+G14
=EOMONTH(A14,1)	=B14+1	=H14	=-C15*($B$4/12)	=MAX(-C15,$B$7-D15)	=D15+E15	=-C15*Prepayment_Rate/12	=C15+E15+G15
=EOMONTH(A15,1)	=B15+1	=H15	=-C16*($B$4/12)	=MAX(-C16,$B$7-D16)	=D16+E16	=-C16*Prepayment_Rate/12	=C16+E16+G16
=EOMONTH(A16,1)	=B16+1	=H16	=-C17*($B$4/12)	=MAX(-C17,$B$7-D17)	=D17+E17	=-C17*Prepayment_Rate/12	=C17+E17+G17
=EOMONTH(A17,1)	=B17+1	=H17	=-C18*($B$4/12)	=MAX(-C18,$B$7-D18)	=D18+E18	=-C18*Prepayment_Rate/12	=C18+E18+G18
=EOMONTH(A18,1)	=B18+1	=H18	=-C19*($B$4/12)	=MAX(-C19,$B$7-D19)	=D19+E19	=-C19*Prepayment_Rate/12	=C19+E19+G19
=EOMONTH(A19,1)	=B19+1	=H19	=-C20*($B$4/12)	=MAX(-C20,$B$7-D20)	=D20+E20	=-C20*Prepayment_Rate/12	=C20+E20+G20
=EOMONTH(A20,1)	=B20+1	=H20	=-C21*($B$4/12)	=MAX(-C21,$B$7-D21)	=D21+E21	=-C21*Prepayment_Rate/12	=C21+E21+G21
=EOMONTH(A21,1)	=B21+1	=H21	=-C22*($B$4/12)	=MAX(-C22,$B$7-D22)	=D22+E22	=-C22*Prepayment_Rate/12	=C22+E22+G22
=EOMONTH(A22,1)	=B22+1	=H22	=-C23*($B$4/12)	=MAX(-C23,$B$7-D23)	=D23+E23	=-C23*Prepayment_Rate/12	=C23+E23+G23
=EOMONTH(A23,1)	=B23+1	=H23	=-C24*($B$4/12)	=MAX(-C24,$B$7-D24)	=D24+E24	=-C24*Prepayment_Rate/12	=C24+E24+G24
=EOMONTH(A24,1)	=B24+1	=H24	=-C25*($B$4/12)	=MAX(-C25,$B$7-D25)	=D25+E25	=-C25*Prepayment_Rate/12	=C25+E25+G25
=EOMONTH(A25,1)	=B25+1	=H25	=-C26*($B$4/12)	=MAX(-C26,$B$7-D26)	=D26+E26	=-C26*Prepayment_Rate/12	=C26+E26+G26
=EOMONTH(A26,1)	=B26+1	=H26	=-C27*($B$4/12)	=MAX(-C27,$B$7-D27)	=D27+E27	=-C27*Prepayment_Rate/12	=C27+E27+G27
=EOMONTH(A27,1)	=B27+1	=H27	=-C28*($B$4/12)	=MAX(-C28,$B$7-D28)	=D28+E28	=-C28*Prepayment_Rate/12	=C28+E28+G28
=EOMONTH(A28,1)	=B28+1	=H28	=-C29*($B$4/12)	=MAX(-C29,$B$7-D29)	=D29+E29	=-C29*Prepayment_Rate/12	=C29+E29+G29
=EOMONTH(A29,1)	=B29+1	=H29	=-C30*($B$4/12)	=MAX(-C30,$B$7-D30)	=D30+E30	=-C30*Prepayment_Rate/12	=C30+E30+G30
=EOMONTH(A30,1)	=B30+1	=H30	=-C31*($B$4/12)	=MAX(-C31,$B$7-D31)	=D31+E31	=-C31*Prepayment_Rate/12	=C31+E31+G31
=EOMONTH(A31,1)	=B31+1	=H31	=-C32*($B$4/12)	=MAX(-C32,$B$7-D32)	=D32+E32	=-C32*Prepayment_Rate/12	=C32+E32+G32
=EOMONTH(A32,1)	=B32+1	=H32	=-C33*($B$4/12)	=MAX(-C33,$B$7-D33)	=D33+E33	=-C33*Prepayment_Rate/12	=C33+E33+G33

Figure 13.30 Cash Flow Completion

Security	2
Underwriter	FNMA
Principal	345,000,000
Interest Rate	5.25%
Term	29.7
Payment	($1,913,047)

Payment Date	Period	BoP	Int. PMT	Princ. PMT	Total PMT	Prepayment	EoP
4/20/2014	0	345,000,000	0	0	0	0	345,000,000
4/30/2014	1	345,000,000	(1,509,375)	(403,672)	(1,913,047)	(575,000)	344,021,328
5/31/2014	2	344,021,328	(1,505,093)	(407,954)	(1,913,047)	(573,369)	343,040,005
6/30/2014	3	343,040,005	(1,500,800)	(412,247)	(1,913,047)	(571,733)	342,056,024
7/31/2014	4	342,056,024	(1,496,495)	(416,552)	(1,913,047)	(570,093)	341,069,378
8/31/2014	5	341,069,378	(1,492,179)	(420,869)	(1,913,047)	(568,449)	340,080,060
9/30/2014	6	340,080,060	(1,487,850)	(425,197)	(1,913,047)	(566,800)	339,088,063
10/31/2014	7	339,088,063	(1,483,510)	(429,537)	(1,913,047)	(565,147)	338,093,379
11/30/2014	8	338,093,379	(1,479,159)	(433,889)	(1,913,047)	(563,489)	337,096,001
12/31/2014	9	337,096,001	(1,474,795)	(438,252)	(1,913,047)	(561,827)	336,095,922
1/31/2015	10	336,095,922	(1,470,420)	(442,628)	(1,913,047)	(560,160)	335,093,135
2/28/2015	11	335,093,135	(1,466,032)	(447,015)	(1,913,047)	(558,489)	334,087,631
3/31/2015	12	334,087,631	(1,461,633)	(451,414)	(1,913,047)	(556,813)	333,079,404
4/30/2015	13	333,079,404	(1,457,222)	(455,825)	(1,913,047)	(555,132)	332,068,447
5/31/2015	14	332,068,447	(1,452,799)	(460,248)	(1,913,047)	(553,447)	331,054,752
6/30/2015	15	331,054,752	(1,448,365)	(464,683)	(1,913,047)	(551,758)	330,038,311
7/31/2015	16	330,038,311	(1,443,918)	(469,130)	(1,913,047)	(550,064)	329,019,117

Figure 13.31 Cash Flows for Additional Assets

Cash flow summary

In a collateralized mortgage obligation, the cash flows of each individual pool of mortgages should be combined to create one theoretical cash flow (Figure 13.32).

This can be done rather simply by creating a new sheet and starting a table with the same headings as were used previously. For clarity, name the new sheet 'CF Total' (Figure 13.33).

The payment dates and periods can be entered just as before, using the analysis date as the payment date for period 0. The data in this table will be the sum of the corresponding data on pages 1, 2, 3, 4, and 5.

The balance at the beginning of period 0, for the entire pool of mortgages, is simply the sum of the balance in periods 1, 2, 3, 4, and 5. Enter this by adding the BoP cells in period 0 from each page or "='1'!C10+'2'!C10+'3'!C10+'4'!C10+'5'!C10" (Figure 13.34).

Notice the outstanding principal at the beginning of period 0 is equal to the total outstanding principal on the Summary page (Figure 13.35). The remaining columns can be filled in

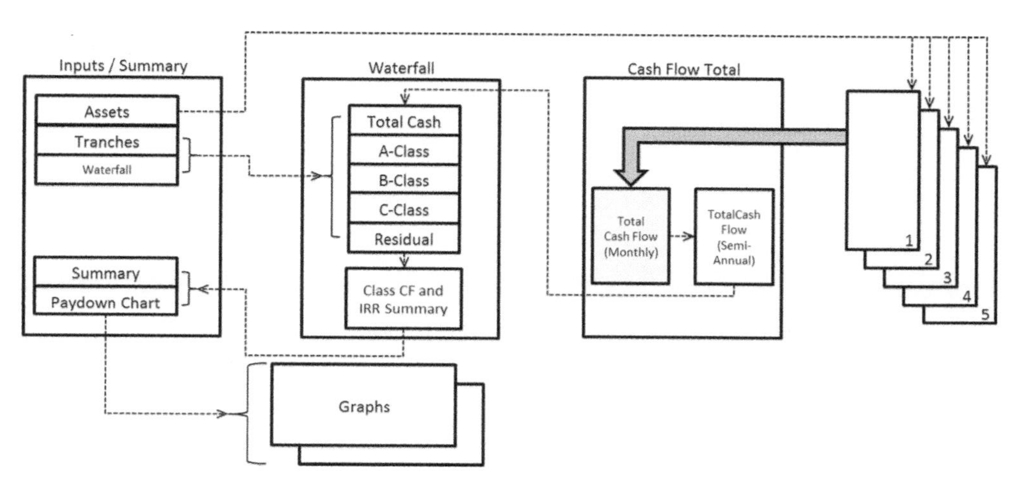

Figure 13.32 Model Flow Diagram (Cash Flow Summary Worksheet)

Total Cash Flows

Payment Date	Period	BoP	Int. PMT	Princ. PMT	Total PMT	Prepayment	EoP
4/20/2014	0						
4/30/2014	1						
5/31/2014	2						
6/30/2014	3						
7/31/2014	4						
8/31/2014	5						
9/30/2014	6						
10/31/2014	7						
11/30/2014	8						
12/31/2014	9						
1/31/2015	10						
2/28/2015	11						
3/31/2015	12						
4/30/2015	13						
5/31/2015	14						
6/30/2015	15						
7/31/2015	16						
8/31/2015	17						
9/30/2015	18						
10/31/2015	19						
11/30/2015	20						
12/31/2015	21						
1/31/2016	22						

Figure 13.33 Cash Flow Summary Page Setup

CMO Model - Excel

C4 =’1’!C10+’2’!C10+’3’!C10+’4’!C10+’5’!C10

Total Cash Flows

Payment Date	Period	BoP	Int. PMT	Princ. PMT	Total PMT	Prepayment	EoP
4/20/2014	0	959,000,000					
4/30/2014	1						
5/31/2014	2						
6/30/2014	3						
7/31/2014	4						
8/31/2014	5						
9/30/2014	6						
10/31/2014	7						
11/30/2014	8						
12/31/2014	9						
1/31/2015	10						
2/28/2015	11						
3/31/2015	12						
4/30/2015	13						
5/31/2015	14						
6/30/2015	15						
7/31/2015	16						
8/31/2015	17						
9/30/2015	18						
10/31/2015	19						
11/30/2015	20						
12/31/2015	21						
1/31/2016	22						

Summary | CF Total | 1 | 2 | 3 | 4 | 5

Figure 13.34 Cash Flow Summary – Beginning of Period Balance

CMO Model - Excel

C5 =’1’!C11+’2’!C11+’3’!C11+’4’!C11+’5’!C11

Total Cash Flows

Payment Date	Period	BoP	Int. PMT	Princ. PMT	Total PMT	Prepayment	EoP
4/20/2014	0	959,000,000	-	-	-	-	959,000,000
4/30/2014	1	959,000,000	(5,285,521)	(962,273)	(6,247,794)	(1,598,333)	956,439,394
5/31/2014	2	956,439,394	(5,271,515)	(976,278)	(6,247,794)	(1,594,066)	953,869,050
6/30/2014	3	953,869,050	(5,257,455)	(990,339)	(6,247,794)	(1,589,782)	951,288,930
7/31/2014	4	951,288,930	(5,243,339)	(1,004,455)	(6,247,794)	(1,585,482)	948,698,993
8/31/2014	5	948,698,993	(5,229,167)	(1,018,626)	(6,247,794)	(1,581,165)	946,099,202
9/30/2014	6	946,099,202	(5,214,940)	(1,032,854)	(6,247,794)	(1,576,832)	943,489,516
10/31/2014	7	943,489,516	(5,200,656)	(1,047,137)	(6,247,794)	(1,572,483)	940,869,896
11/30/2014	8	940,869,896	(5,186,316)	(1,061,477)	(6,247,794)	(1,568,116)	938,240,302
12/31/2014	9	938,240,302	(5,171,920)	(1,075,874)	(6,247,794)	(1,563,734)	935,600,695
1/31/2015	10	935,600,695	(5,157,467)	(1,090,327)	(6,247,794)	(1,559,334)	932,951,033
2/28/2015	11	932,951,033	(5,142,956)	(1,104,837)	(6,247,794)	(1,554,918)	930,291,278
3/31/2015	12	930,291,278	(5,128,389)	(1,119,405)	(6,247,794)	(1,550,485)	927,621,388
4/30/2015	13	927,621,388	(5,113,764)	(1,134,030)	(6,247,794)	(1,546,036)	924,941,322
5/31/2015	14	924,941,322	(5,099,081)	(1,148,712)	(6,247,794)	(1,541,569)	922,251,041
6/30/2015	15	922,251,041	(5,084,341)	(1,163,453)	(6,247,794)	(1,537,085)	919,550,503
7/31/2015	16	919,550,503	(5,069,542)	(1,178,252)	(6,247,794)	(1,532,584)	916,839,667
8/31/2015	17	916,839,667	(5,054,684)	(1,193,109)	(6,247,794)	(1,528,066)	914,118,491
9/30/2015	18	914,118,491	(5,039,768)	(1,208,025)	(6,247,794)	(1,523,531)	911,386,935
10/31/2015	19	911,386,935	(5,024,793)	(1,223,000)	(6,247,794)	(1,518,978)	908,644,956
11/30/2015	20	908,644,956	(5,009,759)	(1,238,034)	(6,247,794)	(1,514,408)	905,892,514
12/31/2015	21	905,892,514	(4,994,666)	(1,253,128)	(6,247,794)	(1,509,821)	903,129,565
1/31/2016	22	903,129,565	(4,979,512)	(1,268,281)	(6,247,794)	(1,505,216)	900,356,068

Summary | CF Total | 1 | 2 | 3 | 4 | 5

Figure 13.35 Cash Flow Summary Table Completion

by simply adding the data in the corresponding column on sheets 1 through 5; or by selecting cell C4 and copying to each cell in the row to cell H4. Notice when copying the formula that the columns update automatically. Double click the bottom right of cell H4 and the chart will automatically complete.

One table indicating monthly cash flows for the entire CMO has been generated.

In collateralized mortgage obligations, however, payments are distributed semiannually. The data must be reconfigured so that the semiannual payments are accurately represented. This will be done by creating a separate table on the same page to the right of cash flow (created earlier).

In this case the first step will be to combine periods 1 through 6 into the first payment period and periods 7 through 12 into the second payment period. To simplify the conversion process, this new table will be created on the same sheet titled 'CF Total' (Figure 13.36).

Begin by changing the width of column I to 2 (Figure 13.37). This is done by selecting the column heading, right-clicking, and selecting 'Column Width . . .', then entering '2'. Again

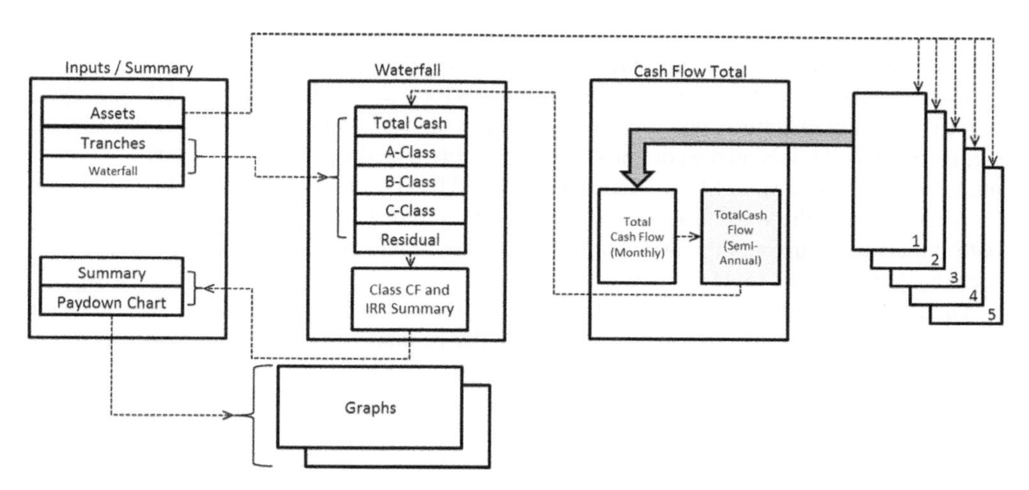

Figure 13.36 Model Flow Diagram (Cash Flow Summary – Conversion to Semiannual Periods)

Total Cash Flows

Payment Date	Period	BoP	Int. PMT	Princ. PMT	Total PMT	Prepayment	EoP
4/20/2014	0	959,000,000	-	-	-	-	959,000,000
4/30/2014	1	959,000,000	(5,285,521)	(962,273)	(6,247,794)	(1,598,333)	956,439,394
5/31/2014	2	956,439,394	(5,271,515)	(976,278)	(6,247,794)	(1,594,066)	953,869,050
6/30/2014	3	953,869,050	(5,257,455)	(990,339)	(6,247,794)	(1,589,782)	951,288,930
7/31/2014	4	951,288,930	(5,243,339)	(1,004,455)	(6,247,794)	(1,585,482)	948,698,993
8/31/2014	5	948,698,993	(5,229,167)	(1,018,626)	(6,247,794)	(1,581,165)	946,099,202
9/30/2014	6	946,099,202	(5,214,940)	(1,032,854)	(6,247,794)	(1,576,832)	943,489,516
10/31/2014	7	943,489,516	(5,200,656)	(1,047,137)	(6,247,794)	(1,572,483)	940,869,896
11/30/2014	8	940,869,896	(5,186,316)	(1,061,477)	(6,247,794)	(1,568,116)	938,240,302
12/31/2014	9	938,240,302	(5,171,920)	(1,075,874)	(6,247,794)	(1,563,734)	935,600,695
1/31/2015	10	935,600,695	(5,157,467)	(1,090,327)	(6,247,794)	(1,559,334)	932,951,033
2/28/2015	11	932,951,033	(5,142,956)	(1,104,837)	(6,247,794)	(1,554,918)	930,291,278
3/31/2015	12	930,291,278	(5,128,389)	(1,119,405)	(6,247,794)	(1,550,485)	927,621,388
4/30/2015	13	927,621,388	(5,113,764)	(1,134,030)	(6,247,794)	(1,546,036)	924,941,322
5/31/2015	14	924,941,322	(5,099,081)	(1,148,712)	(6,247,794)	(1,541,569)	922,251,041
6/30/2015	15	922,251,041	(5,084,341)	(1,163,453)	(6,247,794)	(1,537,085)	919,550,503
7/31/2015	16	919,550,503	(5,069,542)	(1,178,252)	(6,247,794)	(1,532,584)	916,839,667
8/31/2015	17	916,839,667	(5,054,684)	(1,193,109)	(6,247,794)	(1,528,066)	914,118,491
9/30/2015	18	914,118,491	(5,039,768)	(1,208,025)	(6,247,794)	(1,523,531)	911,386,935
10/31/2015	19	911,386,935	(5,024,793)	(1,223,000)	(6,247,794)	(1,518,978)	908,644,956
11/30/2015	20	908,644,956	(5,009,759)	(1,238,034)	(6,247,794)	(1,514,408)	905,892,514
12/31/2015	21	905,892,514	(4,994,666)	(1,253,128)	(6,247,794)	(1,509,821)	903,129,565
1/31/2016	22	903,129,565	(4,979,512)	(1,268,281)	(6,247,794)	(1,505,216)	900,356,068
2/29/2016	23	900,356,068	(4,964,299)	(1,283,494)	(6,247,794)	(1,500,593)	897,571,980
3/31/2016	24	897,571,980	(4,949,026)	(1,298,768)	(6,247,794)	(1,495,953)	894,777,259

Summary | Sheet6 | CF Total | 1 | 2 | 3 | 4 | 5

Figure 13.37 Cash Flow Summary – Conversion to Semiannual Periods

this can be done using keyboard shortcuts. Tab over until a cell in column I is active, then hold down Ctrl + Space Bar simultaneously. This will select the entire column. Shift + F10 will produce the same menu as a right click; then use the arrow keys to select the 'Column Width . . .' option.

The ROUNDUP function will be used to produce a '1' next to periods 1 through 6, a '2' next to periods 7 through 12, and so forth. This indicates that monthly periods 1 through 6 combine to create the first semiannual period. The ROUNDUP function will require two arguments: the number which must be rounded up, and the number of digits to which the number is to be rounded. Since each payment period is 6 months, each period will be divided by 6. In period 1, for instance, this would produce a value of 0.1667. The second argument is to how many digits the value is rounded; by entering '0' the value 0.1667 is rounded up to 1. The formula in cell J4 should read "=ROUNDUP(B4/6,0)" (Figure 13.38).

Cell J4 can then be copied into the remaining cells in column J (Figure 13.39).

The next step is to set up a table for the cash flows for each semiannual period. For analysis of the CMO only cash flows will be necessary; beginning and ending balances are not required. Beginning in cell L3, label the columns of the new table with the following: 'Period', 'Semiannual Payment Date', 'Int. PMT', 'Princ. PMT', 'Total PMT', and 'Prepayment' (Figure 13.40).

When creating the waterfall a period 0 will be required. Thus a 0 period with the analysis date should be set up; and the original principal amount will be entered as the principal payment amount (Figure 13.41).

The periods will be completed just as before. In cell L5 the equation is "=L4+1" (Figure 13.42). This can then be copied to complete the column.

The first payment date can then be entered in cell M5 by simply typing 'First_Payment_Date' (Figure 13.43). Since the first pay date was entered on the Summary page, the date automatically populates in cell M5.

CMO Model - Excel

J4 =ROUNDUP(B4/6,0)

Total Cash Flows

Payment Date	Period	BoP	Int. PMT	Princ. PMT	Total PMT	Prepayment	EoP	J
4/20/2014	0	959,000,000	-	-	-	-	959,000,000	=ROUNDUP(B4/6,0)
4/30/2014	1	959,000,000	(5,285,521)	(962,273)	(6,247,794)	(1,598,333)	956,439,394	
5/31/2014	2	956,439,394	(5,271,515)	(976,278)	(6,247,794)	(1,594,066)	953,869,050	
6/30/2014	3	953,869,050	(5,257,455)	(990,339)	(6,247,794)	(1,589,782)	951,288,930	
7/31/2014	4	951,288,930	(5,243,339)	(1,004,455)	(6,247,794)	(1,585,482)	948,698,993	
8/31/2014	5	948,698,993	(5,229,167)	(1,018,626)	(6,247,794)	(1,581,165)	946,099,202	
9/30/2014	6	946,099,202	(5,214,940)	(1,032,854)	(6,247,794)	(1,576,832)	943,489,516	
10/31/2014	7	943,489,516	(5,200,656)	(1,047,137)	(6,247,794)	(1,572,483)	940,869,896	
11/30/2014	8	940,869,896	(5,186,316)	(1,061,477)	(6,247,794)	(1,568,116)	938,240,302	
12/31/2014	9	938,240,302	(5,171,920)	(1,075,874)	(6,247,794)	(1,563,734)	935,600,695	
1/31/2015	10	935,600,695	(5,157,467)	(1,090,327)	(6,247,794)	(1,559,334)	932,951,033	
2/28/2015	11	932,951,033	(5,142,956)	(1,104,837)	(6,247,794)	(1,554,918)	930,291,278	
3/31/2015	12	930,291,278	(5,128,389)	(1,119,405)	(6,247,794)	(1,550,485)	927,621,388	
4/30/2015	13	927,621,388	(5,113,764)	(1,134,030)	(6,247,794)	(1,546,036)	924,941,322	
5/31/2015	14	924,941,322	(5,099,081)	(1,148,712)	(6,247,794)	(1,541,569)	922,251,041	
6/30/2015	15	922,251,041	(5,084,341)	(1,163,453)	(6,247,794)	(1,537,085)	919,550,503	
7/31/2015	16	919,550,503	(5,069,542)	(1,178,252)	(6,247,794)	(1,532,584)	916,839,667	
8/31/2015	17	916,839,667	(5,054,684)	(1,193,109)	(6,247,794)	(1,528,066)	914,118,491	
9/30/2015	18	914,118,491	(5,039,768)	(1,208,025)	(6,247,794)	(1,523,531)	911,386,935	
10/31/2015	19	911,386,935	(5,024,793)	(1,223,000)	(6,247,794)	(1,518,978)	908,644,956	
11/30/2015	20	908,644,956	(5,009,759)	(1,238,034)	(6,247,794)	(1,514,408)	905,892,514	
12/31/2015	21	905,892,514	(4,994,666)	(1,253,128)	(6,247,794)	(1,509,821)	903,129,565	
1/31/2016	22	903,129,565	(4,979,512)	(1,268,281)	(6,247,794)	(1,505,216)	900,356,068	

Summary | Sheet6 | CF Total | 1 | 2 | 3 | 4 | 5

Figure 13.38 Semiannual Periods

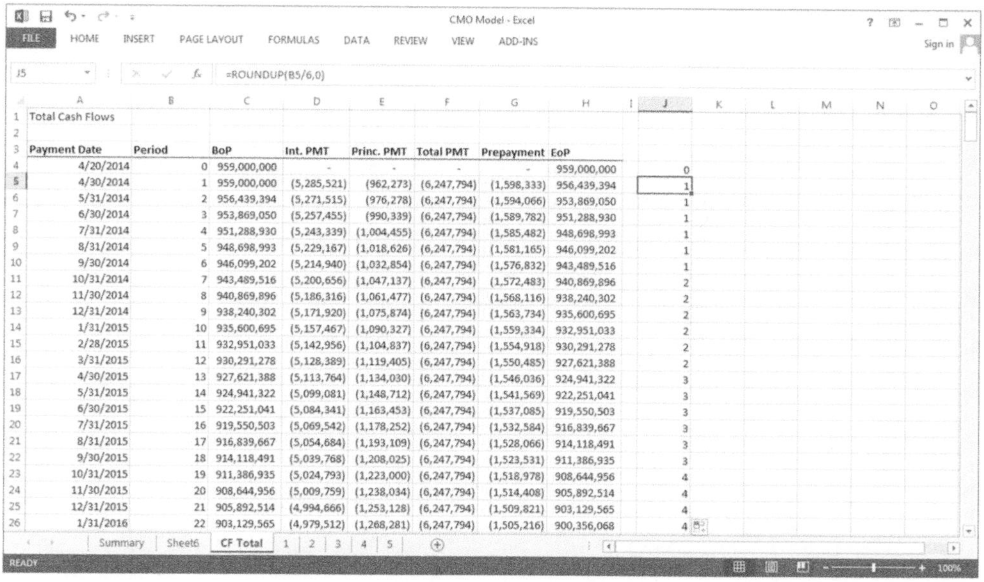

J5: =ROUNDUP(B5/6,0)

Total Cash Flows

Payment Date	Period	BoP	Int. PMT	Princ. PMT	Total PMT	Prepayment	EoP		
4/20/2014	0	959,000,000	-	-	-	-	959,000,000		0
4/30/2014	1	959,000,000	(5,285,521)	(962,273)	(6,247,794)	(1,598,333)	956,439,394		1
5/31/2014	2	956,439,394	(5,271,515)	(976,278)	(6,247,794)	(1,594,066)	953,869,050		1
6/30/2014	3	953,869,050	(5,257,455)	(990,339)	(6,247,794)	(1,589,782)	951,288,930		1
7/31/2014	4	951,288,930	(5,243,339)	(1,004,455)	(6,247,794)	(1,585,482)	948,698,993		1
8/31/2014	5	948,698,993	(5,229,167)	(1,018,626)	(6,247,794)	(1,581,165)	946,099,202		1
9/30/2014	6	946,099,202	(5,214,940)	(1,032,854)	(6,247,794)	(1,576,832)	943,489,516		1
10/31/2014	7	943,489,516	(5,200,656)	(1,047,137)	(6,247,794)	(1,572,483)	940,869,896		2
11/30/2014	8	940,869,896	(5,186,316)	(1,061,477)	(6,247,794)	(1,568,116)	938,240,302		2
12/31/2014	9	938,240,302	(5,171,920)	(1,075,874)	(6,247,794)	(1,563,734)	935,600,695		2
1/31/2015	10	935,600,695	(5,157,467)	(1,090,327)	(6,247,794)	(1,559,334)	932,951,033		2
2/28/2015	11	932,951,033	(5,142,956)	(1,104,837)	(6,247,794)	(1,554,918)	930,291,278		2
3/31/2015	12	930,291,278	(5,128,389)	(1,119,405)	(6,247,794)	(1,550,485)	927,621,388		2
4/30/2015	13	927,621,388	(5,113,764)	(1,134,030)	(6,247,794)	(1,546,036)	924,941,322		3
5/31/2015	14	924,941,322	(5,099,081)	(1,148,712)	(6,247,794)	(1,541,569)	922,251,041		3
6/30/2015	15	922,251,041	(5,084,341)	(1,163,453)	(6,247,794)	(1,537,085)	919,550,503		3
7/31/2015	16	919,550,503	(5,069,542)	(1,178,252)	(6,247,794)	(1,532,584)	916,839,667		3
8/31/2015	17	916,839,667	(5,054,684)	(1,193,109)	(6,247,794)	(1,528,066)	914,118,491		3
9/30/2015	18	914,118,491	(5,039,768)	(1,208,025)	(6,247,794)	(1,523,531)	911,386,935		3
10/31/2015	19	911,386,935	(5,024,793)	(1,223,000)	(6,247,794)	(1,518,978)	908,644,956		4
11/30/2015	20	908,644,956	(5,009,759)	(1,238,034)	(6,247,794)	(1,514,408)	905,892,514		4
12/31/2015	21	905,892,514	(4,994,666)	(1,253,128)	(6,247,794)	(1,509,821)	903,129,565		4
1/31/2016	22	903,129,565	(4,979,512)	(1,268,281)	(6,247,794)	(1,505,216)	900,356,068		4

Figure 13.39 Conversion to Semiannual Periods

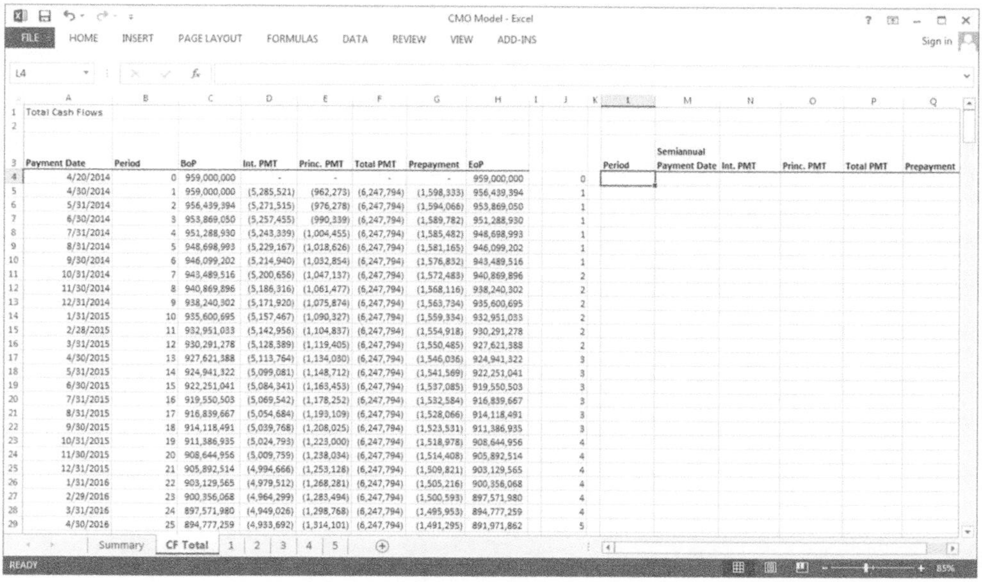

Total Cash Flows

Payment Date	Period	BoP	Int. PMT	Princ. PMT	Total PMT	Prepayment	EoP				Period	Semiannual Payment Date	Int. PMT	Princ. PMT	Total PMT	Prepayment
4/20/2014	0	959,000,000	-	-	-	-	959,000,000		0							
4/30/2014	1	959,000,000	(5,285,521)	(962,273)	(6,247,794)	(1,598,333)	956,439,394		1							
5/31/2014	2	956,439,394	(5,271,515)	(976,278)	(6,247,794)	(1,594,066)	953,869,050		1							
6/30/2014	3	953,869,050	(5,257,455)	(990,339)	(6,247,794)	(1,589,782)	951,288,930		1							
7/31/2014	4	951,288,930	(5,243,339)	(1,004,455)	(6,247,794)	(1,585,482)	948,698,993		1							
8/31/2014	5	948,698,993	(5,229,167)	(1,018,626)	(6,247,794)	(1,581,165)	946,099,202		1							
9/30/2014	6	946,099,202	(5,214,940)	(1,032,854)	(6,247,794)	(1,576,832)	943,489,516		1							
10/31/2014	7	943,489,516	(5,200,656)	(1,047,137)	(6,247,794)	(1,572,483)	940,869,896		2							
11/30/2014	8	940,869,896	(5,186,316)	(1,061,477)	(6,247,794)	(1,568,116)	938,240,302		2							
12/31/2014	9	938,240,302	(5,171,920)	(1,075,874)	(6,247,794)	(1,563,734)	935,600,695		2							
1/31/2015	10	935,600,695	(5,157,467)	(1,090,327)	(6,247,794)	(1,559,334)	932,951,033		2							
2/28/2015	11	932,951,033	(5,142,956)	(1,104,837)	(6,247,794)	(1,554,918)	930,291,278		2							
3/31/2015	12	930,291,278	(5,128,389)	(1,119,405)	(6,247,794)	(1,550,485)	927,621,388		2							
4/30/2015	13	927,621,388	(5,113,764)	(1,134,030)	(6,247,794)	(1,546,036)	924,941,322		3							
5/31/2015	14	924,941,322	(5,099,081)	(1,148,712)	(6,247,794)	(1,541,569)	922,251,041		3							
6/30/2015	15	922,251,041	(5,084,341)	(1,163,453)	(6,247,794)	(1,537,085)	919,550,503		3							
7/31/2015	16	919,550,503	(5,069,542)	(1,178,252)	(6,247,794)	(1,532,584)	916,839,667		3							
8/31/2015	17	916,839,667	(5,054,684)	(1,193,109)	(6,247,794)	(1,528,066)	914,118,491		3							
9/30/2015	18	914,118,491	(5,039,768)	(1,208,025)	(6,247,794)	(1,523,531)	911,386,935		3							
10/31/2015	19	911,386,935	(5,024,793)	(1,223,000)	(6,247,794)	(1,518,978)	908,644,956		4							
11/30/2015	20	908,644,956	(5,009,759)	(1,238,034)	(6,247,794)	(1,514,408)	905,892,514		4							
12/31/2015	21	905,892,514	(4,994,666)	(1,253,128)	(6,247,794)	(1,509,821)	903,129,565		4							
1/31/2016	22	903,129,565	(4,979,512)	(1,268,281)	(6,247,794)	(1,505,216)	900,356,068		4							
2/29/2016	23	900,356,068	(4,964,299)	(1,283,494)	(6,247,794)	(1,500,593)	897,571,980		4							
3/31/2016	24	897,571,980	(4,949,026)	(1,298,768)	(6,247,794)	(1,495,953)	894,777,259		4							
4/30/2016	25	894,777,259	(4,933,692)	(1,314,101)	(6,247,794)	(1,491,295)	891,971,862		5							

Figure 13.40 Conversion to Semiannual Periods

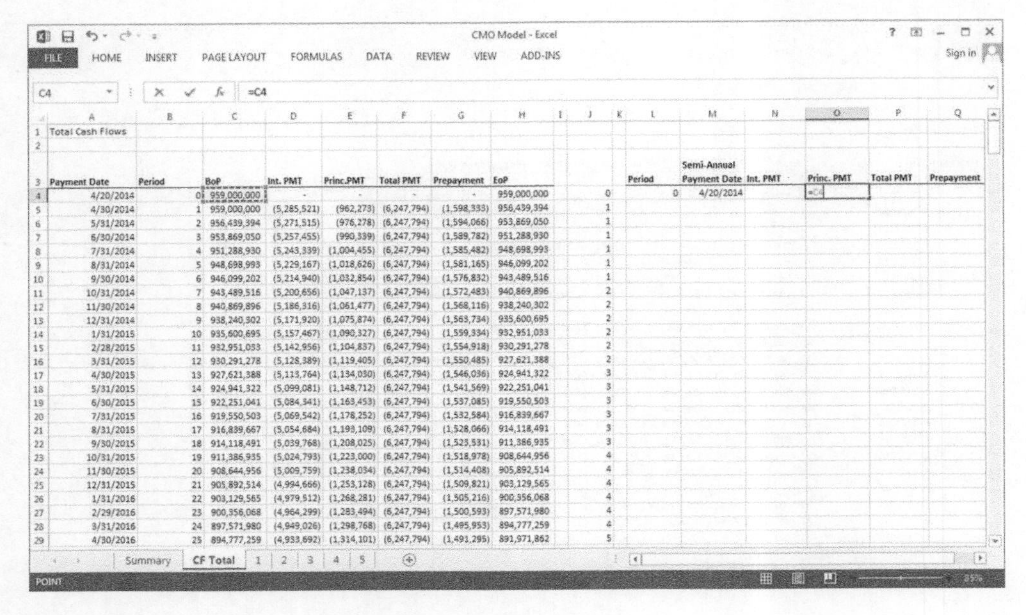

CMO Model - Excel

FILE HOME INSERT PAGE LAYOUT FORMULAS DATA REVIEW VIEW ADD-INS

C4 =C4

	A	B	C	D	E	F	G	H	J	L	M	N	O	P	Q
1	Total Cash Flows														
2															
3	Payment Date	Period	BoP	Int. PMT	Princ.PMT	Total PMT	Prepayment	EoP		Period	Semi-Annual Payment Date	Int. PMT	Princ. PMT	Total PMT	Prepayment
4	4/20/2014	0	959,000,000	-	-	-	-	959,000,000	0	0	4/20/2014		=C4		
5	4/30/2014	1	959,000,000	(5,285,521)	(962,273)	(6,247,794)	(1,598,333)	956,439,394	1						
6	5/31/2014	2	956,439,394	(5,271,515)	(976,278)	(6,247,794)	(1,594,066)	953,869,050	1						
7	6/30/2014	3	953,869,050	(5,257,455)	(990,339)	(6,247,794)	(1,589,782)	951,288,930	1						
8	7/31/2014	4	951,288,930	(5,243,339)	(1,004,455)	(6,247,794)	(1,585,482)	948,698,993	1						
9	8/31/2014	5	948,698,993	(5,229,167)	(1,018,626)	(6,247,794)	(1,581,165)	946,099,202	1						
10	9/30/2014	6	946,099,202	(5,214,940)	(1,032,854)	(6,247,794)	(1,576,832)	943,489,516	1						
11	10/31/2014	7	943,489,516	(5,200,656)	(1,047,137)	(6,247,794)	(1,572,483)	940,869,896	2						
12	11/30/2014	8	940,869,896	(5,186,316)	(1,061,477)	(6,247,794)	(1,568,116)	938,240,302	2						
13	12/31/2014	9	938,240,302	(5,171,920)	(1,075,874)	(6,247,794)	(1,563,734)	935,600,695	2						
14	1/31/2015	10	935,600,695	(5,157,467)	(1,090,327)	(6,247,794)	(1,559,334)	932,951,033	2						
15	2/28/2015	11	932,951,033	(5,142,956)	(1,104,837)	(6,247,794)	(1,554,918)	930,291,278	2						
16	3/31/2015	12	930,291,278	(5,128,389)	(1,119,405)	(6,247,794)	(1,550,485)	927,621,388	2						
17	4/30/2015	13	927,621,388	(5,113,764)	(1,134,030)	(6,247,794)	(1,546,036)	924,941,322	3						
18	5/31/2015	14	924,941,322	(5,099,081)	(1,148,712)	(6,247,794)	(1,541,569)	922,251,041	3						
19	6/30/2015	15	922,251,041	(5,084,341)	(1,163,453)	(6,247,794)	(1,537,085)	919,550,503	3						
20	7/31/2015	16	919,550,503	(5,069,542)	(1,178,252)	(6,247,794)	(1,532,584)	916,839,667	3						
21	8/31/2015	17	916,839,667	(5,054,684)	(1,193,109)	(6,247,794)	(1,528,066)	914,118,491	3						
22	9/30/2015	18	914,118,491	(5,039,768)	(1,208,025)	(6,247,794)	(1,523,531)	911,386,935	3						
23	10/31/2015	19	911,386,935	(5,024,793)	(1,223,000)	(6,247,794)	(1,518,978)	908,644,956	4						
24	11/30/2015	20	908,644,956	(5,009,759)	(1,238,034)	(6,247,794)	(1,514,408)	905,892,514	4						
25	12/31/2015	21	905,892,514	(4,994,666)	(1,253,128)	(6,247,794)	(1,509,821)	903,129,565	4						
26	1/31/2016	22	903,129,565	(4,979,512)	(1,268,281)	(6,247,794)	(1,505,216)	900,356,068	4						
27	2/29/2016	23	900,356,068	(4,964,299)	(1,283,494)	(6,247,794)	(1,500,593)	897,571,980	4						
28	3/31/2016	24	897,571,980	(4,949,026)	(1,298,768)	(6,247,794)	(1,495,953)	894,777,259	4						
29	4/30/2016	25	894,777,259	(4,933,692)	(1,314,101)	(6,247,794)	(1,491,295)	891,971,862	5						

Summary | CF Total | 1 | 2 | 3 | 4 | 5

POINT

Figure 13.41 Principal Payment at Period 0 (Analysis Date)

CMO Model - Excel

FILE HOME INSERT PAGE LAYOUT FORMULAS DATA REVIEW VIEW ADD-INS

L4 =L4+1

	A	B	C	D	E	F	G	H	J	L	M	N	O	P	Q
1	Total Cash Flows														
2															
3	Payment Date	Period	BoP	Int. PMT	Princ.PMT	Total PMT	Prepayment	EoP		Period	Semi-Annual Payment Date	Int. PMT	Princ. PMT	Total PMT	Prepayment
4	4/20/2014	0	959,000,000	-	-	-	-	959,000,000	0	0	4/20/2014		959,000,000		
5	4/30/2014	1	959,000,000	(5,285,521)	(962,273)	(6,247,794)	(1,598,333)	956,439,394	1	=L4+1					
6	5/31/2014	2	956,439,394	(5,271,515)	(976,278)	(6,247,794)	(1,594,066)	953,869,050	1						
7	6/30/2014	3	953,869,050	(5,257,455)	(990,339)	(6,247,794)	(1,589,782)	951,288,930	1						
8	7/31/2014	4	951,288,930	(5,243,339)	(1,004,455)	(6,247,794)	(1,585,482)	948,698,993	1						
9	8/31/2014	5	948,698,993	(5,229,167)	(1,018,626)	(6,247,794)	(1,581,165)	946,099,202	1						
10	9/30/2014	6	946,099,202	(5,214,940)	(1,032,854)	(6,247,794)	(1,576,832)	943,489,516	1						
11	10/31/2014	7	943,489,516	(5,200,656)	(1,047,137)	(6,247,794)	(1,572,483)	940,869,896	2						
12	11/30/2014	8	940,869,896	(5,186,316)	(1,061,477)	(6,247,794)	(1,568,116)	938,240,302	2						
13	12/31/2014	9	938,240,302	(5,171,920)	(1,075,874)	(6,247,794)	(1,563,734)	935,600,695	2						
14	1/31/2015	10	935,600,695	(5,157,467)	(1,090,327)	(6,247,794)	(1,559,334)	932,951,033	2						
15	2/28/2015	11	932,951,033	(5,142,956)	(1,104,837)	(6,247,794)	(1,554,918)	930,291,278	2						
16	3/31/2015	12	930,291,278	(5,128,389)	(1,119,405)	(6,247,794)	(1,550,485)	927,621,388	2						
17	4/30/2015	13	927,621,388	(5,113,764)	(1,134,030)	(6,247,794)	(1,546,036)	924,941,322	3						
18	5/31/2015	14	924,941,322	(5,099,081)	(1,148,712)	(6,247,794)	(1,541,569)	922,251,041	3						
19	6/30/2015	15	922,251,041	(5,084,341)	(1,163,453)	(6,247,794)	(1,537,085)	919,550,503	3						
20	7/31/2015	16	919,550,503	(5,069,542)	(1,178,252)	(6,247,794)	(1,532,584)	916,839,667	3						
21	8/31/2015	17	916,839,667	(5,054,684)	(1,193,109)	(6,247,794)	(1,528,066)	914,118,491	3						
22	9/30/2015	18	914,118,491	(5,039,768)	(1,208,025)	(6,247,794)	(1,523,531)	911,386,935	3						
23	10/31/2015	19	911,386,935	(5,024,793)	(1,223,000)	(6,247,794)	(1,518,978)	908,644,956	4						
24	11/30/2015	20	908,644,956	(5,009,759)	(1,238,034)	(6,247,794)	(1,514,408)	905,892,514	4						
25	12/31/2015	21	905,892,514	(4,994,666)	(1,253,128)	(6,247,794)	(1,509,821)	903,129,565	4						
26	1/31/2016	22	903,129,565	(4,979,512)	(1,268,281)	(6,247,794)	(1,505,216)	900,356,068	4						
27	2/29/2016	23	900,356,068	(4,964,299)	(1,283,494)	(6,247,794)	(1,500,593)	897,571,980	4						
28	3/31/2016	24	897,571,980	(4,949,026)	(1,298,768)	(6,247,794)	(1,495,953)	894,777,259	4						
29	4/30/2016	25	894,777,259	(4,933,692)	(1,314,101)	(6,247,794)	(1,491,295)	891,971,862	5						

Summary | CF Total | 1 | 2 | 3 | 4 | 5

ENTER

Figure 13.42 Semiannual Periods

CMO Model - Excel

IFERROR | =First

Payment Date	Period	BoP	Int. PMT	Princ.PMT	Total PMT	Prepayment	EoP		Period	Semi-Annual Payment Date	Int. PMT	Princ. PMT	Total PMT	Prepayment
4/20/2014	0	959,000,000	-	-	-	-	959,000,000	0	0	4/20/2014		959,000,000		
4/30/2014	1	959,000,000	(5,285,521)	(962,273)	(6,247,794)	(1,598,333)	956,439,394	1	1	=First				
5/31/2014	2	956,439,394	(5,271,515)	(976,278)	(6,247,794)	(1,594,066)	953,869,050	1	2	First_Payment_Date				
6/30/2014	3	953,869,050	(5,257,455)	(990,339)	(6,247,794)	(1,589,782)	951,288,930	1	3					
7/31/2014	4	951,288,930	(5,243,339)	(1,004,455)	(6,247,794)	(1,585,482)	948,698,993	1	4					
8/31/2014	5	948,698,993	(5,229,167)	(1,018,626)	(6,247,794)	(1,581,165)	946,099,202	1	5					
9/30/2014	6	946,099,202	(5,214,940)	(1,032,854)	(6,247,794)	(1,576,832)	943,489,516	1	6					
10/31/2014	7	943,489,516	(5,200,656)	(1,047,137)	(6,247,794)	(1,572,483)	940,869,896	2	7					
11/30/2014	8	940,869,896	(5,186,316)	(1,061,477)	(6,247,794)	(1,568,116)	938,240,302	2	8					
12/31/2014	9	938,240,302	(5,171,920)	(1,075,874)	(6,247,794)	(1,563,734)	935,600,695	2	9					
1/31/2015	10	935,600,695	(5,157,467)	(1,090,327)	(6,247,794)	(1,559,334)	932,951,033	2	10					
2/28/2015	11	932,951,033	(5,142,956)	(1,104,837)	(6,247,794)	(1,554,918)	930,291,278	2	11					
3/31/2015	12	930,291,278	(5,128,389)	(1,119,405)	(6,247,794)	(1,550,485)	927,621,388	2	12					
4/30/2015	13	927,621,388	(5,113,764)	(1,134,030)	(6,247,794)	(1,546,036)	924,941,322	3	13					
5/31/2015	14	924,941,322	(5,099,081)	(1,148,712)	(6,247,794)	(1,541,569)	922,251,041	3	14					
6/30/2015	15	922,251,041	(5,084,341)	(1,163,453)	(6,247,794)	(1,537,085)	919,550,503	3	15					
7/31/2015	16	919,550,503	(5,069,542)	(1,178,252)	(6,247,794)	(1,532,584)	916,839,667	3	16					
8/31/2015	17	916,839,667	(5,054,684)	(1,193,109)	(6,247,794)	(1,528,066)	914,118,491	3	17					
9/30/2015	18	914,118,491	(5,039,768)	(1,208,025)	(6,247,794)	(1,523,531)	911,386,935	3	18					
10/31/2015	19	911,386,935	(5,024,793)	(1,223,000)	(6,247,794)	(1,518,978)	908,644,956	4	19					
11/30/2015	20	908,644,956	(5,009,759)	(1,238,034)	(6,247,794)	(1,514,408)	905,892,514	4	20					
12/31/2015	21	905,892,514	(4,994,666)	(1,253,128)	(6,247,794)	(1,509,821)	903,129,565	4	21					
1/31/2016	22	903,129,565	(4,979,512)	(1,268,281)	(6,247,794)	(1,505,216)	900,356,068	4	22					
2/29/2016	23	900,356,068	(4,964,299)	(1,283,494)	(6,247,794)	(1,500,593)	897,571,980	4	23					
3/31/2016	24	897,571,980	(4,949,026)	(1,298,768)	(6,247,794)	(1,495,953)	894,777,259	4	24					
4/30/2016	25	894,777,259	(4,933,692)	(1,314,101)	(6,247,794)	(1,491,295)	891,971,862	5	25					

Summary | CF Total | 1 | 2 | 3 | 4 | 5

Figure 13.43 First Semiannual Payment

CMO Model - Excel

M6 | =EOMONTH(M5,6)

Payment Date	Period	BoP	Int. PMT	Princ.PMT	Total PMT	Prepayment	EoP		Period	Semi-Annual Payment Date	Int. PMT	Princ. PMT	Total PMT	Prepayment
4/20/2014	0	959,000,000	-	-	-	-	959,000,000	0	0	4/20/2014		959,000,000		
4/30/2014	1	959,000,000	(5,285,521)	(962,273)	(6,247,794)	(1,598,333)	956,439,394	1	1	9/30/2014				
5/31/2014	2	956,439,394	(5,271,515)	(976,278)	(6,247,794)	(1,594,066)	953,869,050	1	2	=EOMONTH(M5,6)				
6/30/2014	3	953,869,050	(5,257,455)	(990,339)	(6,247,794)	(1,589,782)	951,288,930	1	3					
7/31/2014	4	951,288,930	(5,243,339)	(1,004,455)	(6,247,794)	(1,585,482)	948,698,993	1	4					
8/31/2014	5	948,698,993	(5,229,167)	(1,018,626)	(6,247,794)	(1,581,165)	946,099,202	1	5					
9/30/2014	6	946,099,202	(5,214,940)	(1,032,854)	(6,247,794)	(1,576,832)	943,489,516	1	6					
10/31/2014	7	943,489,516	(5,200,656)	(1,047,137)	(6,247,794)	(1,572,483)	940,869,896	2	7					
11/30/2014	8	940,869,896	(5,186,316)	(1,061,477)	(6,247,794)	(1,568,116)	938,240,302	2	8					
12/31/2014	9	938,240,302	(5,171,920)	(1,075,874)	(6,247,794)	(1,563,734)	935,600,695	2	9					
1/31/2015	10	935,600,695	(5,157,467)	(1,090,327)	(6,247,794)	(1,559,334)	932,951,033	2	10					
2/28/2015	11	932,951,033	(5,142,956)	(1,104,837)	(6,247,794)	(1,554,918)	930,291,278	2	11					
3/31/2015	12	930,291,278	(5,128,389)	(1,119,405)	(6,247,794)	(1,550,485)	927,621,388	2	12					
4/30/2015	13	927,621,388	(5,113,764)	(1,134,030)	(6,247,794)	(1,546,036)	924,941,322	3	13					
5/31/2015	14	924,941,322	(5,099,081)	(1,148,712)	(6,247,794)	(1,541,569)	922,251,041	3	14					
6/30/2015	15	922,251,041	(5,084,341)	(1,163,453)	(6,247,794)	(1,537,085)	919,550,503	3	15					
7/31/2015	16	919,550,503	(5,069,542)	(1,178,252)	(6,247,794)	(1,532,584)	916,839,667	3	16					
8/31/2015	17	916,839,667	(5,054,684)	(1,193,109)	(6,247,794)	(1,528,066)	914,118,491	3	17					
9/30/2015	18	914,118,491	(5,039,768)	(1,208,025)	(6,247,794)	(1,523,531)	911,386,935	3	18					
10/31/2015	19	911,386,935	(5,024,793)	(1,223,000)	(6,247,794)	(1,518,978)	908,644,956	4	19					
11/30/2015	20	908,644,956	(5,009,759)	(1,238,034)	(6,247,794)	(1,514,408)	905,892,514	4	20					
12/31/2015	21	905,892,514	(4,994,666)	(1,253,128)	(6,247,794)	(1,509,821)	903,129,565	4	21					
1/31/2016	22	903,129,565	(4,979,512)	(1,268,281)	(6,247,794)	(1,505,216)	900,356,068	4	22					
2/29/2016	23	900,356,068	(4,964,299)	(1,283,494)	(6,247,794)	(1,500,593)	897,571,980	4	23					
3/31/2016	24	897,571,980	(4,949,026)	(1,298,768)	(6,247,794)	(1,495,953)	894,777,259	4	24					
4/30/2016	25	894,777,259	(4,933,692)	(1,314,101)	(6,247,794)	(1,491,295)	891,971,862	5	25					

Summary | CF Total | 1 | 2 | 3 | 4 | 5

Figure 13.44 Semiannual Payment Dates

The remaining months may be entered by using the EOMONTH function as was used previously (Figure 13.44). Remember that the second argument will be a 6 because the payments take place semiannually, or every 6 months. The column can be completed by copying the formula in cell M6 to the remainder of the column.

CMO Model - Excel

D5 =SUMIF($J$5:$J$364,$L5,D$5:D$364

Total Cash Flows

Payment Date	Period	BoP	Int. PMT	Princ.PMT	Total PMT	Prepayment	EoP		Period	Semi-Annual Payment Date	Int. PMT	Princ. PMT	Total PMT	Prepayment
4/20/2014	0	959,000,000	-	-	-	-	959,000,000	0	0	4/20/2014		959,000,000		
4/30/2014	1	959,000,000	(5,285,521)	(962,273)	(6,247,794)	(1,598,333)	956,439,394	1	1	9/30/2014	=SUMIF($J$5:$J$364,$L5,D$5:D$364			
5/31/2014	2	956,439,394	(5,271,515)	(976,278)	(6,247,794)	(1,594,066)	953,869,050	1	2	3/31/2015	SUMIF(range, criteria, [sum_range])			
6/30/2014	3	953,869,050	(5,257,455)	(990,339)	(6,247,794)	(1,589,782)	951,288,930	1	3	9/30/2015				
7/31/2014	4	951,288,930	(5,243,339)	(1,004,455)	(6,247,794)	(1,585,482)	948,698,993	1	4	3/31/2016				
8/31/2014	5	948,698,993	(5,229,167)	(1,018,626)	(6,247,794)	(1,581,165)	946,099,202	1	5	9/30/2016				
9/30/2014	6	946,099,202	(5,214,940)	(1,032,854)	(6,247,794)	(1,576,832)	943,489,516	1	6	3/31/2017				
10/31/2014	7	943,489,516	(5,200,656)	(1,047,137)	(6,247,794)	(1,572,483)	940,869,896	2	7	9/30/2017				
11/30/2014	8	940,869,896	(5,186,316)	(1,061,477)	(6,247,794)	(1,568,116)	938,240,302	2	8	3/31/2018				
12/31/2014	9	938,240,302	(5,171,920)	(1,075,874)	(6,247,794)	(1,563,734)	935,600,695	2	9	9/30/2018				
1/31/2015	10	935,600,695	(5,157,467)	(1,090,327)	(6,247,794)	(1,559,334)	932,951,033	2	10	3/31/2019				
2/28/2015	11	932,951,033	(5,142,956)	(1,104,837)	(6,247,794)	(1,554,918)	930,291,278	2	11	9/30/2019				
3/31/2015	12	930,291,278	(5,128,389)	(1,119,405)	(6,247,794)	(1,550,485)	927,621,388	2	12	3/31/2020				
4/30/2015	13	927,621,388	(5,113,764)	(1,134,030)	(6,247,794)	(1,546,036)	924,941,322	3	13	9/30/2020				
5/31/2015	14	924,941,322	(5,099,081)	(1,148,712)	(6,247,794)	(1,541,569)	922,251,041	3	14	3/31/2021				
6/30/2015	15	922,251,041	(5,084,341)	(1,163,453)	(6,247,794)	(1,537,085)	919,550,503	3	15	9/30/2021				
7/31/2015	16	919,550,503	(5,069,542)	(1,178,252)	(6,247,794)	(1,532,584)	916,839,667	3	16	3/31/2022				
8/31/2015	17	916,839,667	(5,054,684)	(1,193,109)	(6,247,794)	(1,528,066)	914,118,491	3	17	9/30/2022				
9/30/2015	18	914,118,491	(5,039,768)	(1,208,025)	(6,247,794)	(1,523,531)	911,386,935	3	18	3/31/2023				
10/31/2015	19	911,386,935	(5,024,793)	(1,223,000)	(6,247,794)	(1,518,978)	908,644,956	4	19	9/30/2023				
11/30/2015	20	908,644,956	(5,009,759)	(1,238,034)	(6,247,794)	(1,514,408)	905,892,514	4	20	3/31/2024				
12/31/2015	21	905,892,514	(4,994,666)	(1,253,128)	(6,247,794)	(1,509,821)	903,129,565	4	21	9/30/2024				
1/31/2016	22	903,129,565	(4,979,512)	(1,268,281)	(6,247,794)	(1,505,216)	900,356,068	4	22	3/31/2025				
2/29/2016	23	900,356,068	(4,964,299)	(1,283,494)	(6,247,794)	(1,500,593)	897,571,980	4	23	9/30/2025				
3/31/2016	24	897,571,980	(4,949,026)	(1,298,768)	(6,247,794)	(1,495,953)	894,777,259	4	24	3/31/2026				
4/30/2016	25	894,777,259	(4,933,692)	(1,314,101)	(6,247,794)	(1,491,295)	891,971,862	5	25	9/30/2026				

Summary | CF Total | 1 | 2 | 3 | 4 | 5

Figure 13.45 Using SUMIF Function to Determine Semiannual Interest Payments

The interest payment amount to be filled in is the sum of the principal payment amounts from the first 6 months. These will be added automatically using the SUMIF function. The idea is that some particular values will be summed *if* a particular argument is true. There are three arguments to be entered into a SUMIF function. The first argument is the range; this refers to the cells to be evaluated. In this case that would be column J, or cells J5 to J364 (Figure 13.45). Criteria is the second argument. The period listed in the L column is what one is trying to match, so the criteria for period 1 would be $L4. The $ is inserted in front of the L so that the equation may be copied to other columns. The third argument is the Sum_ range, the data that will be summed. In this case that is the data in the 'Princ. PMT' column, or cells D5 through D364. The argument is stating: sum any data in cells D5 through D364, if the corresponding value in cells J5 through J364 is equal to the value in cell L5. The equation in cell N5 is, therefore, "=SUMIF($J$5:$J$364,$L5,D$5:D$364)".

The equation in cell N5 can then be copied to cells O5, P5, and Q5; and those may be copied to the entire row to complete the table (Figure 13.46). Since the direct references were used, the equation should adjust accordingly. That is, the column in the 'Sum_range' argument should update.

Once the table is complete, these numbers will be used to create the waterfall. Select cells L3 through Q64 and name the entire array 'Total_Cash_Flow' (Figure 13.47).

Waterfall

Begin the waterfall by opening a blank sheet, renaming the sheet, and typing the title in cell A1. See Figure 13.48 for the model structure.

Waterfalls are more easily followed when the cash flow is shown horizontally. Thus, the periods will be listed in row 2 instead of a column. Column A will be used for titles, column B will be used for various information, and column C will be used for totals. Start by typing 'Period' in cell C2 and enter a '0' into cell D2 (Figure 13.49). Just as before, cell E2 will be "=D2+1".

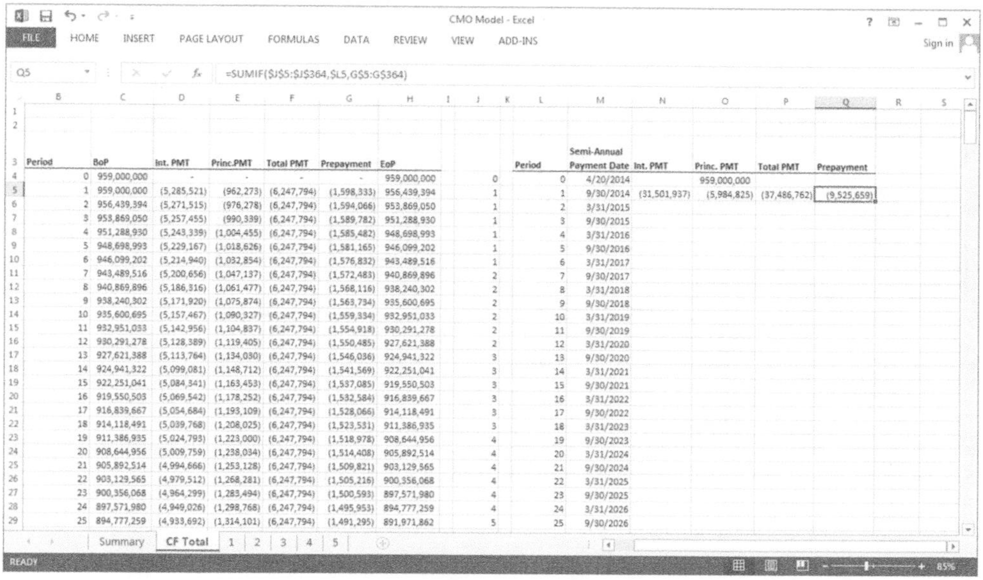

CMO Model - Excel

Q5 =SUMIF($J$5:$J$364,$L5,G$5:G$364)

Period	BoP	Int. PMT	Princ.PMT	Total PMT	Prepayment	EoP	
0	959,000,000	-	-	-	-	959,000,000	0
1	959,000,000	(5,285,521)	(962,273)	(6,247,794)	(1,598,333)	956,439,394	1
2	956,439,394	(5,271,515)	(976,278)	(6,247,794)	(1,594,066)	953,869,050	1
3	953,869,050	(5,257,455)	(990,339)	(6,247,794)	(1,589,782)	951,288,930	1
4	951,288,930	(5,243,339)	(1,004,455)	(6,247,794)	(1,585,482)	948,698,993	1
5	948,698,993	(5,229,167)	(1,018,626)	(6,247,794)	(1,581,165)	946,099,202	1
6	946,099,202	(5,214,940)	(1,032,854)	(6,247,794)	(1,576,832)	943,489,516	1
7	943,489,516	(5,200,656)	(1,047,137)	(6,247,794)	(1,572,483)	940,869,896	2
8	940,869,896	(5,186,316)	(1,061,477)	(6,247,794)	(1,568,116)	938,240,302	2
9	938,240,302	(5,171,920)	(1,075,874)	(6,247,794)	(1,563,734)	935,600,695	2
10	935,600,695	(5,157,467)	(1,090,327)	(6,247,794)	(1,559,334)	932,951,033	2
11	932,951,033	(5,142,956)	(1,104,837)	(6,247,794)	(1,554,918)	930,291,278	2
12	930,291,278	(5,128,389)	(1,119,405)	(6,247,794)	(1,550,485)	927,621,388	2
13	927,621,388	(5,113,764)	(1,134,030)	(6,247,794)	(1,546,036)	924,941,322	3
14	924,941,322	(5,099,081)	(1,148,712)	(6,247,794)	(1,541,569)	922,251,041	3
15	922,251,041	(5,084,341)	(1,163,453)	(6,247,794)	(1,537,085)	919,550,503	3
16	919,550,503	(5,069,542)	(1,178,252)	(6,247,794)	(1,532,584)	916,839,667	3
17	916,839,667	(5,054,684)	(1,193,109)	(6,247,794)	(1,528,066)	914,118,491	3
18	914,118,491	(5,039,768)	(1,208,025)	(6,247,794)	(1,523,531)	911,386,935	3
19	911,386,935	(5,024,793)	(1,223,000)	(6,247,794)	(1,518,978)	908,644,956	4
20	908,644,956	(5,009,759)	(1,238,034)	(6,247,794)	(1,514,408)	905,892,514	4
21	905,892,514	(4,994,666)	(1,253,128)	(6,247,794)	(1,509,821)	903,129,565	4
22	903,129,565	(4,979,512)	(1,268,281)	(6,247,794)	(1,505,216)	900,356,068	4
23	900,356,068	(4,964,299)	(1,283,494)	(6,247,794)	(1,500,593)	897,571,980	4
24	897,571,980	(4,949,026)	(1,298,768)	(6,247,794)	(1,495,953)	894,777,259	4
25	894,777,259	(4,933,692)	(1,314,101)	(6,247,794)	(1,491,295)	891,971,862	5

Period	Semi-Annual Payment Date	Int. PMT	Princ. PMT	Total PMT	Prepayment
0	4/20/2014		959,000,000		
1	9/30/2014	(31,501,937)	(5,984,825)	(37,486,762)	(9,525,659)
2	3/31/2015				
3	9/30/2015				
4	3/31/2016				
5	9/30/2016				
6	3/31/2017				
7	9/30/2017				
8	3/31/2018				
9	9/30/2018				
10	3/31/2019				
11	9/30/2019				
12	3/31/2020				
13	9/30/2020				
14	3/31/2021				
15	9/30/2021				
16	3/31/2022				
17	9/30/2022				
18	3/31/2023				
19	9/30/2023				
20	3/31/2024				
21	9/30/2024				
22	3/31/2025				
23	9/30/2025				
24	3/31/2026				
25	9/30/2026				

Summary | CF Total | 1 | 2 | 3 | 4 | 5

Figure 13.46 Principal Payment, Total Payment, and Prepayment Values for Period 1

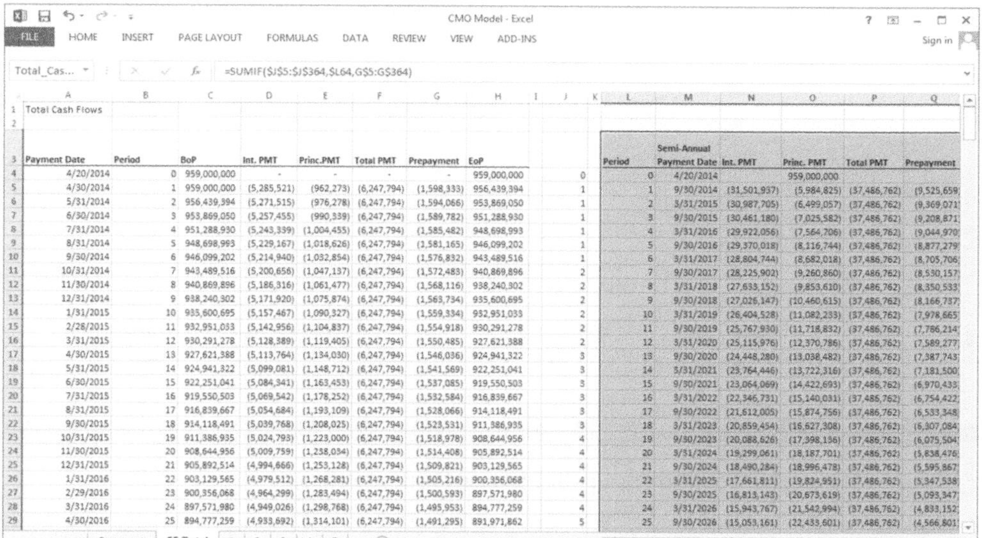

CMO Model - Excel

Total_Cas... =SUMIF($J$5:$J$364,$L64,G$5:G$364)

Total Cash Flows

Payment Date	Period	BoP	Int. PMT	Princ.PMT	Total PMT	Prepayment	EoP	
4/20/2014	0	959,000,000	-	-	-	-	959,000,000	0
4/30/2014	1	959,000,000	(5,285,521)	(962,273)	(6,247,794)	(1,598,333)	956,439,394	1
5/31/2014	2	956,439,394	(5,271,515)	(976,278)	(6,247,794)	(1,594,066)	953,869,050	1
6/30/2014	3	953,869,050	(5,257,455)	(990,339)	(6,247,794)	(1,589,782)	951,288,930	1
7/31/2014	4	951,288,930	(5,243,339)	(1,004,455)	(6,247,794)	(1,585,482)	948,698,993	1
8/31/2014	5	948,698,993	(5,229,167)	(1,018,626)	(6,247,794)	(1,581,165)	946,099,202	1
9/30/2014	6	946,099,202	(5,214,940)	(1,032,854)	(6,247,794)	(1,576,832)	943,489,516	1
10/31/2014	7	943,489,516	(5,200,656)	(1,047,137)	(6,247,794)	(1,572,483)	940,869,896	2
11/30/2014	8	940,869,896	(5,186,316)	(1,061,477)	(6,247,794)	(1,568,116)	938,240,302	2
12/31/2014	9	938,240,302	(5,171,920)	(1,075,874)	(6,247,794)	(1,563,734)	935,600,695	2
1/31/2015	10	935,600,695	(5,157,467)	(1,090,327)	(6,247,794)	(1,559,334)	932,951,033	2
2/28/2015	11	932,951,033	(5,142,956)	(1,104,837)	(6,247,794)	(1,554,918)	930,291,278	2
3/31/2015	12	930,291,278	(5,128,389)	(1,119,405)	(6,247,794)	(1,550,485)	927,621,388	2
4/30/2015	13	927,621,388	(5,113,764)	(1,134,030)	(6,247,794)	(1,546,036)	924,941,322	3
5/31/2015	14	924,941,322	(5,099,081)	(1,148,712)	(6,247,794)	(1,541,569)	922,251,041	3
6/30/2015	15	922,251,041	(5,084,341)	(1,163,453)	(6,247,794)	(1,537,085)	919,550,503	3
7/31/2015	16	919,550,503	(5,069,542)	(1,178,252)	(6,247,794)	(1,532,584)	916,839,667	3
8/31/2015	17	916,839,667	(5,054,684)	(1,193,109)	(6,247,794)	(1,528,066)	914,118,491	3
9/30/2015	18	914,118,491	(5,039,768)	(1,208,025)	(6,247,794)	(1,523,531)	911,386,935	3
10/31/2015	19	911,386,935	(5,024,793)	(1,223,000)	(6,247,794)	(1,518,978)	908,644,956	4
11/30/2015	20	908,644,956	(5,009,759)	(1,238,034)	(6,247,794)	(1,514,408)	905,892,514	4
12/31/2015	21	905,892,514	(4,994,666)	(1,253,128)	(6,247,794)	(1,509,821)	903,129,565	4
1/31/2016	22	903,129,565	(4,979,512)	(1,268,281)	(6,247,794)	(1,505,216)	900,356,068	4
2/29/2016	23	900,356,068	(4,964,299)	(1,283,494)	(6,247,794)	(1,500,593)	897,571,980	4
3/31/2016	24	897,571,980	(4,949,026)	(1,298,768)	(6,247,794)	(1,495,953)	894,777,259	4
4/30/2016	25	894,777,259	(4,933,692)	(1,314,101)	(6,247,794)	(1,491,295)	891,971,862	5

Period	Semi-Annual Payment Date	Int. PMT	Princ. PMT	Total PMT	Prepayment
0	4/20/2014		959,000,000		
1	9/30/2014	(31,501,937)	(5,984,825)	(37,486,762)	(9,525,659
2	3/31/2015	(30,987,705)	(6,499,057)	(37,486,762)	(9,369,071
3	9/30/2015	(30,461,180)	(7,025,582)	(37,486,762)	(9,208,871
4	3/31/2016	(29,922,056)	(7,564,706)	(37,486,762)	(9,044,970
5	9/30/2016	(29,370,018)	(8,116,744)	(37,486,762)	(8,877,279
6	3/31/2017	(28,804,744)	(8,682,018)	(37,486,762)	(8,705,706
7	9/30/2017	(28,225,902)	(9,260,860)	(37,486,762)	(8,530,157
8	3/31/2018	(27,633,152)	(9,853,610)	(37,486,762)	(8,350,533
9	9/30/2018	(27,026,147)	(10,460,615)	(37,486,762)	(8,166,737
10	3/31/2019	(26,404,528)	(11,082,233)	(37,486,762)	(7,978,665
11	9/30/2019	(25,767,930)	(11,718,832)	(37,486,762)	(7,786,214
12	3/31/2020	(25,115,976)	(12,370,786)	(37,486,762)	(7,589,277
13	9/30/2020	(24,448,280)	(13,038,482)	(37,486,762)	(7,387,743
14	3/31/2021	(23,764,446)	(13,722,316)	(37,486,762)	(7,181,500
15	9/30/2021	(23,064,069)	(14,422,693)	(37,486,762)	(6,970,433
16	3/31/2022	(22,346,731)	(15,140,031)	(37,486,762)	(6,754,422
17	9/30/2022	(21,612,005)	(15,874,756)	(37,486,762)	(6,533,348
18	3/31/2023	(20,859,454)	(16,627,308)	(37,486,762)	(6,307,084
19	9/30/2023	(20,088,626)	(17,398,136)	(37,486,762)	(6,075,504
20	3/31/2024	(19,299,061)	(18,187,701)	(37,486,762)	(5,838,476
21	9/30/2024	(18,490,284)	(18,996,478)	(37,486,762)	(5,595,867
22	3/31/2025	(17,661,811)	(19,824,951)	(37,486,762)	(5,347,538
23	9/30/2025	(16,813,143)	(20,673,619)	(37,486,762)	(5,093,347
24	3/31/2026	(15,943,767)	(21,542,994)	(37,486,762)	(4,833,152
25	9/30/2026	(15,053,161)	(22,433,601)	(37,486,762)	(4,566,801

Summary | CF Total | 1 | 2 | 3 | 4 | 5

AVERAGE: (5,935,701) COUNT: 369 SUM: (2,154,659,321)

Figure 13.47 Total Cash Flow Array

This equation can then be copied to the right as far as necessary. In this case it was determined that there will be cash flows for 42 semiannual periods. This could change, however, if the prepayment rate is altered, so it is suggested that the table include 60 periods.

The period end dates will be added in row 3 just below the corresponding period numbers. These could be completed using the EOMONTH function, but in this case a VLOOKUP

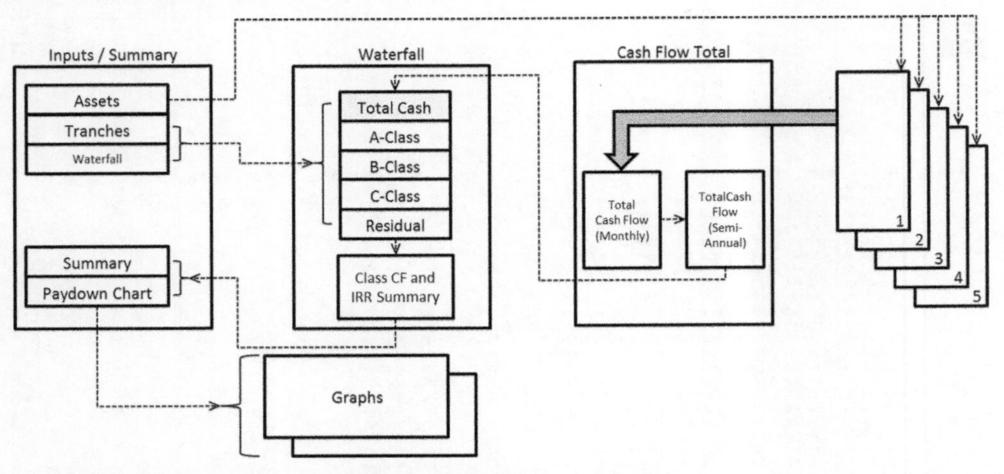

Figure 13.48 Model Flow Diagram (Waterfall Worksheet – Total Cash)

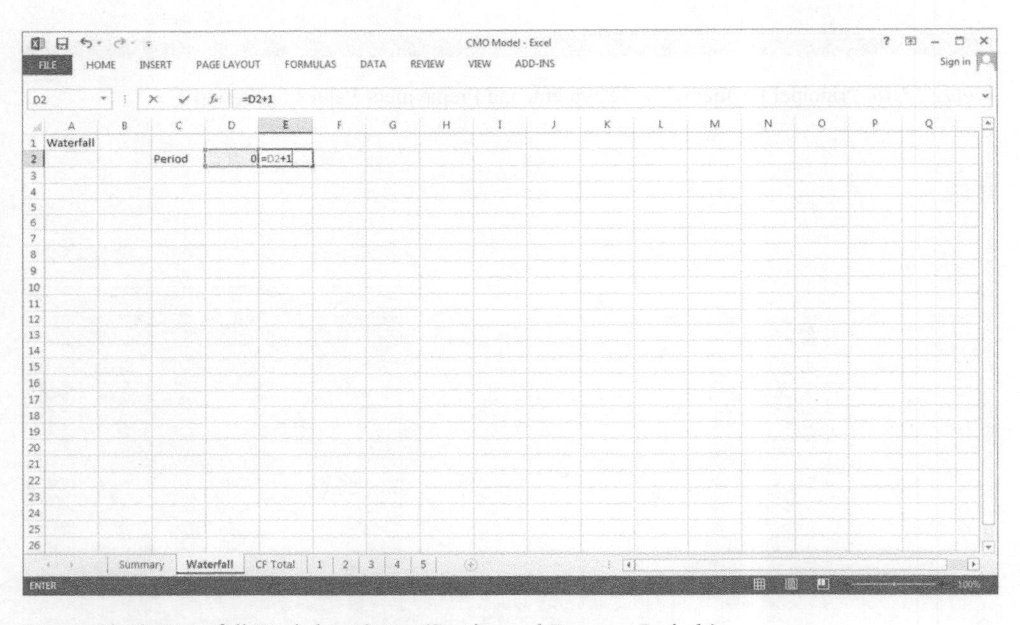

Figure 13.49 Waterfall Worksheet Setup (Semiannual Payment Periods)

function will be used instead. The first argument, the lookup value, will be the period for which the corresponding date is desired (cell D2). The table array will be the previously created Total_Cash_Flow since the numbers to be used here are for the entire CMO. And the third argument is the column with the value to be returned, the period end date. Thus, the equation in cell D3 is "=VLOOKUP(D2,Total_Cash_Flow,2)" (Figure 13.50). Remember that from this point on the periods are semiannual instead of monthly.

Row 3 may be completed by copying cell D3 to the remainder of the row.

Next, include a summary of the cash flows (which has already been created), see Figure 13.48. Label cells A5 through A8 as follows: 'Principal Payments', 'Interest Payments', 'Prepayments', and

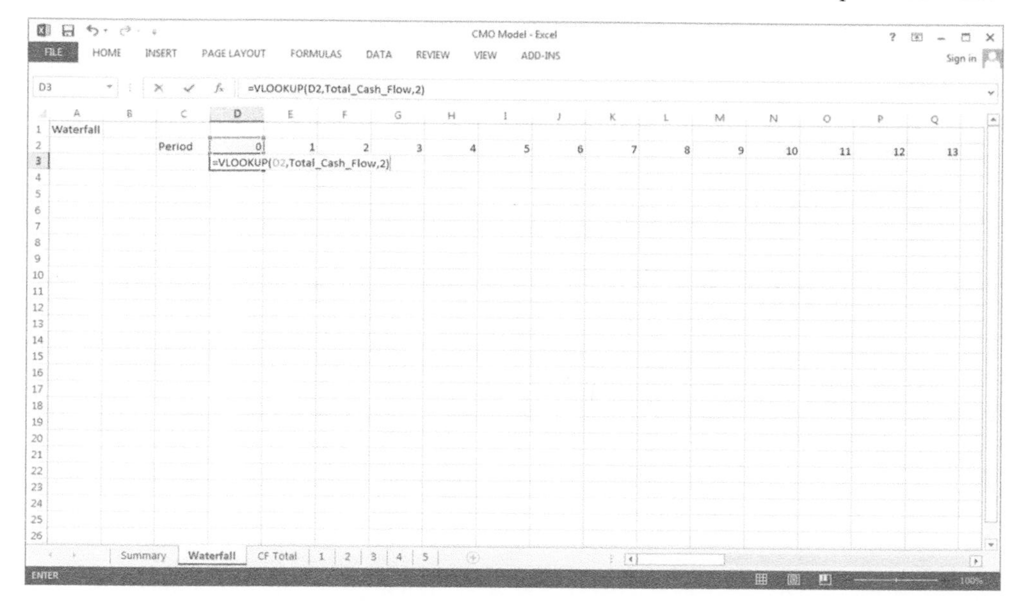

Figure 13.50 Waterfall Worksheet Setup (Semiannual Payment Dates)

'Total Funds Available' (Figure 13.52). The cells in row 8 will be the sum of the corresponding cells in rows 5, 6, and 7. The next step is to indicate where the information from the CF Total sheet will be transposed. This can be done in several ways, but this model will proceed using the VLOOKUP function. The first argument will be the period number. That is the value for which the principal payment amount should be returned (i.e. row 2). The second argument is the array in which the value will be found; thus the previously created Total_Cash_Flow. And the third argument is the column within the selected array where the value resides. In the case of principal payment the column would be 4; column 3 will be the interest payments, and column 6 will be the prepayments.

These column designations can be entered into cells B5, B6, and B7 so that one formula may be created and then copied to the rows below (see Figure 13.52).

Cell D5 will contain the principal payment for period 0. This can be entered using the VLOOKUP function and the following formula: "=-VLOOKUP(D$2,Total_Cash_Flow,$B5)" (Figure 13.53).

The placement of the absolute references allows the formula to be copied to the right while maintaining the reference to cell B5 (Figure 13.54). It may also be copied down while maintaining the reference to the period in cell D2. Notice the "-" sign in front of the equation. Since the payments are representing cash in, or deposits, these numbers will be positive. This is not true at period 0, however. At period 0 cash is being spent, so the numbers returned will be negative.

Next type 'IRR' into cell A9 (Figure 13.55). Cell B9 is where the IRR for the overall cash flow will be calculated.

There are two IRR formulas which can be used: IRR and XIRR. The IRR function only considers one variable, the cash flow values. Each cash flow is discounted at the end of each period and the periods are to be regular. The XIRR function evaluates a second variable, the period end dates. The cash flows are discounted on a daily basis and the function uses the

Period	Semi-Annual Payment Date	Int. PMT	Princ. PMT	Total PMT	Prepayment
0	4/20/2014		959,000,000		
1	9/30/2014	(31,501,937)	(5,984,825)	(37,486,762)	(9,525,659)
2	3/31/2015	(30,987,705)	(6,499,057)	(37,486,762)	(9,369,071)
3	9/30/2015	(30,461,180)	(7,025,582)	(37,486,762)	(9,208,871)
4	3/31/2016	(29,922,056)	(7,564,706)	(37,486,762)	(9,044,970)
5	9/30/2016	(29,370,018)	(8,116,744)	(37,486,762)	(8,877,279)
6	3/31/2017	(28,804,744)	(8,682,018)	(37,486,762)	(8,705,706)
7	9/30/2017	(28,225,902)	(9,260,860)	(37,486,762)	(8,530,157)
8	3/31/2018	(27,633,152)	(9,853,610)	(37,486,762)	(8,350,533)
9	9/30/2018	(27,026,147)	(10,460,615)	(37,486,762)	(8,166,737)
10	3/31/2019	(26,404,528)	(11,082,233)	(37,486,762)	(7,978,665)
11	9/30/2019	(25,767,930)	(11,718,832)	(37,486,762)	(7,786,214)
12	3/31/2020	(25,115,976)	(12,370,786)	(37,486,762)	(7,589,277)
13	9/30/2020	(24,448,280)	(13,038,482)	(37,486,762)	(7,387,743)
14	3/31/2021	(23,764,446)	(13,722,316)	(37,486,762)	(7,181,500)
15	9/30/2021	(23,064,069)	(14,422,693)	(37,486,762)	(6,970,433)
16	3/31/2022	(22,346,731)	(15,140,031)	(37,486,762)	(6,754,422)
17	9/30/2022	(21,612,005)	(15,874,756)	(37,486,762)	(6,533,348)
18	3/31/2023	(20,859,454)	(16,627,308)	(37,486,762)	(6,307,084)
19	9/30/2023	(20,088,626)	(17,398,136)	(37,486,762)	(6,075,504)
20	3/31/2024	(19,299,061)	(18,187,701)	(37,486,762)	(5,838,476)
21	9/30/2024	(18,490,284)	(18,996,478)	(37,486,762)	(5,595,867)
22	3/31/2025	(17,661,811)	(19,824,951)	(37,486,762)	(5,347,538)
23	9/30/2025	(16,813,143)	(20,673,619)	(37,486,762)	(5,093,347)
24	3/31/2026	(15,943,767)	(21,542,994)	(37,486,762)	(4,833,152)
25	9/30/2026	(15,053,161)	(22,433,601)	(37,486,762)	(4,566,801)
26	3/31/2027	(14,140,785)	(23,345,977)	(37,486,762)	(4,294,145)
27	9/30/2027	(13,206,087)	(24,280,675)	(37,486,762)	(4,015,025)

Figure 13.51 Column References for VLOOKUP Function

Waterfall															
		Period	0	1	2	3	4	5	6	7	8	9	10	11	12
			4/20/14	9/30/14	3/31/15	9/30/15	3/31/16	9/30/16	3/31/17	9/30/17	3/31/18	9/30/18	3/31/19	9/30/19	3/31/20
Principal Payments	4														
Interest Payments	3														
Prepayments	6														
Total Funds Available															

Figure 13.52 Column References for VLOOKUP Function

period end dates to consider the irregular intervals. Since the irregular intervals are considered, the XIRR is regarded as a more accurate value and will be used in this model.

Enter "=XIRR" into cell B9. The first array to be entered is the values for which a return will be calculated; in this case D8:BL8. Since the XIRR formula uses dates, the next array is

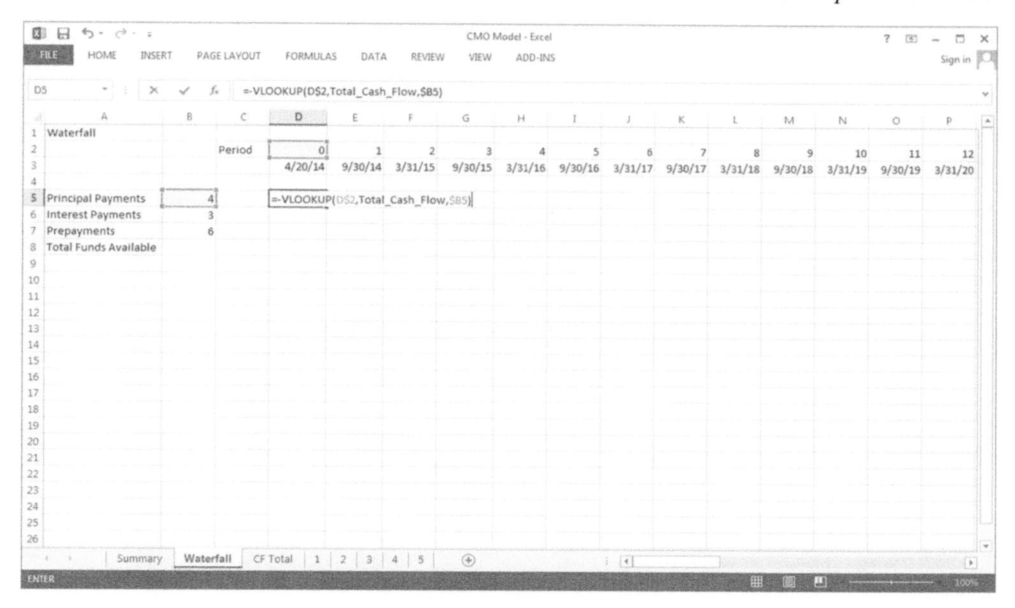

Figure 13.53 Using VLOOKUP Function to Display Payment Values

CMO Model - Excel

D7 =VLOOKUP(D$2,Total_Cash_Flow,$B7)

	A	B	C	D	E	F	G	H	I	J	K	L	M	
1	Waterfall													
2			Period	0	1	2	3	4	5	6	7	8	9	
3				4/20/14	9/30/14	3/31/15	9/30/15	3/31/16	9/30/16	3/31/17	9/30/17	3/31/18	9/30/18	
4														
5	Principal Payments	4		(959,000,000)	5,984,825	6,499,057	7,025,582	7,564,706	8,116,744	8,682,018	9,260,860	9,853,610	10,460,615	11
6	Interest Payments	3		-	31,501,937	30,987,705	30,461,180	29,922,056	29,370,018	28,804,744	28,225,902	27,633,152	27,026,147	26
7	Prepayments	6		-	9,525,659	9,369,071	9,208,871	9,044,970	8,877,279	8,705,706	8,530,157	8,350,533	8,166,737	7
8	Total Funds Available													

Summary | Waterfall | CF Total | 1 | 2 | 3 | 4 | 5

READY

Figure 13.54 Complete Semiannual Total Cash Summary

the dates for which the return should be calculated. In this case it is the period end dates, or the dates listed in row 3. The formula in cell B9 should now read "=XIRR(D8:BL8,$D$3:$BL$3)". Notice the direct references are used so the formulas may be copied.

The next step is to show how the cash is used to pay down the tranches. Each class will be set up based upon its individual characteristics. The model structure is shown in Figure 13.56.

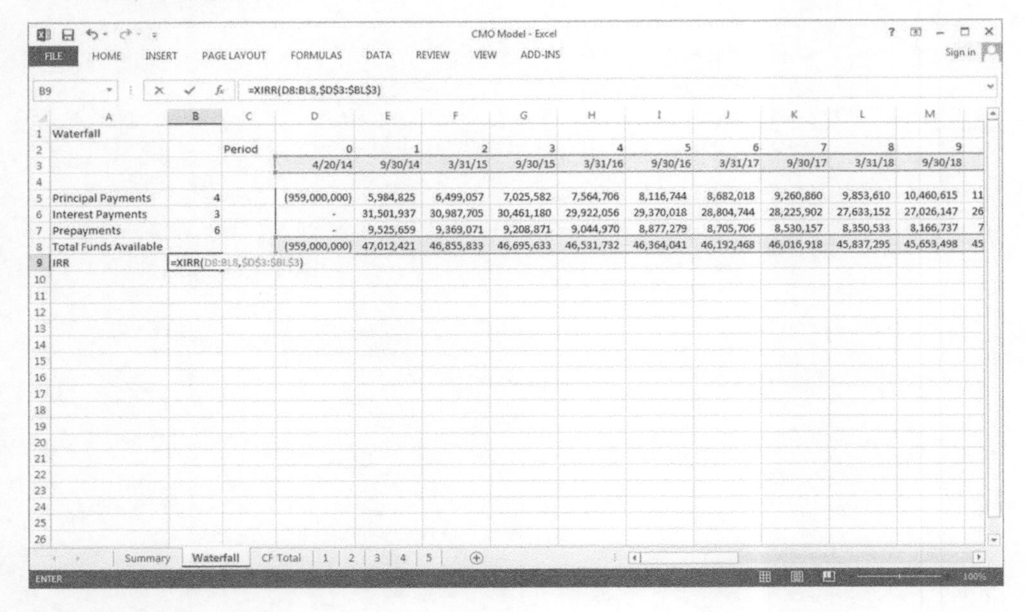

	A	B	C	D	E	F	G	H	I	J	K	L	M	
1	Waterfall													
2			Period	0	1	2	3	4	5	6	7	8	9	
3				4/20/14	9/30/14	3/31/15	9/30/15	3/31/16	9/30/16	3/31/17	9/30/17	3/31/18	9/30/18	
4														
5	Principal Payments	4		(959,000,000)	5,984,825	6,499,057	7,025,582	7,564,706	8,116,744	8,682,018	9,260,860	9,853,610	10,460,615	11
6	Interest Payments	3		-	31,501,937	30,987,705	30,461,180	29,922,056	29,370,018	28,804,744	28,225,902	27,633,152	27,026,147	26
7	Prepayments	6		-	9,525,659	9,369,071	9,208,871	9,044,970	8,877,279	8,705,706	8,530,157	8,350,533	8,166,737	7
8	Total Funds Available			(959,000,000)	47,012,421	46,855,833	46,695,633	46,531,732	46,364,041	46,192,468	46,016,918	45,837,295	45,653,498	45
9	IRR	=XIRR(D8:BL8,$D$3:$BL$3)												

Figure 13.55 XIRR Function

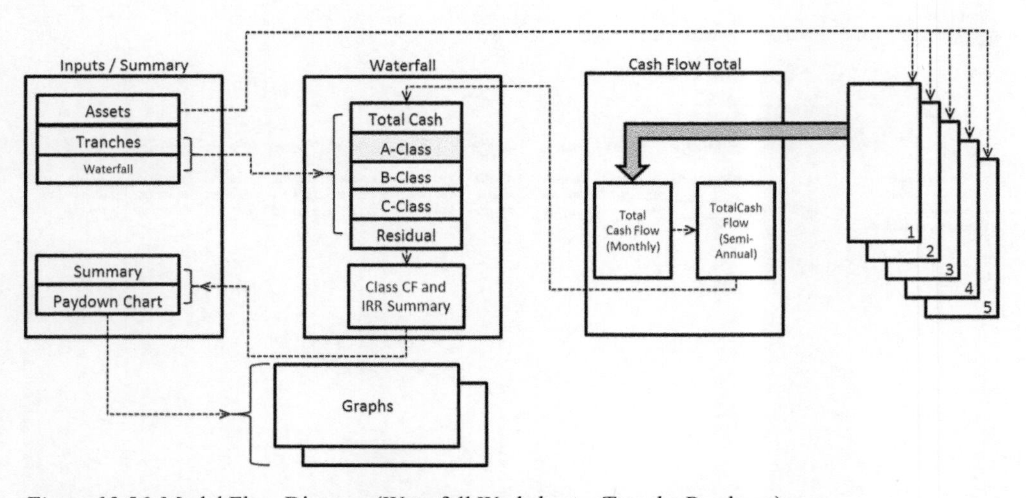

Figure 13.56 Model Flow Diagram (Waterfall Worksheet – Tranche Paydown)

Because it is wise to leave a blank row where more information could be inserted at a later time, skip row 10 and enter 'A' in cell A11 (Figure 13.57). The next few rows will be used to show the cash flow for Class A.

Row 13 will be used to show the Principal Balance at the beginning of each period for Class A. Begin by labeling A13 'Beginning Principal'. Because Column B is used for various information, the beginning principal amount for Class A can be entered here. This value was entered in the Summary page and labeled, so this can be entered by simply typing "=ClassA_Principal".

CMO Model - Excel

=ClassA_Principal

	A	B	C	D	E	F	G	H	I	J	K	L	M	
1	Waterfall													
2			Period	0	1	2	3	4	5	6	7	8	9	
3				4/20/14	9/30/14	3/31/15	9/30/15	3/31/16	9/30/16	3/31/17	9/30/17	3/31/18	9/30/18	
4														
5	Principal Payments	4		(959,000,000)	5,984,825	6,499,057	7,025,582	7,564,706	8,116,744	8,682,018	9,260,860	9,853,610	10,460,615	11
6	Interest Payments	3		-	31,501,937	30,987,705	30,461,180	29,922,056	29,370,018	28,804,744	28,225,902	27,633,152	27,026,147	26
7	Prepayments	6		-	9,525,659	9,369,071	9,208,871	9,044,970	8,877,279	8,705,706	8,530,157	8,350,533	8,166,737	7
8	Total Funds Available			(959,000,000)	47,012,421	46,855,833	46,695,633	46,531,732	46,364,041	46,192,468	46,016,918	45,837,295	45,653,498	45
9	IRR	6.66%												
10														
11	A													
12														
13	Beginning Principal	=ClassA_Principal												

Summary | Waterfall | CF Total | 1 | 2 | 3 | 4 | 5

Figure 13.57 Tranche Setup – Beginning Principal Balance

CMO Model - Excel

=ClassA_Interest

	A	B	C	D	E	F	G	H	I	J	K	L	M
1	Waterfall												
2			Period	0	1	2	3	4	5	6	7	8	9
3				4/20/14	9/30/14	3/31/15	9/30/15	3/31/16	9/30/16	3/31/17	9/30/17	3/31/18	9/30/18
4													
5	Principal Payments	4		(959,000,000)	5,984,825	6,499,057	7,025,582	7,564,706	8,116,744	8,682,018	9,260,860	9,853,610	10,460,615
6	Interest Payments	3		-	31,501,937	30,987,705	30,461,180	29,922,056	29,370,018	28,804,744	28,225,902	27,633,152	27,026,147
7	Prepayments	6		-	9,525,659	9,369,071	9,208,871	9,044,970	8,877,279	8,705,706	8,530,157	8,350,533	8,166,737
8	Total Funds Available			(959,000,000)	47,012,421	46,855,833	46,695,633	46,531,732	46,364,041	46,192,468	46,016,918	45,837,295	45,653,498
9	IRR	6.66%											
10													
11	A												
12													
13	Beginning Principal	455,000,000											
14													
15	Interest Due	=ClassA_Interest											

Summary | Waterfall | CF Total | 1 | 2 | 3 | 4 | 5

Figure 13.58 Tranche Setup – Interest

Row 15 will be used to calculate interest due each period; therefore, label row 15 'Interest Due' (Figure 13.58). Because Column B is used for various information, the interest rate for Class A can be entered here. Since the interest rate was entered in the Summary page and labeled, this can be entered by simply typing "=ClassA_Interest".

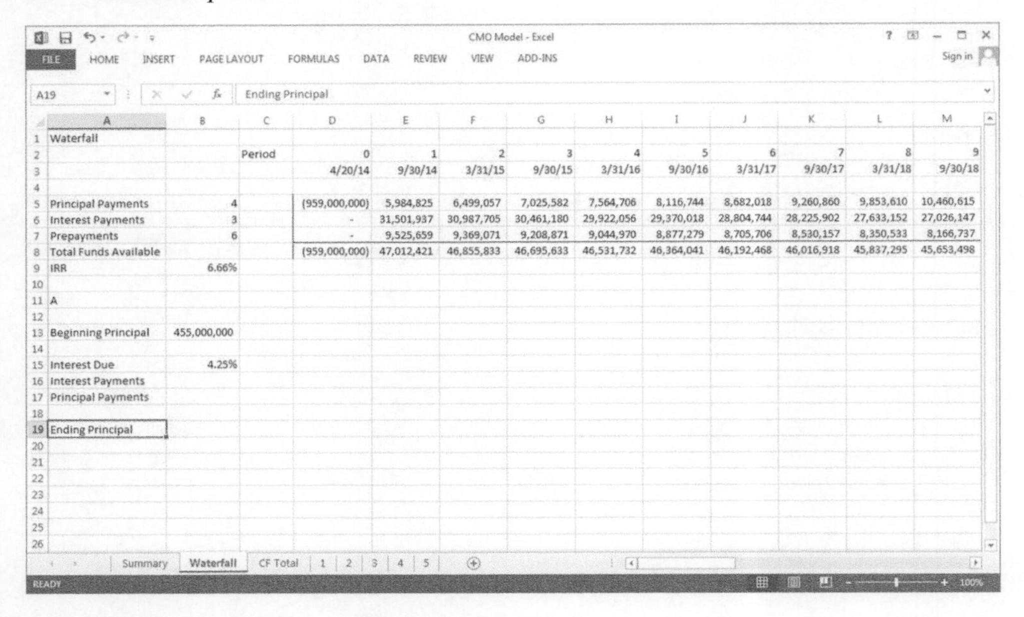

	A	B	C	D	E	F	G	H	I	J	K	L	M
1	Waterfall												
2			Period	0	1	2	3	4	5	6	7	8	9
3				4/20/14	9/30/14	3/31/15	9/30/15	3/31/16	9/30/16	3/31/17	9/30/17	3/31/18	9/30/18
4													
5	Principal Payments	4		(959,000,000)	5,984,825	6,499,057	7,025,582	7,564,706	8,116,744	8,682,018	9,260,860	9,853,610	10,460,615
6	Interest Payments	3		-	31,501,937	30,987,705	30,461,180	29,922,056	29,370,018	28,804,744	28,225,902	27,633,152	27,026,147
7	Prepayments	6		-	9,525,659	9,369,071	9,208,871	9,044,970	8,877,279	8,705,706	8,530,157	8,350,533	8,166,737
8	Total Funds Available			(959,000,000)	47,012,421	46,855,833	46,695,633	46,531,732	46,364,041	46,192,468	46,016,918	45,837,295	45,653,498
9	IRR	6.66%											
10													
11	A												
12													
13	Beginning Principal	455,000,000											
14													
15	Interest Due	4.25%											
16	Interest Payments												
17	Principal Payments												
18													
19	Ending Principal												

Figure 13.59 Tranche Setup – Ending Principal

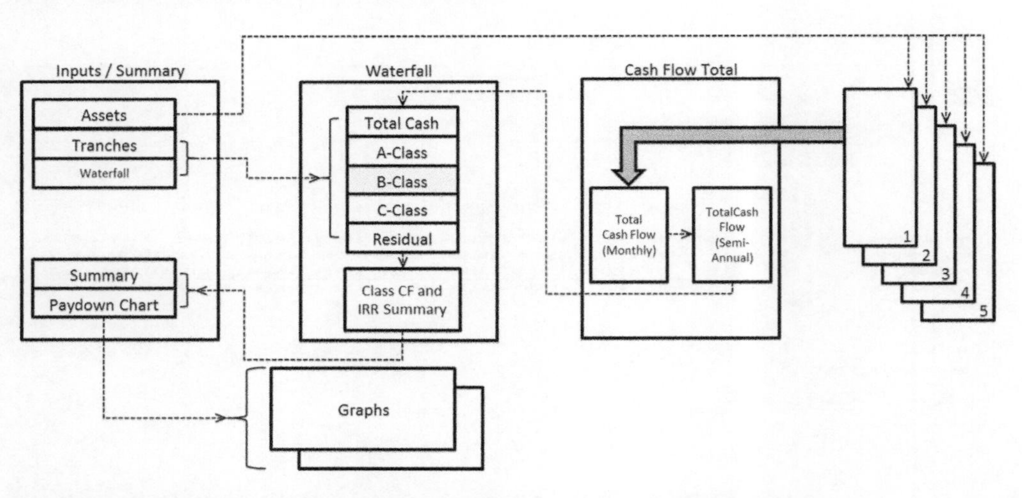

Figure 13.60 Model Flow Diagram (Waterfall Worksheet – Tranche Paydown)

Row 16 will be used for interest payments, and row 17 will be used for principal payments. Skip a row and then enter 'Ending Principal' in cell A19 (Figure 13.59).

The junior class will then be set up. See Figure 13.60 for model structure.

Again, skip a row and type 'B' in cell A21 (Figure 13.61). The next few rows will show the cash distributions to Class B. Set up the Beginning Principal just as was done for Class A. In Class B, no payments are received until Class A is completely retired. In this instance, interest will accrue and be added to the principal. Label cell A25 with 'Interest Accrued' and insert the interest rate in column B as was done in Class A.

Since accrued interest is added to the principal in Class B, there is no need to disinguish between interest and principal payments. The next line, therefore, will be labeled 'Payments'. Skip a row and label cell A28 'Ending Principal'.

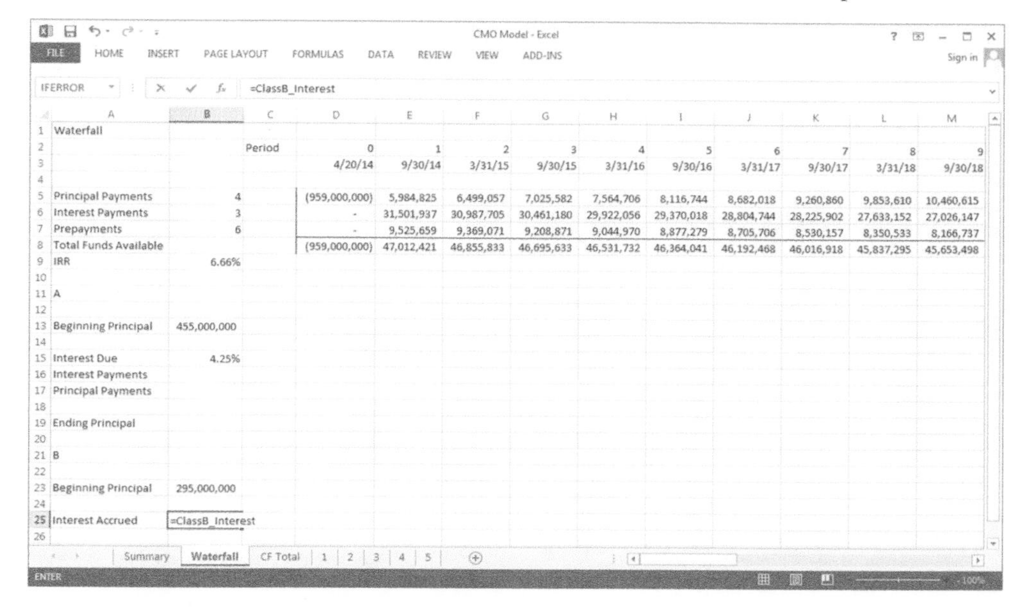

	A	B	C	D	E	F	G	H	I	J	K	L	M
1	Waterfall												
2			Period	0	1	2	3	4	5	6	7	8	9
3				4/20/14	9/30/14	3/31/15	9/30/15	3/31/16	9/30/16	3/31/17	9/30/17	3/31/18	9/30/18
4													
5	Principal Payments	4		(959,000,000)	5,984,825	6,499,057	7,025,582	7,564,706	8,116,744	8,682,018	9,260,860	9,853,610	10,460,615
6	Interest Payments	3		-	31,501,937	30,987,705	30,461,180	29,922,056	29,370,018	28,804,744	28,225,902	27,633,152	27,026,147
7	Prepayments	6		-	9,525,659	9,369,071	9,208,871	9,044,970	8,877,279	8,705,706	8,530,157	8,350,533	8,166,737
8	Total Funds Available			(959,000,000)	47,012,421	46,855,833	46,695,633	46,531,732	46,364,041	46,192,468	46,016,918	45,837,295	45,653,498
9	IRR	6.66%											
10													
11	A												
12													
13	Beginning Principal	455,000,000											
14													
15	Interest Due	4.25%											
16	Interest Payments												
17	Principal Payments												
18													
19	Ending Principal												
20													
21	B												
22													
23	Beginning Principal	295,000,000											
24													
25	Interest Accrued	=ClassB_Interest											
26													

Figure 13.61 Junior Tranche Setup

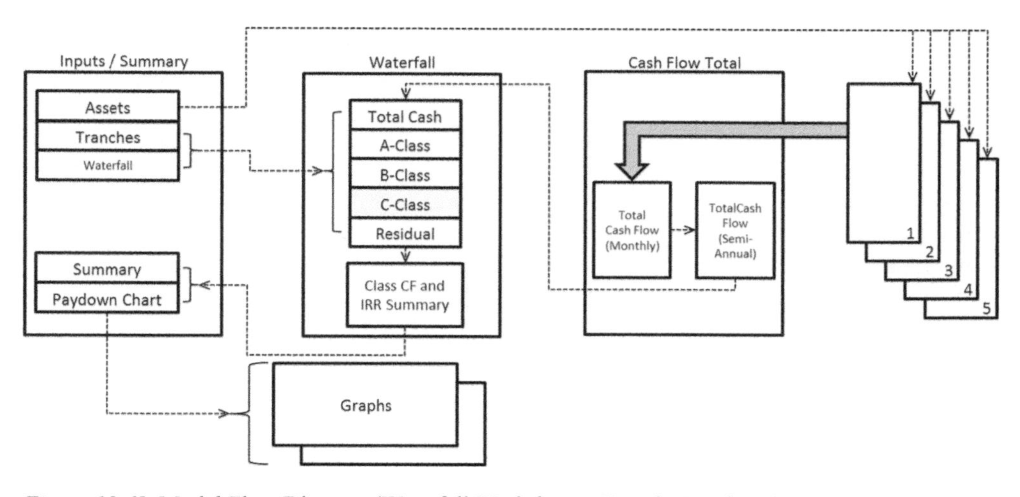

Figure 13.62 Model Flow Diagram (Waterfall Worksheet – Tranche Paydown)

Class C will be set up just as Class A was. See Figure 13.62 for model structure.

Label cell A30 with "C" and then add the Beginning Principal and Interest Due amounts just as before (Figure 13.63).

Complete the setup for Class C by labeling the rows for Interest Payments, Principal Payments, and Ending Principal. At this point it will help to add some formatting.

Since period 0 is unique, the values will be entered manually. The balance for each class at the end of period 0 is equal to the outstanding principal amount. Cell D19, forexample, is "=B13" (Figure 13.64).

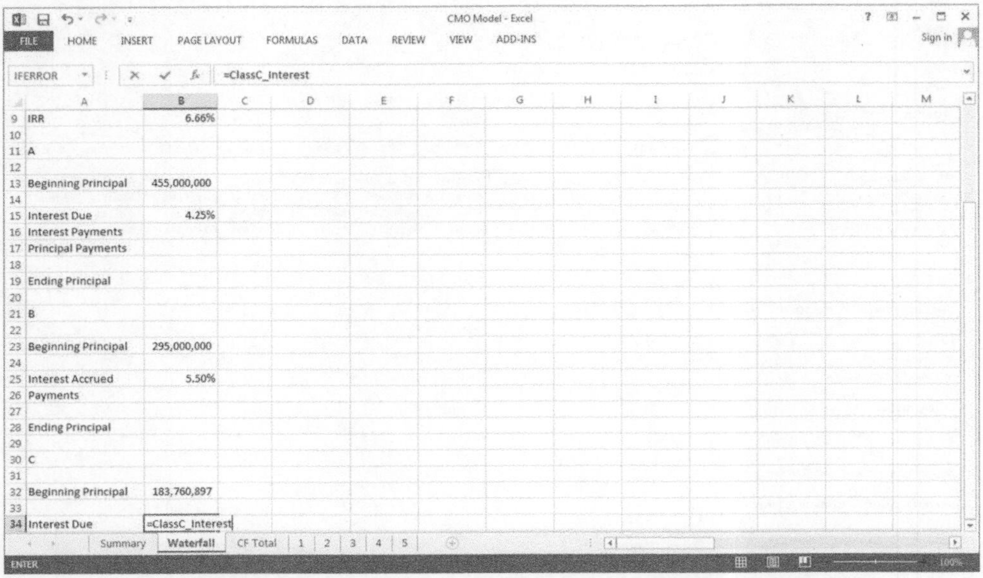

CMO Model - Excel

FILE HOME INSERT PAGE LAYOUT FORMULAS DATA REVIEW VIEW ADD-INS

IFERROR | =ClassC_Interest

	A	B
9	IRR	6.66%
10		
11	A	
12		
13	Beginning Principal	455,000,000
14		
15	Interest Due	4.25%
16	Interest Payments	
17	Principal Payments	
18		
19	Ending Principal	
20		
21	B	
22		
23	Beginning Principal	295,000,000
24		
25	Interest Accrued	5.50%
26	Payments	
27		
28	Ending Principal	
29		
30	C	
31		
32	Beginning Principal	183,760,897
33		
34	Interest Due	=ClassC_Interest

Summary | Waterfall | CF Total | 1 | 2 | 3 | 4 | 5

ENTER

Figure 13.63 Tranche C Setup

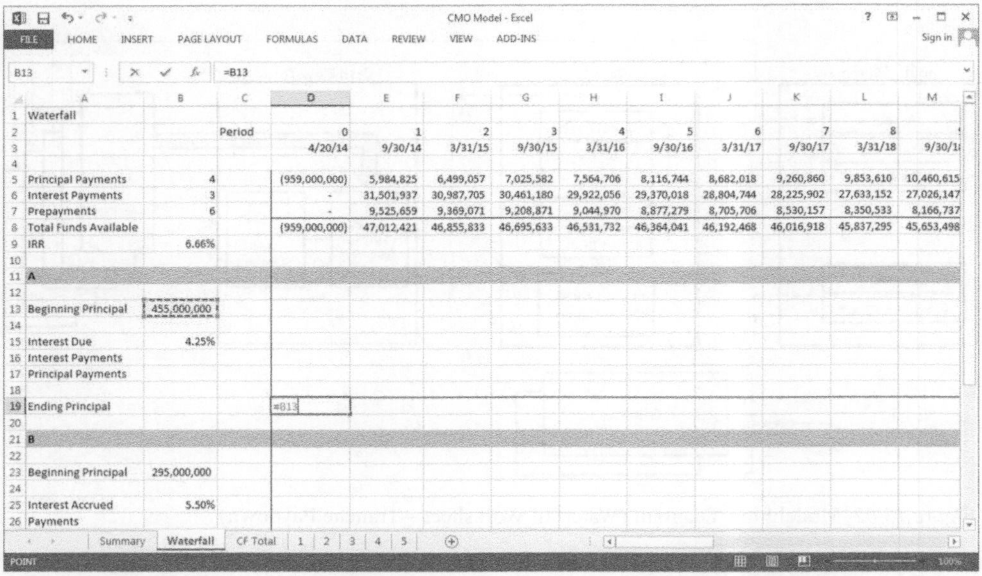

CMO Model - Excel

FILE HOME INSERT PAGE LAYOUT FORMULAS DATA REVIEW VIEW ADD-INS

B13 | =B13

	A	B	C	D	E	F	G	H	I	J	K	L	M
1	Waterfall												
2			Period	0	1	2	3	4	5	6	7	8	
3				4/20/14	9/30/14	3/31/15	9/30/15	3/31/16	9/30/16	3/31/17	9/30/17	3/31/18	9/30/1
4													
5	Principal Payments	4		(959,000,000)	5,984,825	6,499,057	7,025,582	7,564,706	8,116,744	8,682,018	9,260,860	9,853,610	10,460,615
6	Interest Payments	3		-	31,501,937	30,987,705	30,461,180	29,922,056	29,370,018	28,804,744	28,225,902	27,633,152	27,026,147
7	Prepayments	6		-	9,525,659	9,369,071	9,208,871	9,044,970	8,877,279	8,705,706	8,530,157	8,350,533	8,166,737
8	Total Funds Available			(959,000,000)	47,012,421	46,855,833	46,695,633	46,531,732	46,364,041	46,192,468	46,016,918	45,837,295	45,653,498
9	IRR	6.66%											
10													
11	A												
12													
13	Beginning Principal	455,000,000											
14													
15	Interest Due	4.25%											
16	Interest Payments												
17	Principal Payments												
18													
19	Ending Principal			=B13									
20													
21	B												
22													
23	Beginning Principal	295,000,000											
24													
25	Interest Accrued	5.50%											
26	Payments												

Summary | Waterfall | CF Total | 1 | 2 | 3 | 4 | 5

POINT

Figure 13.64 Ending Principal Amount at Period 0

Calculations for cash flow will begin in period 1. Begin by entering the Beginning Principal Balance for each Class at the beginning of period 1. Recall that the beginning balance is equal to the balance at the end of the preceding period. Thus, E13 is "=D19" (Figure 13.65). If no direct references are used then the formula can be copied into the corresponding cells for Class B and C correctly.

The interest rates can then be calculated for each class rather simply. The formula will be the interest rate (as shown in column B) times the beginning principal balance. It is important to remember, however, that the rates are given as annual rates, and in this instance the payments are made semiannually. Thus the formula will be "=($B15/Pays_per_Yr)*E13" (Figure 13.66). Note that the number of payments per year was entered on the Summary page and the cell was named 'Pays_per_Yr'. Again, notice the location of the direct references and how the formula may be copied into the corresponding cell for the junior classes.

At this point, actual calculations can begin. Using the diagram in Figure 13.67, it is noted that interest on class C is to be paid before any other cash is distributed.

CMO Model - Excel

FILE HOME INSERT PAGE LAYOUT FORMULAS DATA REVIEW VIEW ADD-INS Sign in

D19 =D19

	A	B	C	D	E	F	G	H	I	J	K	L	M
1	Waterfall												
2			Period	0	1	2	3	4	5	6	7	8	
3				4/20/14	9/30/14	3/31/15	9/30/15	3/31/16	9/30/16	3/31/17	9/30/17	3/31/18	9/30/1
4													
5	Principal Payments	4		(959,000,000)	5,984,825	6,499,057	7,025,582	7,564,706	8,116,744	8,682,018	9,260,860	9,853,610	10,460,615
6	Interest Payments	3		-	31,501,937	30,987,705	30,461,180	29,922,056	29,370,018	28,804,744	28,225,902	27,633,152	27,026,147
7	Prepayments	6		-	9,525,659	9,369,071	9,208,871	9,044,970	8,877,279	8,705,706	8,530,157	8,350,533	8,166,737
8	Total Funds Available			(959,000,000)	47,012,421	46,855,833	46,695,633	46,531,732	46,364,041	46,192,468	46,016,918	45,837,295	45,653,498
9	IRR	6.66%											
10													
11	A												
12													
13	Beginning Principal	455,000,000			=D19								
14													
15	Interest Due	4.25%											
16	Interest Payments												
17	Principal Payments												
18													
19	Ending Principal			455,000,000									
20													
21	B												
22													
23	Beginning Principal	295,000,000											
24													
25	Interest Accrued	5.50%											
26	Payments												

Summary | Waterfall | CF Total | 1 | 2 | 3 | 4 | 5

POINT 100%

Figure 13.65 Principal Balance at Beginning of Period 1

CMO Model - Excel

FILE HOME INSERT PAGE LAYOUT FORMULAS DATA REVIEW VIEW ADD-INS Sign in

E13 =($B15/Pays_per_Year)*E13

	A	B	C	D	E	F	G	H	I	J	K	L	M
7	Prepayments	6		-	9,525,659	9,369,071	9,208,871	9,044,970	8,877,279	8,705,706	8,530,157	8,350,533	8,166,737
8	Total Funds Available			(959,000,000)	47,012,421	46,855,833	46,695,633	46,531,732	46,364,041	46,192,468	46,016,918	45,837,295	45,653,498
9	IRR	6.66%											
10													
11	A												
12													
13	Beginning Principal	455,000,000			455,000,000								
14													
15	Interest Due	4.25%			=($B15/Pays_per_Year)*E13								
16	Interest Payments												
17	Principal Payments												
18													
19	Ending Principal			455,000,000									
20													
21	B												
22													
23	Beginning Principal	295,000,000			295,000,000								
24													
25	Interest Accrued	5.50%											
26	Payments												
27													
28	Ending Principal			295,000,000									
29													
30	C												
31													
32	Beginning Principal	183,760,897			183,760,897								

Summary | Waterfall | CF Total | 1 | 2 | 3 | 4 | 5

POINT 100%

Figure 13.66 Interest Due at Period 1

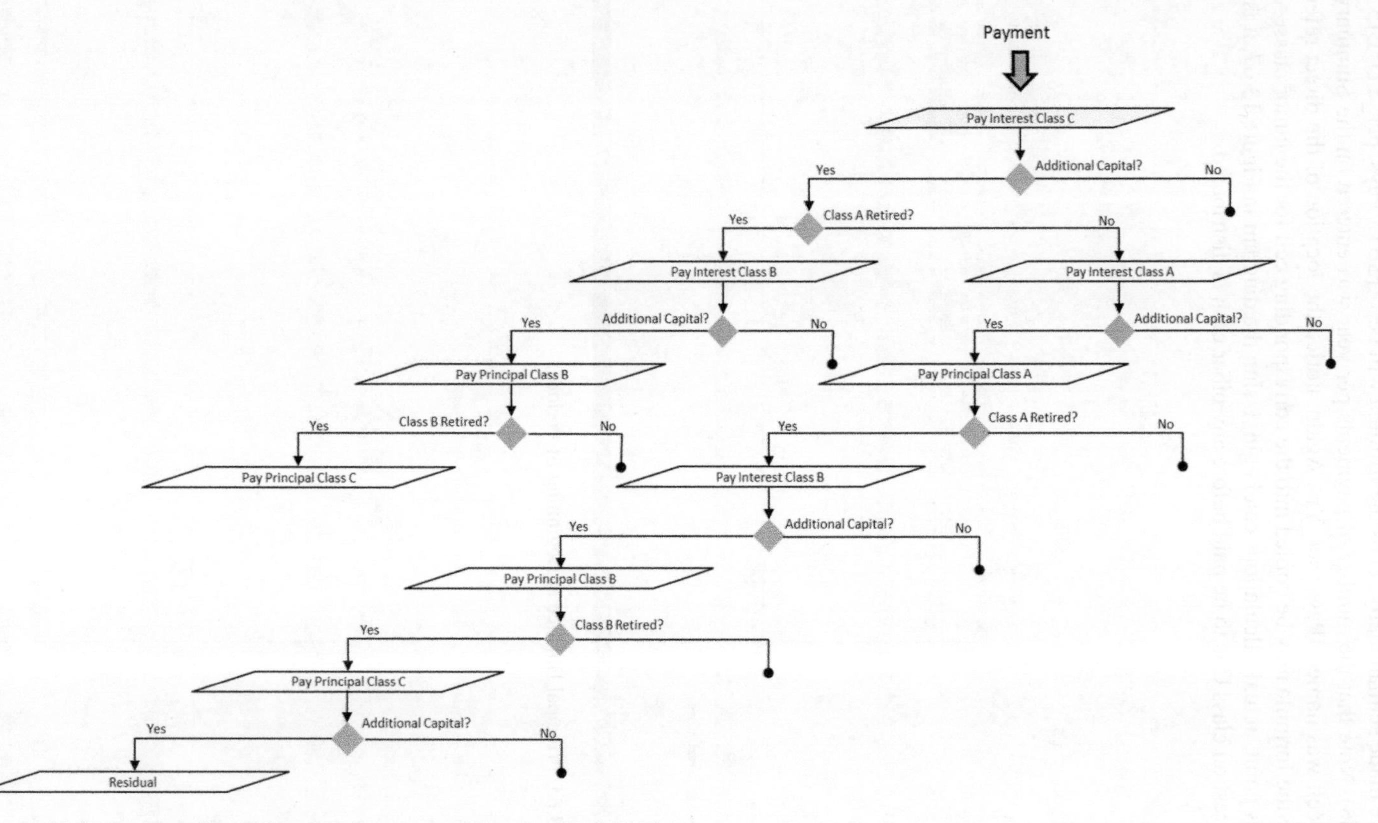

Figure 13.67 Payment Flow Diagram – Interest on Class C to Be Paid First

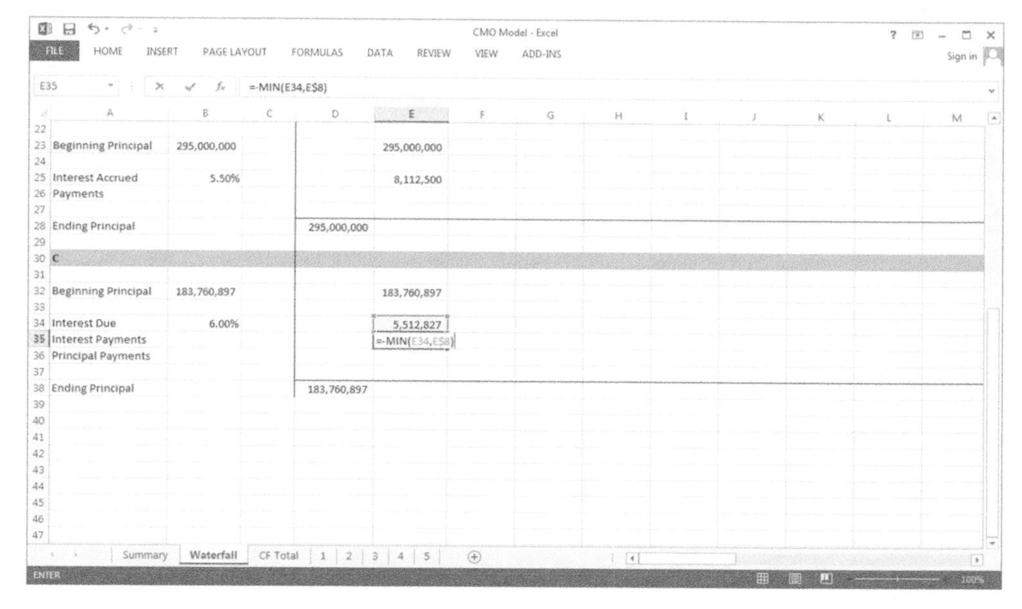

Figure 13.68 Class C Interest Payments Made at the End of Period 1

Therefore, begin in cell E35 (Figure 13.68). The total funds available is listed in E8, and the amount that is supposed to be paid as interest to Class C is noted in E34. The minimum value will be paid and whatever is remaining will be used to pay the next according to priority. The amount due may be less than the cash available so the amount due cannot be automatically entered into cell E35. A minimum formula will be used so that the smallest of the two values considered will be entered. A '-' will also be used in anticipation of the future formula for the ending principal balance. Enter into cell E35 the following: "=-MIN(E8,E$34)".

Since the total cash available is greater than the amount due as interest on Class C, there will be some cash left over to make further distributions.

According to Figure 13.67, after interest is paid on Class C, and if there is additional capital, the interest on Class A is to be addressed. The next formula, therefore, will be in row 16. The interest due is shown in row 15; however, there may not be enough cash remaining to cover the full amount due, or there may be surplus. The formula used in E16 will build on the formula in E35. Again a minimum value will be determined, and a '-' symbol will be used in anticipation of formulas to be entered at a later time. The interest payment is going to be equal to the minimum value of one of the following: (a) the amount due or (b) the total cash available for distribution minus what was already paid to cover the interest on Class C. The formula, therefore, is "=-MIN(E15,E$8+E$35)" (Figure 13.69). Again, notice the position of the absolute references. These are used so that the formula may be copied.

The next payment to be made is against the principal in Class A. The amount due is shown in cell E13. The payment will be made after interest has been paid to Class C and to Class A. The formula, again using a '-' symbol and a minimum function, will be the minimum value of the cash remaining after higher-priority distributions are paid or the total principal amount due. The formula thus is "=-MIN(E13,E$8+E$35+E$16)" (Figure 13.70).

The next distribution to be made is against the principal in Class B. Notice that Class B is different in that interest is accrued and then added to the principal amount. This continues until Class A is retired. At that point cash distributions will be used to pay down the principal

CMO Model - Excel

FILE HOME INSERT PAGE LAYOUT FORMULAS DATA REVIEW VIEW ADD-INS Sign in

IFERROR | =-MIN(E15,E$8+E$35)

	A	B	C	D	E
10					
11	A				
12					
13	Beginning Principal	455,000,000			455,000,000
14					
15	Interest Due	4.25%			9,668,750
16	Interest Payments				=-MIN(E15,E$8+E$35)
17	Principal Payments				MIN(number1, [number2], [number3], ...)
18					
19	Ending Principal			455,000,000	
20					
21	B				
22					
23	Beginning Principal	295,000,000			295,000,000
24					
25	Interest Accrued	5.50%			8,112,500
26	Payments				
27					
28	Ending Principal			295,000,000	
29					
30	C				
31					
32	Beginning Principal	183,760,897			183,760,897
33					
34	Interest Due	6.00%			5,512,827
35	Interest Payments				(5,512,827)

Summary | Waterfall | CF Total | 1 | 2 | 3 | 4 | 5

EDIT 100%

Figure 13.69 Class A Interest Payments Made after Class C Interest Payments Made

CMO Model - Excel

FILE HOME INSERT PAGE LAYOUT FORMULAS DATA REVIEW VIEW ADD-INS Sign in

E16 | =-MIN(E13,E$8+E$35+E$16)

	A	B	C	D	E
10					
11	A				
12					
13	Beginning Principal	455,000,000			455,000,000
14					
15	Interest Due	4.25%			9,668,750
16	Interest Payments				(9,668,750)
17	Principal Payments				=-MIN(E13,E$8+E$35+E$16)
18					MIN(number1, [number2], [number3], ...)
19	Ending Principal			455,000,000	
20					
21	B				
22					
23	Beginning Principal	295,000,000			295,000,000
24					
25	Interest Accrued	5.50%			8,112,500
26	Payments				
27					
28	Ending Principal			295,000,000	
29					
30	C				
31					
32	Beginning Principal	183,760,897			183,760,897
33					
34	Interest Due	6.00%			5,512,827
35	Interest Payments				(5,512,827)

Summary | Waterfall | CF Total | 1 | 2 | 3 | 4 | 5

POINT 100%

Figure 13.70 Class A Principal Payments

amount of Class B. The formula will again be entered as a negative number, and the value will be the minimum of two numbers. One value to consider is the beginning principal amount plus the interest accrued during the period. The other value to consider is the total cash available for distribution minus the amount used to pay interest on Class C, minus the amount paid as interest in Class A, minus the amount paid as principal in Class A. The formula, therefore, is "=-MIN(E23+E25,E8+E35+E16+E17)" (Figure 13.71).

CMO Model - Excel

IFERROR | =-MIN(E23+E25,E$8+E$35+E$16+E$17)

	A	B	C	D	E
10					
11	A				
12					
13	Beginning Principal	455,000,000			455,000,000
14					
15	Interest Due	4.25%			9,668,750
16	Interest Payments				(9,668,750)
17	Principal Payments				(31,830,844)
18					
19	Ending Principal			455,000,000	
20					
21	B				
22					
23	Beginning Principal	295,000,000			295,000,000
24					
25	Interest Accrued	5.50%			8,112,500
26	Payments				=-MIN(E23+E25,E$8+E$35+E$16+E$17)
27					MIN(number1, [number2], [number3], ...)
28	Ending Principal			295,000,000	
29					
30	C				
31					
32	Beginning Principal	183,760,897			183,760,897
33					
34	Interest Due	6.00%			5,512,827
35	Interest Payments				(5,512,827)

Summary | Waterfall | CF Total | 1 | 2 | 3 | 4 | 5

Figure 13.71 Class B Principal Payments

CMO Model - Excel

IFERROR | =-MIN(E32,E$8+E$35+E$16+E$17+E$26)

	A	B	C	D	E
15	Interest Due	4.25%			9,668,750
16	Interest Payments				(9,668,750)
17	Principal Payments				(31,830,844)
18					
19	Ending Principal			455,000,000	
20					
21	B				
22					
23	Beginning Principal	295,000,000			295,000,000
24					
25	Interest Accrued	5.50%			8,112,500
26	Payments				-
27					
28	Ending Principal			295,000,000	
29					
30	C				
31					
32	Beginning Principal	183,760,897			183,760,897
33					
34	Interest Due	6.00%			5,512,827
35	Interest Payments				(5,512,827)
36	Principal Payments				=-MIN(E32,E$8+E$35+E$16+E$17+E$26)
37					MIN(number1, [number2], [number3], ...)
38	Ending Principal			183,760,897	
39					
40					

Summary | Waterfall | CF Total | 1 | 2 | 3 | 4 | 5

Figure 13.72 Class C Principal Payments

Notice in this example that there is not enough cash available during the first period to pay anything into Class B, the value is therefore '0'.

The final payment to be made is to the principal due in Class C. This formula is "=-MIN (E32,E$8+E$35+E$16+E$17+E$26)" (Figure 13.72).

At this point formulas have been entered for all cash distributions. Finally, the ending balances can be calculated. This is fairly simple for Class A and C. The formulas for Class A and C are simply the original principal amount minus any payments made against that principal (Figure 13.73).

Notice that the principal went down in Class A, but it remained the same in Class C. As a senior tranche, Class A will be paid earlier.

Remember that in Class B the interest was accrued and added to the principal amount. The formula for Ending Balance in Class B is the initial balance plus interest accrued minus any payments made (Figure 13.74).

Note that the principal increased in Class B.

CMO Model - Excel

E17 =E13+E17

	A	B	C	D	E	F	G	H	I	J	K	L	M
1	Waterfall												
2			Period	0	1	2	3	4	5	6	7	8	
3				4/20/14	9/30/14	3/31/15	9/30/15	3/31/16	9/30/16	3/31/17	9/30/17	3/31/18	9/30/1
4													
5	Principal Payments	4		(959,000,000)	5,984,825	6,499,057	7,025,582	7,564,706	8,116,744	8,682,018	9,260,860	9,853,610	10,460,615
6	Interest Payments	3		-	31,501,937	30,987,705	30,461,180	29,922,056	29,370,018	28,804,744	28,225,902	27,633,152	27,026,147
7	Prepayments	6		-	9,525,659	9,369,071	9,208,871	9,044,970	8,877,279	8,705,706	8,530,157	8,350,533	8,166,737
8	Total Funds Available			(959,000,000)	47,012,421	46,855,833	46,695,633	46,531,732	46,364,041	46,192,468	46,016,918	45,837,295	45,653,498
9	IRR	6.66%											
10													
11	A												
12													
13	Beginning Principal	455,000,000			455,000,000								
14													
15	Interest Due	4.25%			9,668,750								
16	Interest Payments				(9,668,750)								
17	Principal Payments				(31,830,844)								
18													
19	Ending Principal			455,000,000	=E13+E17								
20													
21	B												
22													
23	Beginning Principal	295,000,000			295,000,000								
24													
25	Interest Accrued	5.50%			8,112,500								
26	Payments				-								

Summary | Waterfall | CF Total | 1 | 2 | 3 | 4 | 5

POINT

Figure 13.73 End of Period Balances for Class A and C

CMO Model - Excel

IFERROR =E23+E25+E26

	A	B	C	D	E	F	G	H	I	J	K	L	M
8	Total Funds Available			(959,000,000)	47,012,421	46,855,833	46,695,633	46,531,732	46,364,041	46,192,468	46,016,918	45,837,295	45,653,498
9	IRR	6.66%											
10													
11	A												
12													
13	Beginning Principal	455,000,000			455,000,000								
14													
15	Interest Due	4.25%			9,668,750								
16	Interest Payments				(9,668,750)								
17	Principal Payments				(31,830,844)								
18													
19	Ending Principal			455,000,000	423,169,156								
20													
21	B												
22													
23	Beginning Principal	295,000,000			295,000,000								
24													
25	Interest Accrued	5.50%			8,112,500								
26	Payments				-								
27													
28	Ending Principal			295,000,000	=E23+E25−E26								
29													
30	C												
31													
32	Beginning Principal	183,760,897			183,760,897								
33													

Summary | Waterfall | CF Total | 1 | 2 | 3 | 4 | 5

EDIT

Figure 13.74 Principal Balance at End of Period for Class B

CMO Model - Excel

G32 =F38

	A	B	C	D	E	F	G	H	I	J	K	L	
11	A												
12													
13	Beginning Principal	455,000,000			455,000,000	423,169,156	390,818,494	357,940,582	324,527,914	290,572,918	256,067,951	221,005,304	18
14													
15	Interest Due	4.25%			9,668,750	8,992,345	8,304,893	7,606,237	6,896,218	6,174,675	5,441,444	4,696,363	
16	Interest Payments				(9,668,750)	(8,992,345)	(8,304,893)	(7,606,237)	(6,896,218)	(6,174,675)	(5,441,444)	(4,696,363)	(
17	Principal Payments				(31,830,844)	(32,350,662)	(32,877,913)	(33,412,668)	(33,954,996)	(34,504,967)	(35,062,648)	(35,628,105)	(3
18													
19	Ending Principal			455,000,000	423,169,156	390,818,494	357,940,582	324,527,914	290,572,918	256,067,951	221,005,304	185,377,198	14
20													
21	B												
22													
23	Beginning Principal	295,000,000			295,000,000	303,112,500	311,448,094	320,012,916	328,813,272	337,855,636	347,146,666	356,693,200	36
24													
25	Interest Accrued	5.50%			8,112,500	8,335,594	8,564,823	8,800,355	9,042,365	9,291,030	9,546,533	9,809,063	1
26	Payments				-	-	-	-	-	-	-	-	
27													
28	Ending Principal			295,000,000	303,112,500	311,448,094	320,012,916	328,813,272	337,855,636	347,146,666	356,693,200	366,502,263	37
29													
30	C												
31													
32	Beginning Principal	183,760,897			183,760,897	183,760,897	183,760,897	183,760,897	183,760,897	183,760,897	183,760,897	183,760,897	18
33													
34	Interest Due	6.00%			5,512,827	5,512,827	5,512,827	5,512,827	5,512,827	5,512,827	5,512,827	5,512,827	
35	Interest Payments				(5,512,827)	(5,512,827)	(5,512,827)	(5,512,827)	(5,512,827)	(5,512,827)	(5,512,827)	(5,512,827)	(
36	Principal Payments				-	-	-	-	-	-	-	-	

Summary | Waterfall | CF Total | 1 | 2 | 3 | 4 | 5

READY

Figure 13.75 Tranche Paydown Values Completed

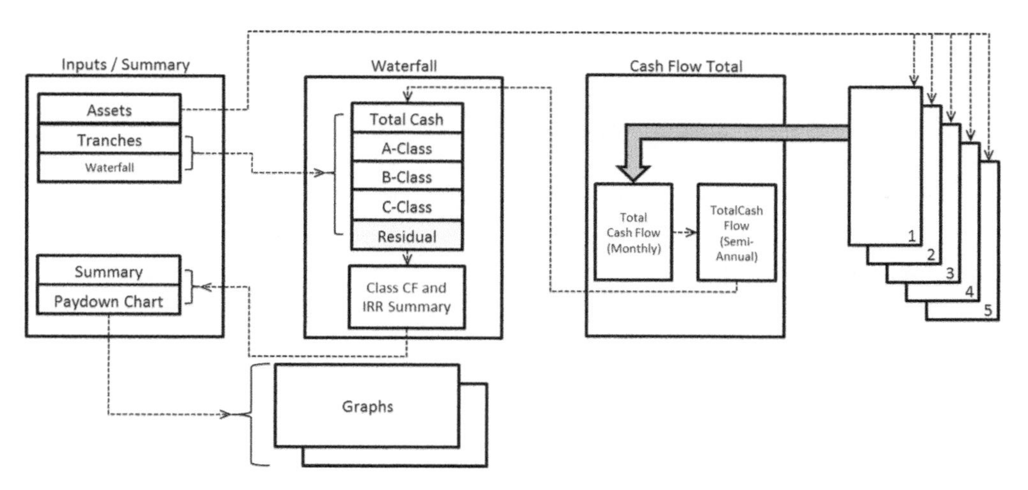

Figure 13.76 Model Flow Diagram (Waterfall Worksheet – Residual Payments)

At this point the cash flows may be completed by copying the formulas in cells E13 through E38 to the right until Period 60 (Figure 13.75).

At this point the residual dollars, which are absorbed by Anderson Insurance Corporation, can be determined. See Figure 13.76 for model structure.

The residual cash is any leftover amounts that will be earned by Anderson Insurance Corporation. This is a straight calculation of Total Funds Available minus any payments made into the tranches. Begin by labeling A40 'Residual' (Figure 13.77). The calculation at period 0 is the amount of money expended by Anderson Insurance Corporation. The total value of – i.e. the amount spent to acquire – the five pools is shown in cell D8. Note that the row represents total cash flow for the period; at time 0 there is a negative cash flow indicating CF_0. The residual income is the difference between what it cost to acquire the assets

CMO Model - Excel

FILE HOME INSERT PAGE LAYOUT FORMULAS DATA REVIEW VIEW ADD-INS

D8 =D8+Total_Price_to_Public

	A	B	C	D	E	F	G	H	I	J	K	L
23	Beginning Principal	295,000,000			295,000,000	303,112,500	311,448,094	320,012,916	328,813,272	337,855,636	347,146,666	356,693,200
24												
25	Interest Accrued	5.50%			8,112,500	8,335,594	8,564,823	8,800,355	9,042,365	9,291,030	9,546,533	9,809,063
26	Payments				-	-	-	-	-	-	-	-
27												
28	Ending Principal			295,000,000	303,112,500	311,448,094	320,012,916	328,813,272	337,855,636	347,146,666	356,693,200	366,502,263
29												
30	C											
31												
32	Beginning Principal	183,760,897			183,760,897	183,760,897	183,760,897	183,760,897	183,760,897	183,760,897	183,760,897	183,760,897
33												
34	Interest Due	6.00%			5,512,827	5,512,827	5,512,827	5,512,827	5,512,827	5,512,827	5,512,827	5,512,827
35	Interest Payments				(5,512,827)	(5,512,827)	(5,512,827)	(5,512,827)	(5,512,827)	(5,512,827)	(5,512,827)	(5,512,827)
36	Principal Payments				-	-	-	-	-	-	-	-
37												
38	Ending Principal			183,760,897	183,760,897	183,760,897	183,760,897	183,760,897	183,760,897	183,760,897	183,760,897	183,760,897
39												
40	Residual											
41				=D8+Total_Price_to_Public								

Summary | Waterfall | CF Total | 1 | 2 | 3 | 4 | 5

ENTER

Figure 13.77 Residual Payments Setup

CMO Model - Excel

FILE HOME INSERT PAGE LAYOUT FORMULAS DATA REVIEW VIEW ADD-INS

E36 =E8+E35+E16+E17+E26+E36

	A	B	C	D	E	F	G	H	I	J	K	L
21	B											
22												
23	Beginning Principal	295,000,000			295,000,000	303,112,500	311,448,094	320,012,916	328,813,272	337,855,636	347,146,666	356,693,200
24												
25	Interest Accrued	5.50%			8,112,500	8,335,594	8,564,823	8,800,355	9,042,365	9,291,030	9,546,533	9,809,063
26	Payments				-	-	-	-	-	-	-	-
27												
28	Ending Principal			295,000,000	303,112,500	311,448,094	320,012,916	328,813,272	337,855,636	347,146,666	356,693,200	366,502,263
29												
30	C											
31												
32	Beginning Principal	183,760,897			183,760,897	183,760,897	183,760,897	183,760,897	183,760,897	183,760,897	183,760,897	183,760,897
33												
34	Interest Due	6.00%			5,512,827	5,512,827	5,512,827	5,512,827	5,512,827	5,512,827	5,512,827	5,512,827
35	Interest Payments				(5,512,827)	(5,512,827)	(5,512,827)	(5,512,827)	(5,512,827)	(5,512,827)	(5,512,827)	(5,512,827)
36	Principal Payments				-	-	-	-	-	-	-	-
37												
38	Ending Principal			183,760,897	183,760,897	183,760,897	183,760,897	183,760,897	183,760,897	183,760,897	183,760,897	183,760,897
39												
40	Residual											
41				(149,075,734)	=E8+E35+E16+E17+E26+E36							

Summary | Waterfall | CF Total | 1 | 2 | 3 | 4 | 5

POINT

Figure 13.78 Residual Payments Equation

(listed in D8) and what amount of those assets was sold to investors. The value will be the difference between D8 and the cell on the Summary page which represents what was sold to the public. This cell was previously labeled 'Total_Price_to_Public'. The formula in cell D41 is, therefore, "=D8+Total_Price_to_Public".

The values in periods 1 through 60 are simply the total cash minus any payments made. The formula for E41 is "=E8+E35+E16+E17+E26+E36" (Figure 13.78). Remember to add the numbers since payments were listed as negative values. This formula can then be copied to the remaining cells in the row.

Notice in this example that there will be no residual cash flow until much farther into the term.

Now label the entire cash flow, by selecting cells A2:BL41, with 'Waterfall'.

At this point the analyst will consider what is being analyzed. If an investment is going to be made, an investor will be interested in the profit and the return. Therefore, the next step is to use what has already been calculated to illustrate these values (Figure 13.79).

The next four lines will be used to summarize the cash flows into each class, including Residual (Figure 13.80). Label cells A43:A46 'CF to Class A', 'CF to Class B', 'CF to Class C', and 'CF to Residual', respectively.

At period 0 the values in all cases will be negative and will be equal to the initial expenditure, in this case the price paid by the public. This may be filled in by using '=' or by typing

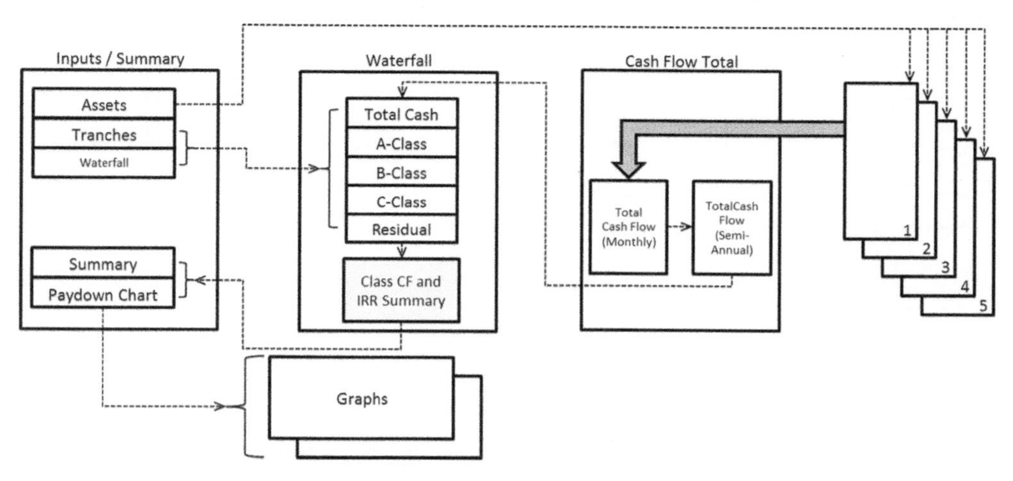

Figure 13.79 Model Flow Diagram (Waterfall Worksheet – Class Summary)

	A	B	C	D	E	F	G	H	I	J	K	L
24												
25	Interest Accrued	5.50%			8,112,500	8,335,594	8,564,823	8,800,355	9,042,365	9,291,030	9,546,533	9,809,063
26	Payments				-	-	-	-	-	-	-	-
27												
28	Ending Principal			295,000,000	303,112,500	311,448,094	320,012,916	328,813,272	337,855,636	347,146,666	356,693,200	366,502,263
29												
30	C											
31												
32	Beginning Principal	183,760,897			183,760,897	183,760,897	183,760,897	183,760,897	183,760,897	183,760,897	183,760,897	183,760,897
33												
34	Interest Due	6.00%			5,512,827	5,512,827	5,512,827	5,512,827	5,512,827	5,512,827	5,512,827	5,512,827
35	Interest Payments				(5,512,827)	(5,512,827)	(5,512,827)	(5,512,827)	(5,512,827)	(5,512,827)	(5,512,827)	(5,512,827)
36	Principal Payments				-	-	-	-	-	-	-	-
37												
38	Ending Principal			183,760,897	183,760,897	183,760,897	183,760,897	183,760,897	183,760,897	183,760,897	183,760,897	183,760,897
39												
40	Residual											
41				(149,075,734)	-	-	-	-	-	-	-	-
42												
43	CF to Class A											
44	CF to Class B											
45	CF to Class C											
46	CF to Residual											
47												
48												
49												

Figure 13.80 Class Summary Setup

in 'ClassA_Price_to_Public' since the cells were named previously (Figure 13.81). Note that the residual amount is equal to the Total Outstanding Principal minus the price paid by the public. The difference is the amount paid by Anderson Insurance Corporation.

To complete the rows, simply add the expenditures for each class. Class A, for example, in period 1 will be the sum of the interest payments and principal payments. This is a positive number since this represents 'cash in' so be sure to use '-' signs where appropriate. The formula for E43, therefore, is "=-E16-E17" (Figure 13.82). The formula can then be copied to the right to complete the row.

CMO Model - Excel

IFERROR | =-Total_Outstanding_Principal+Total_Price_to_Public

	A	B	C	D	E	F	G	H	I	J	K	L
27												
28	Ending Principal			295,000,000	303,112,500	311,448,094	320,012,916	328,813,272	337,855,636	347,146,666	356,693,200	366,502,263
29												
30	C											
31												
32	Beginning Principal	183,760,897			183,760,897	183,760,897	183,760,897	183,760,897	183,760,897	183,760,897	183,760,897	183,760,897
33												
34	Interest Due	6.00%			5,512,827	5,512,827	5,512,827	5,512,827	5,512,827	5,512,827	5,512,827	5,512,827
35	Interest Payments				(5,512,827)	(5,512,827)	(5,512,827)	(5,512,827)	(5,512,827)	(5,512,827)	(5,512,827)	(5,512,827)
36	Principal Payments				-	-	-	-	-	-	-	-
37												
38	Ending Principal			183,760,897	183,760,897	183,760,897	183,760,897	183,760,897	183,760,897	183,760,897	183,760,897	183,760,897
39												
40	Residual											
41				(149,075,734)	-	-	-	-	-	-	-	-
42												
43	CF to Class A			(398,807,500)								
44	CF to Class B			(233,787,500)								
45	CF to Class C			(177,329,266)								
46	CF to Residual			=-Total_Outstanding_Principal+Total_Price_to_Public								

Summary | Waterfall | CF Total | 1 | 2 | 3 | 4 | 5

Figure 13.81 Summary of Tranche Prices Paid Initially

CMO Model - Excel

E43 | =-E16-E17

	A	B	C	D	E	F	G	H	I	J	K	L	M	N
14														
15	Interest Due	4.25%			9,668,750	8,992,345	8,304,893	7,606,237	6,896,218	6,174,675	5,441,444	4,696,363	3,939,265	3,169,98
16	Interest Payments				(9,668,750)	(8,992,345)	(8,304,893)	(7,606,237)	(6,896,218)	(6,174,675)	(5,441,444)	(4,696,363)	(3,939,265)	(3,169,98
17	Principal Payments				(31,830,844)	(32,350,662)	(32,877,913)	(33,412,668)	(33,954,996)	(34,504,967)	(35,062,648)	(35,628,105)	(36,201,406)	(36,782,61
18														
19	Ending Principal			455,000,000	423,169,156	390,818,494	357,940,582	324,527,914	290,572,918	256,067,951	221,005,304	185,377,198	149,175,792	112,393,17
20														
21	B													
22														
23	Beginning Principal	295,000,000			295,000,000	303,112,500	311,448,094	320,012,916	328,813,272	337,855,636	347,146,666	356,693,200	366,502,263	376,581,07
24														
25	Interest Accrued	5.50%			8,112,500	8,335,594	8,564,823	8,800,355	9,042,365	9,291,030	9,546,533	9,809,063	10,078,812	10,355,98
26	Payments				-	-	-	-	-	-	-	-	-	-
27														
28	Ending Principal			295,000,000	303,112,500	311,448,094	320,012,916	328,813,272	337,855,636	347,146,666	356,693,200	366,502,263	376,581,075	386,937,05
29														
30	C													
31														
32	Beginning Principal	183,760,897			183,760,897	183,760,897	183,760,897	183,760,897	183,760,897	183,760,897	183,760,897	183,760,897	183,760,897	183,760,89
33														
34	Interest Due	6.00%			5,512,827	5,512,827	5,512,827	5,512,827	5,512,827	5,512,827	5,512,827	5,512,827	5,512,827	5,512,82
35	Interest Payments				(5,512,827)	(5,512,827)	(5,512,827)	(5,512,827)	(5,512,827)	(5,512,827)	(5,512,827)	(5,512,827)	(5,512,827)	(5,512,82
36	Principal Payments				-	-	-	-	-	-	-	-	-	-
37														
38	Ending Principal			183,760,897	183,760,897	183,760,897	183,760,897	183,760,897	183,760,897	183,760,897	183,760,897	183,760,897	183,760,897	183,760,89
39														
40	Residual													
41				(149,075,734)	-	-	-	-	-	-	-	-	-	-
42														
43	CF to Class A			(398,807,500)	41,499,594									
44	CF to Class B			(233,787,500)										

Summary | Waterfall | CF Total | 1 | 2 | 3 | 4 | 5

Figure 13.82 Summary of Payments Made into Class A Tranche

CMO Model - Excel

IFERROR | =-E26

	A	B	C	D	E	F	G	H	I	J	K	L	M	N
16	Interest Payments				(9,668,750)	(8,992,345)	(8,304,893)	(7,606,237)	(6,896,218)	(6,174,675)	(5,441,444)	(4,696,363)	(3,939,265)	(3,169,98
17	Principal Payments				(31,830,844)	(32,350,662)	(32,877,913)	(33,412,668)	(33,954,996)	(34,504,967)	(35,062,648)	(35,628,105)	(36,201,406)	(36,782,61
18														
19	Ending Principal			455,000,000	423,169,156	390,818,494	357,940,582	324,527,914	290,572,918	256,067,951	221,005,304	185,377,198	149,175,792	112,393,17
20														
21	B													
22														
23	Beginning Principal	295,000,000			295,000,000	303,112,500	311,448,094	320,012,916	328,813,272	337,855,636	347,146,666	356,693,200	366,502,263	376,581,07
24														
25	Interest Accrued	5.50%			8,112,500	8,335,594	8,564,823	8,800,355	9,042,365	9,291,030	9,546,533	9,809,063	10,078,812	10,355,98
26	Payments				-	-	-	-	-	-	-	-	-	-
27														
28	Ending Principal			295,000,000	303,112,500	311,448,094	320,012,916	328,813,272	337,855,636	347,146,666	356,693,200	366,502,263	376,581,075	386,937,05
29														
30	C													
31														
32	Beginning Principal	183,760,897			183,760,897	183,760,897	183,760,897	183,760,897	183,760,897	183,760,897	183,760,897	183,760,897	183,760,897	183,760,89
33														
34	Interest Due	6.00%			5,512,827	5,512,827	5,512,827	5,512,827	5,512,827	5,512,827	5,512,827	5,512,827	5,512,827	5,512,82
35	Interest Payments				(5,512,827)	(5,512,827)	(5,512,827)	(5,512,827)	(5,512,827)	(5,512,827)	(5,512,827)	(5,512,827)	(5,512,827)	(5,512,82
36	Principal Payments				-	-	-	-	-	-	-	-	-	-
37														
38	Ending Principal			183,760,897	183,760,897	183,760,897	183,760,897	183,760,897	183,760,897	183,760,897	183,760,897	183,760,897	183,760,897	183,760,89
39														
40	Residual													
41				(149,075,734)	-	-	-	-	-	-	-	-	-	-
42														
43	CF to Class A			(398,807,500)	41,499,594									
44	CF to Class B			(233,787,500)	=-E26									
45	CF to Class C			(177,329,266)										
46	CF to Residual			(149,075,734)										

Summary | Waterfall | CF Total | 1 | 2 | 3 | 4 | 5

Figure 13.83 Summary of Payments Made into Class B Tranche

CMO Model - Excel

E45 | =-E35-E36

	A	B	C	D	E	F	G	H	I	J	K	L	M	N
17	Principal Payments				(31,830,844)	(32,350,662)	(32,877,913)	(33,412,668)	(33,954,996)	(34,504,967)	(35,062,648)	(35,628,105)	(36,201,406)	(36,782,61
18														
19	Ending Principal			455,000,000	423,169,156	390,818,494	357,940,582	324,527,914	290,572,918	256,067,951	221,005,304	185,377,198	149,175,792	112,393,17
20														
21	B													
22														
23	Beginning Principal	295,000,000			295,000,000	303,112,500	311,448,094	320,012,916	328,813,272	337,855,636	347,146,666	356,693,200	366,502,263	376,581,07
24														
25	Interest Accrued	5.50%			8,112,500	8,335,594	8,564,823	8,800,355	9,042,365	9,291,030	9,546,533	9,809,063	10,078,812	10,355,98
26	Payments				-	-	-	-	-	-	-	-	-	-
27														
28	Ending Principal			295,000,000	303,112,500	311,448,094	320,012,916	328,813,272	337,855,636	347,146,666	356,693,200	366,502,263	376,581,075	386,937,05
29														
30	C													
31														
32	Beginning Principal	183,760,897			183,760,897	183,760,897	183,760,897	183,760,897	183,760,897	183,760,897	183,760,897	183,760,897	183,760,897	183,760,89
33														
34	Interest Due	6.00%			5,512,827	5,512,827	5,512,827	5,512,827	5,512,827	5,512,827	5,512,827	5,512,827	5,512,827	5,512,82
35	Interest Payments				(5,512,827)	(5,512,827)	(5,512,827)	(5,512,827)	(5,512,827)	(5,512,827)	(5,512,827)	(5,512,827)	(5,512,827)	(5,512,82
36	Principal Payments				-	-	-	-	-	-	-	-	-	-
37														
38	Ending Principal			183,760,897	183,760,897	183,760,897	183,760,897	183,760,897	183,760,897	183,760,897	183,760,897	183,760,897	183,760,897	183,760,89
39														
40	Residual													
41				(149,075,734)	-	-	-	-	-	-	-	-	-	-
42														
43	CF to Class A			(398,807,500)	41,499,594									
44	CF to Class B			(233,787,500)	-									
45	CF to Class C			(177,329,266)	5,512,827									
46	CF to Residual			(149,075,734)										

Summary | Waterfall | CF Total | 1 | 2 | 3 | 4 | 5

Figure 13.84 Summary of Payments Made into Class C Tranche

For Class B the total cash received will be equal to the value in row 26, since the payments were lumped together as one, instead of split into interest and principal (Figure 13.83). The equation for cell E44 is simply "=-E26".

The cash flows for Class C will, again, be equivalent to the sum of the interest payments and the principal payments (Figure 13.84).

The line for Residual cash flow is simply equal to the values shown in row 41 (Figure 13.85).

CMO Model - Excel

FILE HOME INSERT PAGE LAYOUT FORMULAS DATA REVIEW VIEW ADD-INS

E41 =E41

	A	B	C	D	E	F	G	H	I	J	K	L	M	N
17	Principal Payments				(31,830,844)	(32,350,662)	(32,877,913)	(33,412,668)	(33,954,996)	(34,504,967)	(35,062,648)	(35,628,105)	(36,201,406)	(36,782,61
18														
19	Ending Principal			455,000,000	423,169,156	390,818,494	357,940,582	324,527,914	290,572,918	256,067,951	221,005,304	185,377,198	149,175,792	112,393,17
20														
21	B													
22														
23	Beginning Principal	295,000,000			295,000,000	303,112,500	311,448,094	320,012,916	328,813,272	337,855,636	347,146,666	356,693,200	366,502,263	376,581,07
24														
25	Interest Accrued	5.50%			8,112,500	8,335,594	8,564,823	8,800,355	9,042,365	9,291,030	9,546,533	9,809,063	10,078,812	10,355,98
26	Payments				-	-	-	-	-	-	-	-	-	-
27														
28	Ending Principal			295,000,000	303,112,500	311,448,094	320,012,916	328,813,272	337,855,636	347,146,666	356,693,200	366,502,263	376,581,075	386,937,05
29														
30	C													
31														
32	Beginning Principal	183,760,897			183,760,897	183,760,897	183,760,897	183,760,897	183,760,897	183,760,897	183,760,897	183,760,897	183,760,897	183,760,89
33														
34	Interest Due	6.00%			5,512,827	5,512,827	5,512,827	5,512,827	5,512,827	5,512,827	5,512,827	5,512,827	5,512,827	5,512,82
35	Interest Payments				(5,512,827)	(5,512,827)	(5,512,827)	(5,512,827)	(5,512,827)	(5,512,827)	(5,512,827)	(5,512,827)	(5,512,827)	(5,512,82
36	Principal Payments				-	-	-	-	-	-	-	-	-	-
37														
38	Ending Principal			183,760,897	183,760,897	183,760,897	183,760,897	183,760,897	183,760,897	183,760,897	183,760,897	183,760,897	183,760,897	183,760,89
39														
40	Residual													
41				(149,075,734)	-	-	-	-	-	-	-	-	-	-
42														
43	CF to Class A			(398,807,500)	41,499,594									
44	CF to Class B			(233,787,500)	-									
45	CF to Class C			(177,329,266)	5,512,827									
46	CF to Residual			(149,075,734)	=E41									
47														

Summary | Waterfall | CF Total | 1 | 2 | 3 | 4 | 5

POINT

Figure 13.85 Summary of Residual Payments

CMO Model - Excel

FILE HOME INSERT PAGE LAYOUT FORMULAS DATA REVIEW VIEW ADD-INS

IFERROR =SUM(D43:BL43)

	A	B	C	D	E	F	G	H	I	J	K	L	M	
18														
19	Ending Principal			455,000,000	423,169,156	390,818,494	357,940,582	324,527,914	290,572,918	256,067,951	221,005,304	185,377,198	149,175,792	11
20														
21	B													
22														
23	Beginning Principal	295,000,000			295,000,000	303,112,500	311,448,094	320,012,916	328,813,272	337,855,636	347,146,666	356,693,200	366,502,263	37
24														
25	Interest Accrued	5.50%			8,112,500	8,335,594	8,564,823	8,800,355	9,042,365	9,291,030	9,546,533	9,809,063	10,078,812	1
26	Payments				-	-	-	-	-	-	-	-	-	
27														
28	Ending Principal			295,000,000	303,112,500	311,448,094	320,012,916	328,813,272	337,855,636	347,146,666	356,693,200	366,502,263	376,581,075	38
29														
30	C													
31														
32	Beginning Principal	183,760,897			183,760,897	183,760,897	183,760,897	183,760,897	183,760,897	183,760,897	183,760,897	183,760,897	183,760,897	18
33														
34	Interest Due	6.00%			5,512,827	5,512,827	5,512,827	5,512,827	5,512,827	5,512,827	5,512,827	5,512,827	5,512,827	
35	Interest Payments				(5,512,827)	(5,512,827)	(5,512,827)	(5,512,827)	(5,512,827)	(5,512,827)	(5,512,827)	(5,512,827)	(5,512,827)	(
36	Principal Payments				-	-	-	-	-	-	-	-	-	
37														
38	Ending Principal			183,760,897	183,760,897	183,760,897	183,760,897	183,760,897	183,760,897	183,760,897	183,760,897	183,760,897	183,760,897	18
39														
40	Residual													
41				(149,075,734)	-	-	-	-	-	-	-	-	-	
42			Profit											
43	CF to Class A		=SUM(D43:BL43)	(398,807,500)	41,499,594	41,343,006	41,182,806	41,018,905	40,851,214	40,679,641	40,504,091	40,324,468	40,140,671	3
44	CF to Class B			(233,787,500)	-	-	-	-	-	-	-	-	-	
45	CF to Class C			(177,329,266)	5,512,827	5,512,827	5,512,827	5,512,827	5,512,827	5,512,827	5,512,827	5,512,827	5,512,827	
46	CF to Residual			(149,075,734)	-	-	-	-	-	-	-	-	-	
47														
48														

Summary | Waterfall | CF Total | 1 | 2 | 3 | 4 | 5

EDIT

Figure 13.86 Profit Determination

The next step is to determine if there is any profit in each class and, if so, how much. Do this simply by adding all the cash flow values and totaling them in column C (Figure 13.86). The equation for C43 is thus "=SUM(D43:BL43)".

This value includes the initial investment (price of the asset); therefore, a positive number would indicate that a profit was made.

Next, the return will be calculated. The XIRR formula will again be used. Begin by entering "=XIRR" into cell B43, followed by an open parenthesis (Figure 13.87). The first array to be entered is the values for which the return is being calculated. In this case, the analyst will select the cash flows for all periods, beginning with period 0.

The next array is the dates upon which these cash flows are recognized (Figure 13.88). This will be the dates in row 3.

CMO Model - Excel

=XIRR(D43:BL43

	A	B	C	D	E	F	G	H	I	J	K	L	M
18													
19	Ending Principal			455,000,000	423,169,156	390,818,494	357,940,582	324,527,914	290,572,918	256,067,951	221,005,304	185,377,198	149,175,792
20													
21	B												
22													
23	Beginning Principal	295,000,000			295,000,000	303,112,500	311,448,094	320,012,916	328,813,272	337,855,636	347,146,666	356,693,200	366,502,263
24													
25	Interest Accrued	5.50%			8,112,500	8,335,594	8,564,823	8,800,355	9,042,365	9,291,030	9,546,533	9,809,063	10,078,812
26	Payments				-	-	-	-	-	-	-	-	-
27													
28	Ending Principal			295,000,000	303,112,500	311,448,094	320,012,916	328,813,272	337,855,636	347,146,666	356,693,200	366,502,263	376,581,075
29													
30	C												
31													
32	Beginning Principal	183,760,897			183,760,897	183,760,897	183,760,897	183,760,897	183,760,897	183,760,897	183,760,897	183,760,897	183,760,897
33													
34	Interest Due	6.00%			5,512,827	5,512,827	5,512,827	5,512,827	5,512,827	5,512,827	5,512,827	5,512,827	5,512,827
35	Interest Payments				(5,512,827)	(5,512,827)	(5,512,827)	(5,512,827)	(5,512,827)	(5,512,827)	(5,512,827)	(5,512,827)	(5,512,827)
36	Principal Payments				-	-	-	-	-	-	-	-	-
37													
38	Ending Principal			183,760,897	183,760,897	183,760,897	183,760,897	183,760,897	183,760,897	183,760,897	183,760,897	183,760,897	183,760,897
39													
40	Residual												
41				(149,075,734)	-	-	-	-	-	-	-	-	-
42		IRR	Profit										
43	CF to Class A	=XIRR(D43:BL43		(398,807,500)	41,499,594	41,343,006	41,182,806	41,018,905	40,851,214	40,679,641	40,504,091	40,324,468	40,140,671
44	CF to Class B	XIRR(values, dates, [guess])	378	(233,787,500)	-	-	-	-	-	-	-	-	-
45	CF to Class C		168,264,325	(177,329,266)	5,512,827	5,512,827	5,512,827	5,512,827	5,512,827	5,512,827	5,512,827	5,512,827	5,512,827
46	CF to Residual		140,341,349	(149,075,734)	-	-	-	-	-	-	-	-	-
47													
48													

Summary | Waterfall | CF Total | 1 | 2 | 3 | 4 | 5

Figure 13.87 Return on Class A

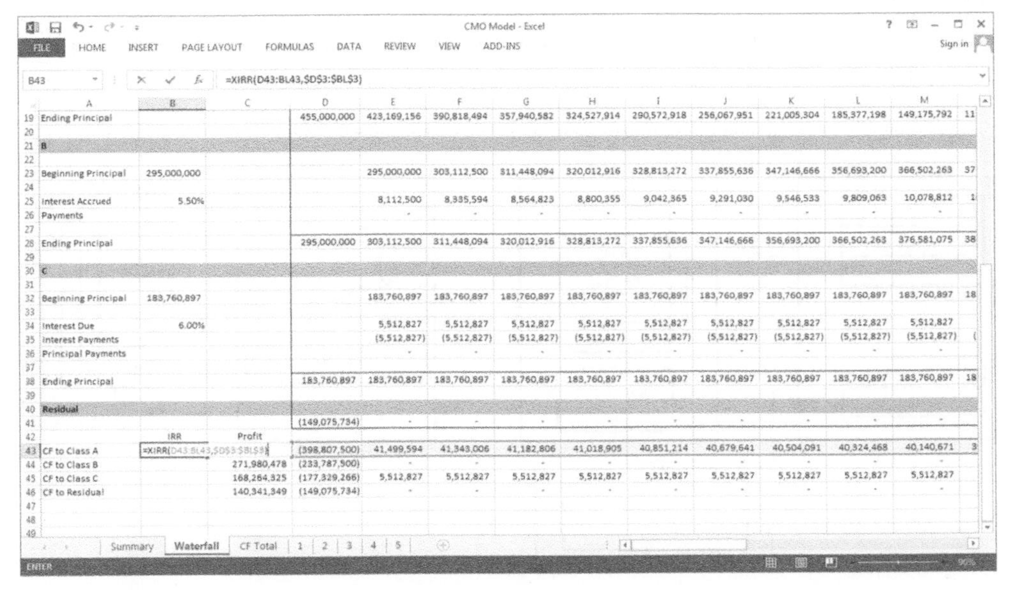

CMO Model - Excel

B43 =XIRR(D43:BL43,$D$3:$BL$3)

	A	B	C	D	E	F	G	H	I	J	K	L	M
19	Ending Principal			455,000,000	423,169,156	390,818,494	357,940,582	324,527,914	290,572,918	256,067,951	221,005,304	185,377,198	149,175,792
20													
21	B												
22													
23	Beginning Principal	295,000,000			295,000,000	303,112,500	311,448,094	320,012,916	328,813,272	337,855,636	347,146,666	356,693,200	366,502,263
24													
25	Interest Accrued	5.50%			8,112,500	8,335,594	8,564,823	8,800,355	9,042,365	9,291,030	9,546,533	9,809,063	10,078,812
26	Payments				-	-	-	-	-	-	-	-	-
27													
28	Ending Principal			295,000,000	303,112,500	311,448,094	320,012,916	328,813,272	337,855,636	347,146,666	356,693,200	366,502,263	376,581,075
29													
30	C												
31													
32	Beginning Principal	183,760,897			183,760,897	183,760,897	183,760,897	183,760,897	183,760,897	183,760,897	183,760,897	183,760,897	183,760,897
33													
34	Interest Due	6.00%			5,512,827	5,512,827	5,512,827	5,512,827	5,512,827	5,512,827	5,512,827	5,512,827	5,512,827
35	Interest Payments				(5,512,827)	(5,512,827)	(5,512,827)	(5,512,827)	(5,512,827)	(5,512,827)	(5,512,827)	(5,512,827)	(5,512,827)
36	Principal Payments				-	-	-	-	-	-	-	-	-
37													
38	Ending Principal			183,760,897	183,760,897	183,760,897	183,760,897	183,760,897	183,760,897	183,760,897	183,760,897	183,760,897	183,760,897
39													
40	Residual												
41				(149,075,734)	-	-	-	-	-	-	-	-	-
42		IRR	Profit										
43	CF to Class A	=XIRR(D43:BL43,$D$3:$BL$3)		(398,807,500)	41,499,594	41,343,006	41,182,806	41,018,905	40,851,214	40,679,641	40,504,091	40,324,468	40,140,671
44	CF to Class B		271,980,478	(233,787,500)	-	-	-	-	-	-	-	-	-
45	CF to Class C		168,264,325	(177,329,266)	5,512,827	5,512,827	5,512,827	5,512,827	5,512,827	5,512,827	5,512,827	5,512,827	5,512,827
46	CF to Residual		140,341,349	(149,075,734)	-	-	-	-	-	-	-	-	-
47													
48													
49													

Summary | Waterfall | CF Total | 1 | 2 | 3 | 4 | 5

Figure 13.88 Return on Class A

B46 | =XIRR(D46:BL46,$D$3:$BL$3)

	A	B	C	D	E	F	G	H	I	J	K	L	M
19	Ending Principal			455,000,000	423,169,156	390,818,494	357,940,582	324,527,914	290,572,918	256,067,951	221,005,304	185,377,198	149,175,792
20													
21	B												
22													
23	Beginning Principal	295,000,000			295,000,000	303,112,500	311,448,094	320,012,916	328,813,272	337,855,636	347,146,666	356,693,200	366,502,263
24													
25	Interest Accrued	5.50%			8,112,500	8,335,594	8,564,823	8,800,355	9,042,365	9,291,030	9,546,533	9,809,063	10,078,812
26	Payments				-	-	-	-	-	-	-	-	-
27													
28	Ending Principal			295,000,000	303,112,500	311,448,094	320,012,916	328,813,272	337,855,636	347,146,666	356,693,200	366,502,263	376,581,075
29													
30	C												
31													
32	Beginning Principal	183,760,897			183,760,897	183,760,897	183,760,897	183,760,897	183,760,897	183,760,897	183,760,897	183,760,897	183,760,897
33													
34	Interest Due	6.00%			5,512,827	5,512,827	5,512,827	5,512,827	5,512,827	5,512,827	5,512,827	5,512,827	5,512,827
35	Interest Payments				(5,512,827)	(5,512,827)	(5,512,827)	(5,512,827)	(5,512,827)	(5,512,827)	(5,512,827)	(5,512,827)	(5,512,827)
36	Principal Payments				-	-	-	-	-	-	-	-	-
37													
38	Ending Principal			183,760,897	183,760,897	183,760,897	183,760,897	183,760,897	183,760,897	183,760,897	183,760,897	183,760,897	183,760,897
39													
40	Residual												
41				(149,075,734)	-	-	-	-	-	-	-	-	-
42		IRR	Profit										
43	CF to Class A	8.77%	125,852,598	(398,807,500)	41,499,594	41,343,006	41,182,806	41,018,905	40,851,214	40,679,641	40,504,091	40,324,468	40,140,671
44	CF to Class B	8.15%	271,980,478	(233,787,500)	-	-	-	-	-	-	-	-	-
45	CF to Class C	6.50%	168,264,325	(177,329,266)	5,512,827	5,512,827	5,512,827	5,512,827	5,512,827	5,512,827	5,512,827	5,512,827	5,512,827
46	CF to Residual	3.81%	140,341,349	(149,075,734)	-	-	-	-	-	-	-	-	-
47													
48													
49													

Summary | Waterfall | CF Total | 1 | 2 | 3 | 4 | 5

Figure 13.89 Return on Each Tranche

Notice the use of direct references, which will allow the formula to be copied to the junior classes (Figure 13.89).

For future use the analyst should now name each cell indicating return and profit. Name cell B43 'ClassA_IRR', and name cell C43 'ClassA_Profit'. Remember that no spaces can be used when naming cells.

Summary

The IRR and Profit will be used to show a summary on the Summary page. See Figure 13.90 for model structure.

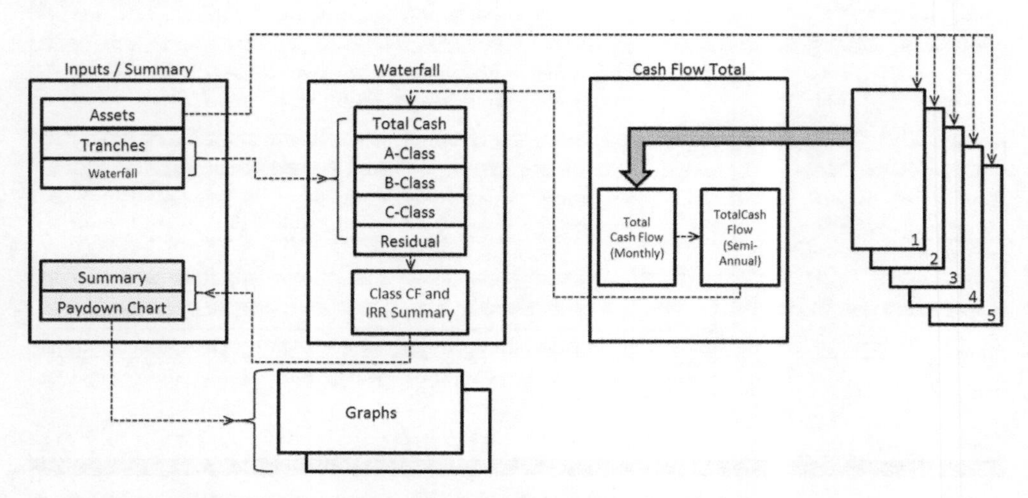

Figure 13.90 Model Flow Diagram (Inputs/Summary Worksheet – Summary)

Return to the Summary page and create the table shown in Figure 13.91.

The analyst will now create a data table to indicate the IRRs if the prepayment rate is changed. The first column in the table, cells B25:B29, will be the actual IRR (Figure 13.92). The next four columns will be the various prepayment rates considered.

CMO Model - Excel

A25 | Prepayment

	A	B	C	D	E	F	G
6	2	FNMA	345,000,000	5.25%	29.7	319,800,000	92.70%
7	3	GNMA	315,000,000	7.50%	28.6	315,000,000	100.00%
8	4	Private Issuers	2,750,000	8%	30	2,715,000	98.73%
9	5	FNMA	295,000,000	7.25%	27.1	295,000,000	100.00%
10			959,000,000	6.61%	28.54	933,760,897	97.37%

Tranches

Class	Original Principal Amount	Interest Rate	Price to Public (%)	Price to Public ($)
A	455,000,000	4.25%	87.65%	398,807,500
B	295,000,000	5.50%	79.25%	233,787,500
C	183,760,897	6%	96.50%	177,329,266
Total	933,760,897			809,924,266

Waterfall Assumptions

Payments per Year:		2
First Payment:		9/30/2014
Prepayment Rate:		2%

IRR Summary

Prepayment					
A					
B					
C					
Residual					

Summary | Waterfall | CF Total | 1 | 2 | 3 | 4 | 5

Figure 13.91 Summary Setup

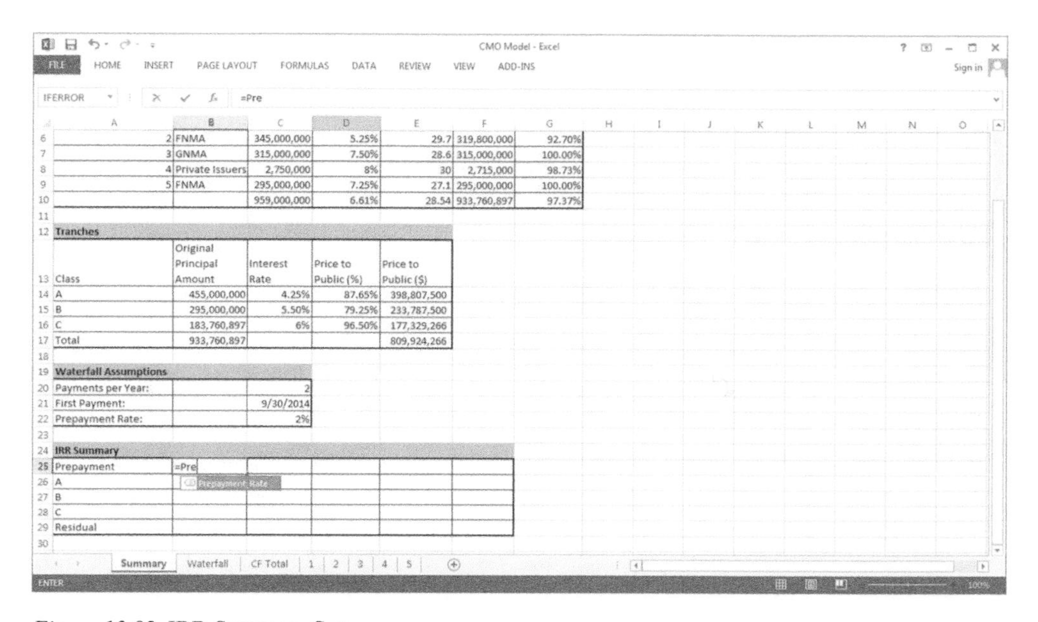

Figure 13.92 IRR Summary Setup

Begin by entering the prepayment rate in cell B26. The prepayment rate was already named so the analyst can simply type "=Prepayment_Rate".

Cells B27:B30 are simply the IRRs calculated on the Waterfall sheet (Figure 13.93). Since these cells were also named, the analyst can simply type in the name of the cell. For instance, cell B26 is "=ClassA_IRR".

Once these values are entered, try changing the prepayment rate, in cell C22 (Figure 13.94).

CMO Model - Excel

IFERROR | =Residual_IRR

	A	B	C	D	E	F	G
6	2	FNMA	345,000,000	5.25%	29.7	319,800,000	92.70%
7	3	GNMA	315,000,000	7.50%	28.6	315,000,000	100.00%
8	4	Private Issuers	2,750,000	8%	30	2,715,000	98.73%
9	5	FNMA	295,000,000	7.25%	27.1	295,000,000	100.00%
10			959,000,000	6.61%	28.54	933,760,897	97.37%
11							
12	Tranches						
13	Class	Original Principal Amount	Interest Rate	Price to Public (%)	Price to Public ($)		
14	A	455,000,000	4.25%	87.65%	398,807,500		
15	B	295,000,000	5.50%	79.25%	233,787,500		
16	C	183,760,897	6%	96.50%	177,329,266		
17	Total	933,760,897			809,924,266		
18							
19	Waterfall Assumptions						
20	Payments per Year:		2				
21	First Payment:		9/30/2014				
22	Prepayment Rate:		2%				
23							
24	IRR Summary						
25	Prepayment	2%					
26	A	8.77%					
27	B	8.15%					
28	C	6.50%					
29	Residual	=Residual_IRR					
30							

Summary | Waterfall | CF Total | 1 | 2 | 3 | 4 | 5

Figure 13.93 IRR Summary Values

CMO Model - Excel

Prepayme... | 4%

	A	B	C	D	E	F	G
6	2	FNMA	345,000,000	5.25%	29.7	319,800,000	92.70%
7	3	GNMA	315,000,000	7.50%	28.6	315,000,000	100.00%
8	4	Private Issuers	2,750,000	8%	30	2,715,000	98.73%
9	5	FNMA	295,000,000	7.25%	27.1	295,000,000	100.00%
10			959,000,000	6.61%	28.54	933,760,897	97.37%
11							
12	Tranches						
13	Class	Original Principal Amount	Interest Rate	Price to Public (%)	Price to Public ($)		
14	A	455,000,000	4.25%	87.65%	398,807,500		
15	B	295,000,000	5.50%	79.25%	233,787,500		
16	C	183,760,897	6%	96.50%	177,329,266		
17	Total	933,760,897			809,924,266		
18							
19	Waterfall Assumptions						
20	Payments per Year:		2				
21	First Payment:		9/30/2014				
22	Prepayment Rate:		4%				
23							
24	IRR Summary						
25	Prepayment	4%					
26	A	9.73%					
27	B	8.75%					
28	C	6.56%					
29	Residual	2.09%					
30							

Summary | Waterfall | CF Total | 1 | 2 | 3 | 4 | 5

Figure 13.94 IRR Summary Values

Notice how the values changed accordingly. Now add 2%, 4%, 5%, and 6% in cells C25, D25, E25, and F25 respectively (Figure 13.95). These numbers will be the prepayment rates analyzed. That is, the IRR for Class B when the prepayment rate is equal to 4% will be indicated in cell D27 once the data table is completed.

Begin by selecting cells B25:F29 (Figure 13.96).

CMO Model - Excel

FILE HOME INSERT PAGE LAYOUT FORMULAS DATA REVIEW VIEW ADD-INS

D27

	A	B	C	D	E	F	G
6		2 FNMA	345,000,000	5.25%	29.7	319,800,000	92.70%
7		3 GNMA	315,000,000	7.50%	28.6	315,000,000	100.00%
8		4 Private Issuers	2,750,000	8%	30	2,715,000	98.73%
9		5 FNMA	295,000,000	7.25%	27.1	295,000,000	100.00%
10			959,000,000	6.61%	28.54	933,760,897	97.37%
11							
12	Tranches						
13	Class	Original Principal Amount	Interest Rate	Price to Public (%)	Price to Public ($)		
14	A	455,000,000	4.25%	87.65%	398,807,500		
15	B	295,000,000	5.50%	79.25%	233,787,500		
16	C	183,760,897	6%	96.50%	177,329,266		
17	Total	933,760,897			809,924,266		
18							
19	Waterfall Assumptions						
20	Payments per Year:		2				
21	First Payment:		9/30/2014				
22	Prepayment Rate:		4%				
23							
24	IRR Summary						
25	Prepayment	4%	2%	4%	5%	6%	
26	A	9.73%					
27	B	8.75%					
28	C	6.56%					
29	Residual	2.09%					
30							

Summary Waterfall CF Total 1 2 3 4 5

READY

Figure 13.95 Data Table Setup

CMO Model - Excel

FILE HOME INSERT PAGE LAYOUT FORMULAS DATA REVIEW VIEW ADD-INS

B25 =Prepayment_Rate

	A	B	C	D	E	F	G
6		2 FNMA	345,000,000	5.25%	29.7	319,800,000	92.70%
7		3 GNMA	315,000,000	7.50%	28.6	315,000,000	100.00%
8		4 Private Issuers	2,750,000	8%	30	2,715,000	98.73%
9		5 FNMA	295,000,000	7.25%	27.1	295,000,000	100.00%
10			959,000,000	6.61%	28.54	933,760,897	97.37%
11							
12	Tranches						
13	Class	Original Principal Amount	Interest Rate	Price to Public (%)	Price to Public ($)		
14	A	455,000,000	4.25%	87.65%	398,807,500		
15	B	295,000,000	5.50%	79.25%	233,787,500		
16	C	183,760,897	6%	96.50%	177,329,266		
17	Total	933,760,897			809,924,266		
18							
19	Waterfall Assumptions						
20	Payments per Year:		2				
21	First Payment:		9/30/2014				
22	Prepayment Rate:		4%				
23							
24	IRR Summary						
25	Prepayment	4%	2%	4%	5%	6%	
26	A	9.73%					
27	B	8.75%					
28	C	6.56%					
29	Residual	2.09%					
30							

Summary Waterfall CF Total 1 2 3 4 5

READY AVERAGE: 5% COUNT: 9 SUM: 48%

Figure 13.96 Variable Data Table Designation

CMO Model - Excel

FILE HOME INSERT PAGE LAYOUT FORMULAS DATA REVIEW VIEW ADD-INS

B25 =Prepayment_Rate

	A	B	C	D	E	F	G
6	2	FNMA	345,000,000	5.25%	29.7	319,800,000	92.70%
7	3	GNMA	315,000,000	7.50%	28.6	315,000,000	100.00%
8	4	Private Issuers	2,750,000	8%	30	2,715,000	98.73%
9	5	FNMA	295,000,000	7.25%	27.1	295,000,000	100.00%
10			959,000,000	6.61%	28.54	933,760,897	97.37%

12 Tranches

	Class	Original Principal Amount	Interest Rate	Price to Public (%)	Price to Public ($)
14	A	455,000,000	4.25%	87.65%	398,807,500
15	B	295,000,000	5.50%	79.25%	233,787,500
16	C	183,760,897	6%	96.50%	177,329,266
17	Total	933,760,897			809,924,266

19 Waterfall Assumptions

20	Payments per Year:		2
21	First Payment:		9/30/2014
22	Prepayment Rate:		4%

24 IRR Summary

25	Prepayment	4%	2%	4%	5%	6%
26	A	9.73%				
27	B	8.75%				
28	C	6.56%				
29	Residual	2.09%				

Summary Waterfall CF Total 1 2 3 4 5

READY AVERAGE: 5% COUNT: 9 SUM: 48%

Figure 13.97 What-If Analysis

CMO Model - Excel

FILE HOME INSERT PAGE LAYOUT FORMULAS DATA REVIEW VIEW ADD-INS

From Access From Web From Text From Other Sources Existing Connections Refresh All Connections Properties Edit Links Sort Filter Clear Reapply Advanced Text to Columns Flash Fill Remove Duplicates Data Validation Consolidate What-If Analysis Group Ungroup Subtotal

Get External Data Connections Sort & Filter Data Tools Outline

Scenario Manager... Goal Seek... Data Table...

Data Table
See the results of multiple inputs at the same time.

	A	B	C	D	E	F	G
7	3	GNMA	315,000,000	7.50%	28.6	315,000,000	100.00%
8	4	Private Issuers	2,750,000	8%	30	2,715,000	98.73%
9	5	FNMA	295,000,000	7.25%	27.1	295,000,000	100.00%
10			959,000,000	6.61%	28.54	933,760,897	97.37%

12 Tranches

	Class	Original Principal Amount	Interest Rate	Price to Public (%)	Price to Public ($)
14	A	455,000,000	4.25%	87.65%	398,807,500
15	B	295,000,000	5.50%	79.25%	233,787,500
16	C	183,760,897	6%	96.50%	177,329,266
17	Total	933,760,897			809,924,266

19 Waterfall Assumptions

20	Payments per Year:		2
21	First Payment:		9/30/2014
22	Prepayment Rate:		4%

24 IRR Summary

25	Prepayment	4%	2%	4%	5%	6%
26	A	9.73%				
27	B	8.75%				
28	C	6.56%				
29	Residual	2.09%				

Summary Waterfall CF Total 1 2 3 4 5

READY AVERAGE: 5% COUNT: 9 SUM: 48%

PAGE 16 OF 16 32 WORDS

10:31 PM 4/19/2014

Figure 13.98 Data Table

These cells will make up the variable data table. The analysis will use the equations entered in the far left, column B, and the variables entered on top, row 25 (Figure 13.97).

While the cells are selected, select the 'Data' tab, then select 'What-If Analysis' (Figure 13.98). From there a pull-down menu will open; select 'Data Table'.

A separate box will pop up with two options: Row input cell and Column input cell. (See Figure 13.99.)

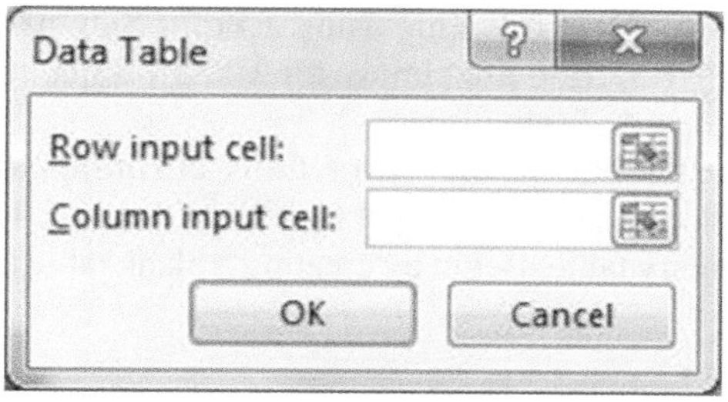

Figure 13.99 Data Table

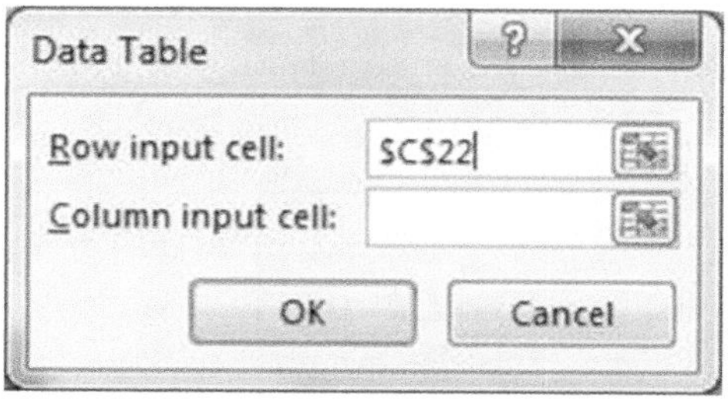

Figure 13.100 Data Table

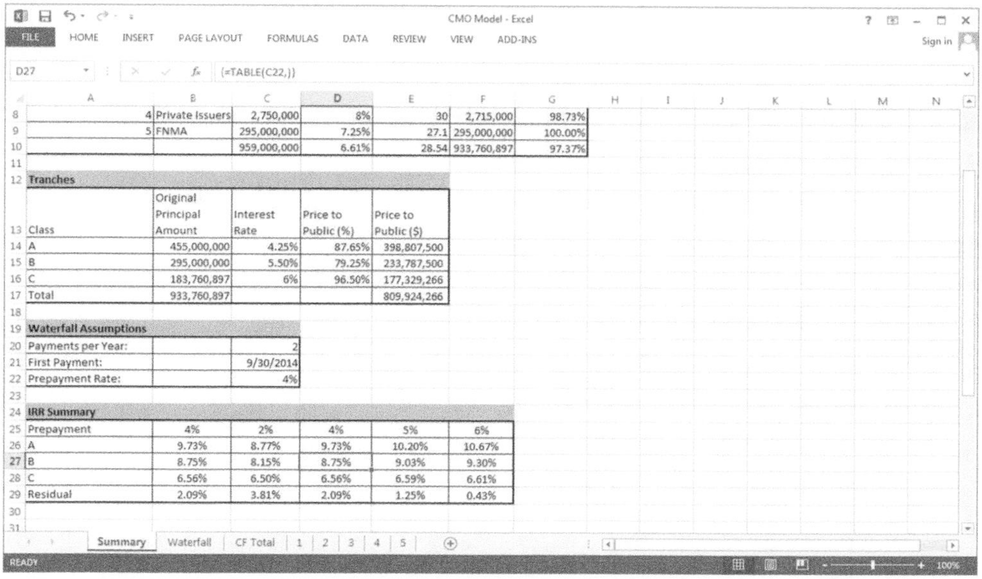

Figure 13.101 Data Table Completion

This is where the analyst indicates which cell will be changed. In this instance cell C22, the prepayment rate, will change (Figure 13.100). The value of the equations listed in the first column (cells B26:B29) will be given based upon the prepayment rate when it is equal to the values shown in the top row, cells C25:F25. Because the data table is using values listed in a row, the Row input cell will be filled in and the Column input cell will remain blank.

Select 'OK' and the table will complete automatically (Figure 13.101).

Next, do the exact same thing, this time using a Profit Summary (Figure 13.102). The equations entered in cells B33:B36 are simply the Class Profits, not the returns; thus, the value for cell B33 is "=ClassA_Profit".

The analyst can now determine the different returns and profits for each.

The next step will be to graph the principal as it is paid down. This can be done by first creating a paydown summary table. Begin by creating a blank table as seen in Figure 13.103.

CMO Model - Excel

C34 {=TABLE(C22,)}

	A	B	C	D	E	F
17	Total	933,760,897			809,924,266	
18						
19	**Waterfall Assumptions**					
20	Payments per Year:		2			
21	First Payment:		9/30/2014			
22	Prepayment Rate:		4%			
23						
24	**IRR Summary**					
25	Prepayment	4%	2%	4%	5%	6%
26	A	9.73%	8.77%	9.73%	10.20%	10.67%
27	B	8.75%	8.15%	8.75%	9.03%	9.30%
28	C	6.56%	6.50%	6.56%	6.59%	6.61%
29	Residual	2.09%	3.81%	2.09%	1.25%	0.43%
30						
31	**Profit Summary**					
32	Prepayment	4%	2%	4%	5%	6%
33	A	112,991,316	125,852,598	112,991,316	108,345,109	104,409,153
34	B	224,013,977	271,980,478	224,013,977	208,171,058	195,510,160
35	C	139,899,210	168,264,325	139,899,210	130,059,374	121,922,402
36	Residual	52,532,684	140,341,349	52,532,684	27,287,753	8,228,965

Summary | Waterfall | CF Total | 1 | 2 | 3 | 4 | 5

Figure 13.102 Profit Summary

CMO Model - Excel

B63

	A	B	C	D
38	**Paydown Summary (Outstanding Principal)**			
39	HLOOKUP Ref:			
40	Period	A	B	C
41	0			
42	1			
43	2			
44	3			
45	4			
46	5			
47	6			
48	7			
49	8			
50	9			
51	10			
52	11			
53	12			
54	13			
55	14			
56	15			
57	16			
58	17			
59	18			
60	19			
61	20			
62	21			
63	22			

Summary | Waterfall | CF Total | 1 | 2 | 3 | 4 | 5

Figure 13.103 Paydown Summary Table Setup

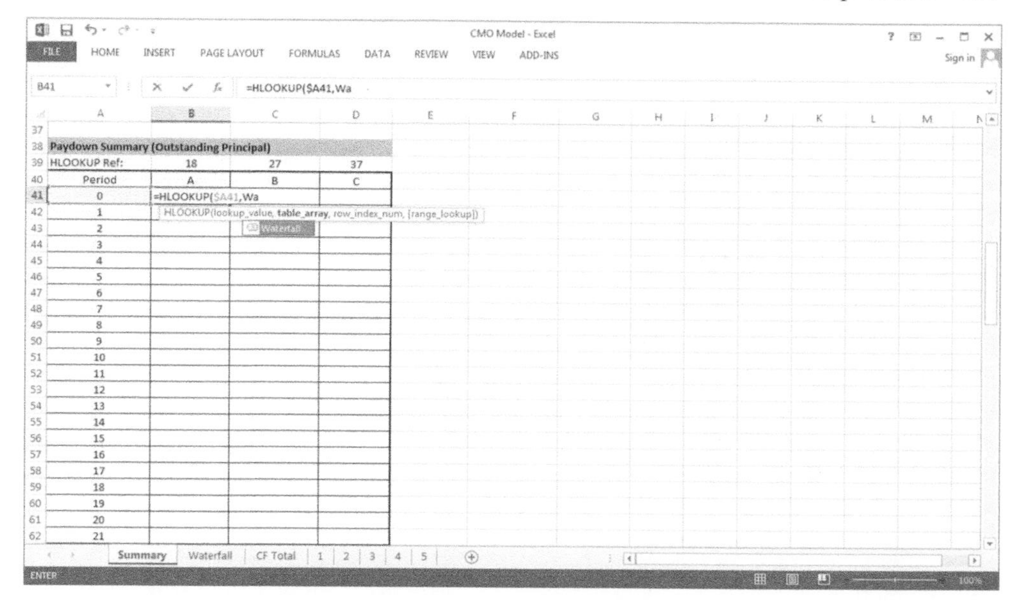

Figure 13.104 HLOOKUP Function

Notice the HLOOKUP reference line on row 39. The VLOOKUP, or vertical lookup, function has been used several times previously. In this chart the HLOOKUP, or horizontal lookup, function will be used. To determine the 'HLOOKUP Ref' for Class A, count how many lines into the Waterfall array the principal balance is. Since the array starts at line 2, this is easy and is simply line 18. For Class B and C, the lookup value will be lines 27 and 37, respectively.

When entering the formula, be sure to use direct references where possible so that the equation may be copied horizontally and vertically. Begin a horizontal lookup the same way as a vertical lookup, by entering "=HLOOKUP" followed by an open parenthesis (Figure 13.104). The first argument required is the lookup value. This is the value which must be found and for which a corresponding value will be returned. In this example the value will be period 0, or cell A41. By inserting a '$' in front of the 'A', the analyst may copy the formula horizontally and the 'A' will remain. Without it the 'A' would change to a 'B' in keeping with the pattern. By neglecting to insert a '$' in front of the '41', the equation may be copied vertically and the referenced row will change in accordance with the corresponding row. The next argument is the table array in which the value must be found. Since the array was already named the analyst can simply begin typing 'Waterfall'.

The final argument is the row from which the value must be returned once the corresponding period number is located. In this case it will be the number listed on row 39. Enter the argument as B39. By entering a '$' in front of the '39' but not the 'B' the analyst may copy the equation horizontally and vertically.

Thus the equation for A43 is "=-HLOOKUP($A41,Waterfall, B$39)". Notice the '-' in front of the HLOOKUP. Because a graph is being created of the principal due values, one wants to be sure this is entered correctly.

	A	B	C	D
38	Paydown Summary (Outstanding Principal)			
39	HLOOKUP Ref:	18	27	37
40	Period	A	B	C
41	0	455,000,000	295,000,000	183,760,897
42	1	413,723,360	303,112,500	183,760,897
43	2	372,074,070	311,448,094	183,760,897
44	3	330,050,778	320,012,916	183,760,897
45	4	287,652,202	328,813,272	183,760,897
46	5	244,877,136	337,855,636	183,760,897
47	6	201,724,448	347,146,666	183,760,897
48	7	158,193,088	356,693,200	183,760,897
49	8	114,282,088	366,502,263	183,760,897
50	9	69,990,568	376,581,075	183,760,897
51	10	25,317,740	386,937,055	183,760,897
52	11	-	377,840,730	183,760,897
53	12	-	343,213,329	183,760,897
54	13	-	308,216,123	183,760,897
55	14	-	272,846,695	183,760,897
56	15	-	237,102,679	183,760,897
57	16	-	200,981,766	183,760,897
58	17	-	164,481,703	183,760,897
59	18	-	127,600,303	183,760,897
60	19	-	90,335,442	183,760,897
61	20	-	52,685,070	183,760,897
62	21	-	14,647,207	183,760,897
63	22	-	-	159,980,852

Figure 13.105 Paydown Summary Table Completion

The equation may now be copied to columns C and D (Figure 13.105).

For the next period the analyst can copy the equations from above, but be sure to remove the '-' symbol. The equations can then be copied vertically.

Analysis

Now that all data has been summarized, the analyst may begin the analysis.

The first step will be to create a graph showing the principal balances at the end of each period (Figure 13.106).

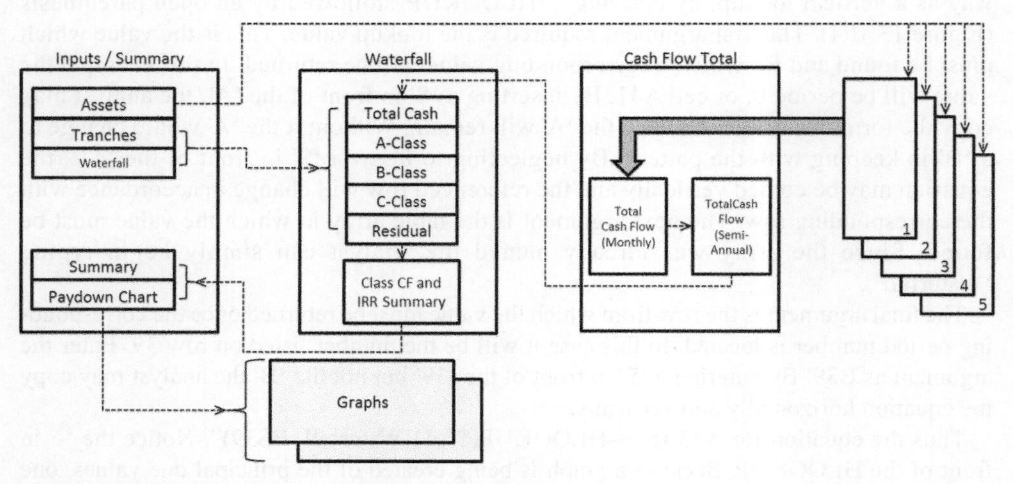

Figure 13.106 Model Flow Diagram (Output – Graphs)

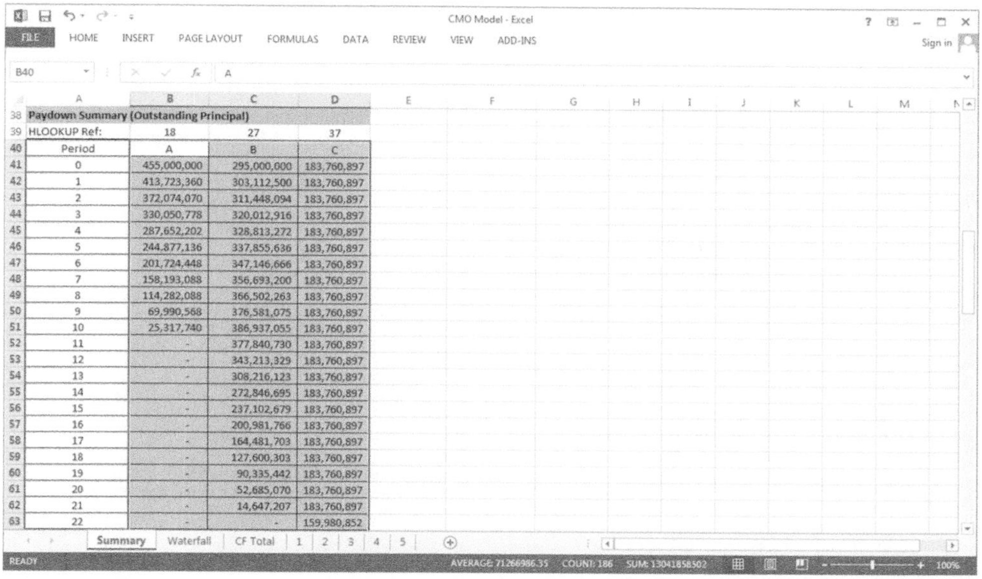

Paydown Summary (Outstanding Principal)			
HLOOKUP Ref:	18	27	37
Period	A	B	C
0	455,000,000	295,000,000	183,760,897
1	413,723,360	303,112,500	183,760,897
2	372,074,070	311,448,094	183,760,897
3	330,050,778	320,012,916	183,760,897
4	287,652,202	328,813,272	183,760,897
5	244,877,136	337,855,636	183,760,897
6	201,724,448	347,146,666	183,760,897
7	158,193,088	356,693,200	183,760,897
8	114,282,088	366,502,263	183,760,897
9	69,990,568	376,581,075	183,760,897
10	25,317,740	386,937,055	183,760,897
11	-	377,840,730	183,760,897
12	-	343,213,329	183,760,897
13	-	308,216,123	183,760,897
14	-	272,846,695	183,760,897
15	-	237,102,679	183,760,897
16	-	200,981,766	183,760,897
17	-	164,481,703	183,760,897
18	-	127,600,303	183,760,897
19	-	90,335,442	183,760,897
20	-	52,685,070	183,760,897
21	-	14,647,207	183,760,897
22	-	-	159,980,852

Figure 13.107 Data Solution for Graph Output

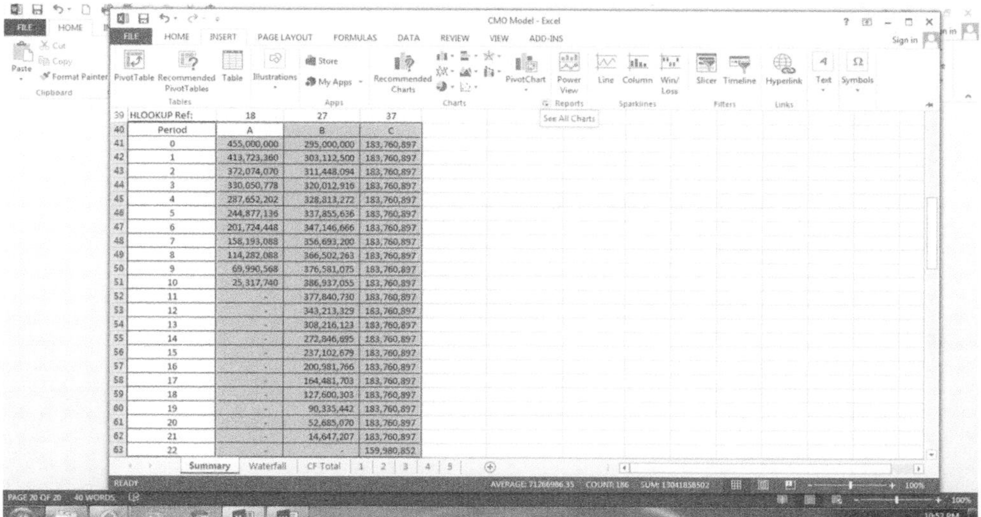

HLOOKUP Ref:	18	27	37
Period	A	B	C
0	455,000,000	295,000,000	183,760,897
1	413,723,360	303,112,500	183,760,897
2	372,074,070	311,448,094	183,760,897
3	330,050,778	320,012,916	183,760,897
4	287,652,202	328,813,272	183,760,897
5	244,877,136	337,855,636	183,760,897
6	201,724,448	347,146,666	183,760,897
7	158,193,088	356,693,200	183,760,897
8	114,282,088	366,502,263	183,760,897
9	69,990,568	376,581,075	183,760,897
10	25,317,740	386,937,055	183,760,897
11	-	377,840,730	183,760,897
12	-	343,213,329	183,760,897
13	-	308,216,123	183,760,897
14	-	272,846,695	183,760,897
15	-	237,102,679	183,760,897
16	-	200,981,766	183,760,897
17	-	164,481,703	183,760,897
18	-	127,600,303	183,760,897
19	-	90,335,442	183,760,897
20	-	52,685,070	183,760,897
21	-	14,647,207	183,760,897
22	-	-	159,980,852

Figure 13.108 Chart Solution

Begin by selecting all the data that will be used to create the graph. In this example, that would be cells B40 through D101 (Figure 13.107). Once the cells are selected, click on the 'Insert' tab.

Within the 'Charts' option, select the arrow on the bottom right (Figure 13.108).

This will open a separate menu for charts: 'Insert Chart' (Figure 13.109).

Scroll until the desired chart is located, select the chart type, and click 'OK'. In this example a line chart will be used first. The chart in Figure 13.110 will appear.

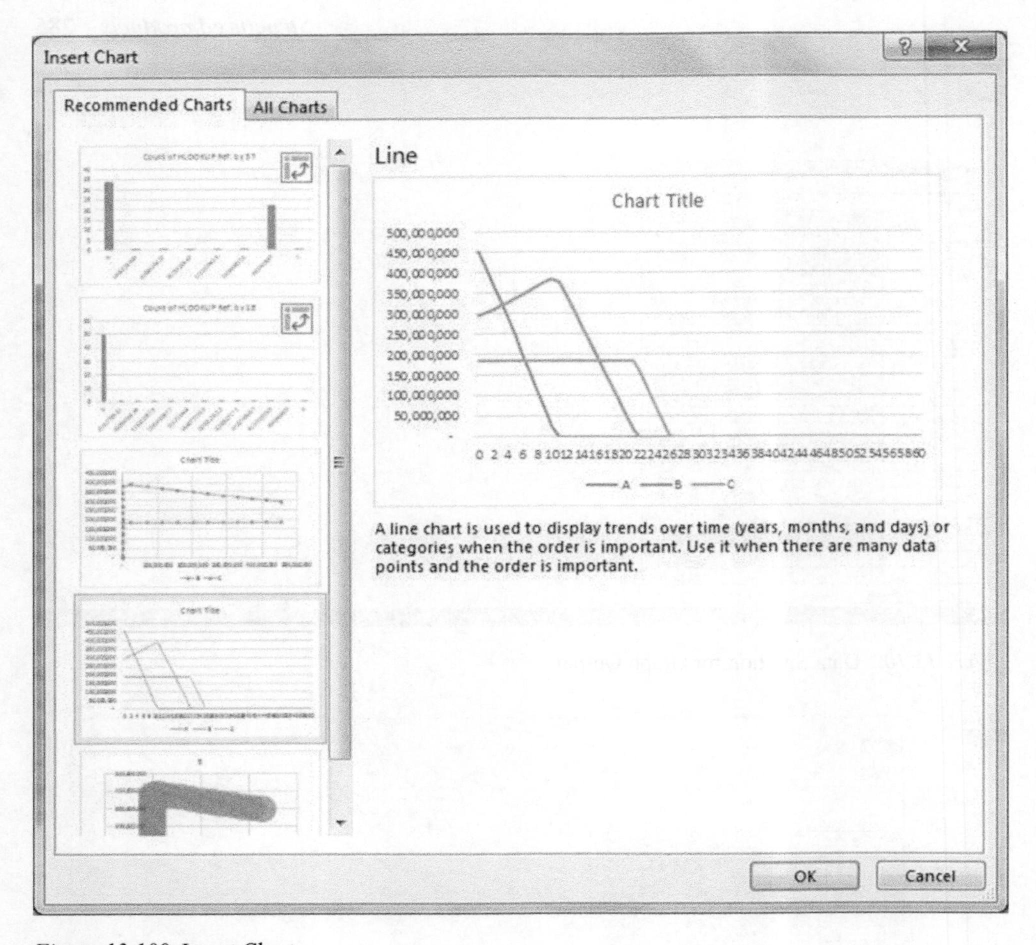

Figure 13.109 Insert Chart

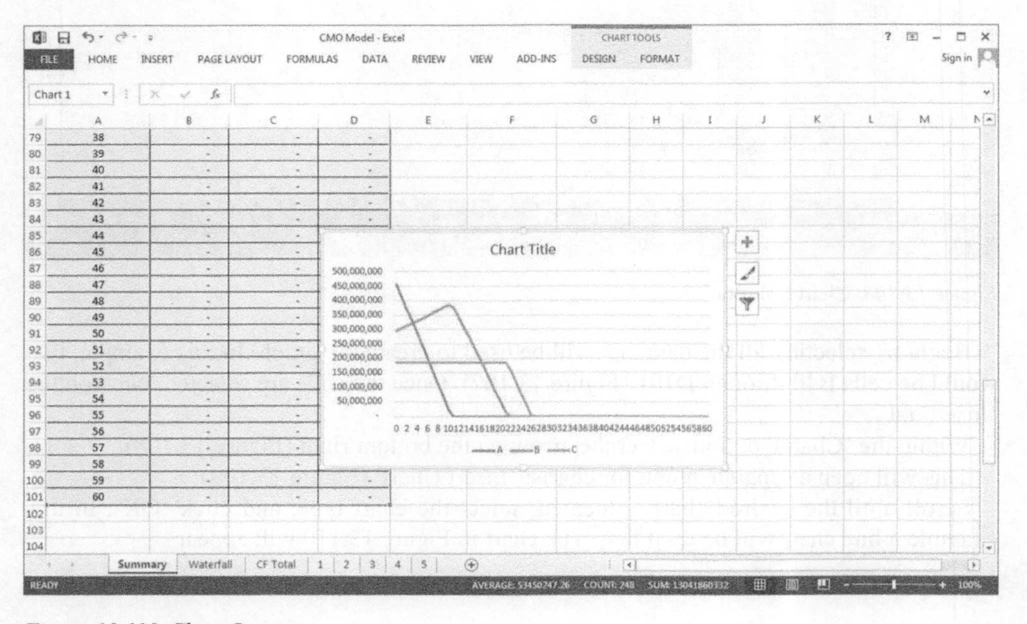

Figure 13.110 Chart Output

Notice that a legend for the data was created automatically. The next step is to format the graph by first adding titles. Select the '+' sign to the right of the graph and notice that an additional new box appears (Figure 13.111). Check the box next to 'Axis Titles'.

Two text boxes will appear. (See Figure 13.112.)

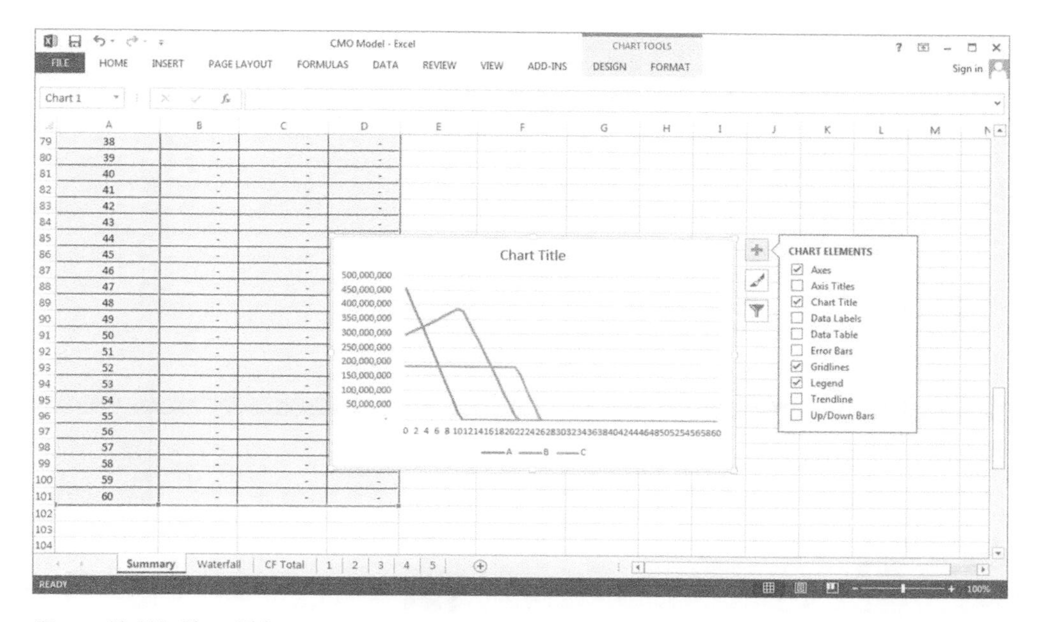

Figure 13.111 Chart Titles

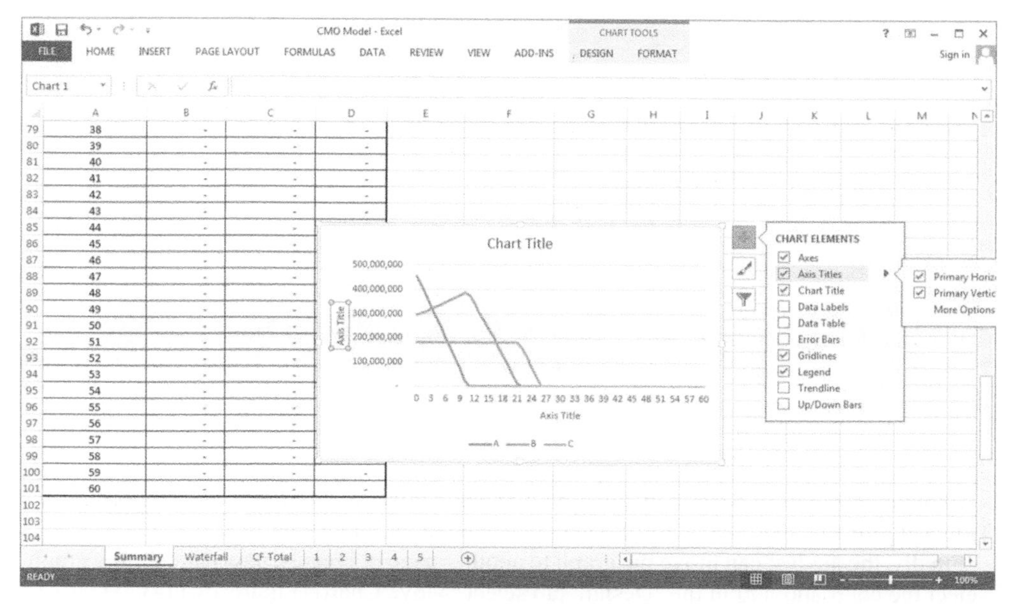

Figure 13.112 Chart Customization

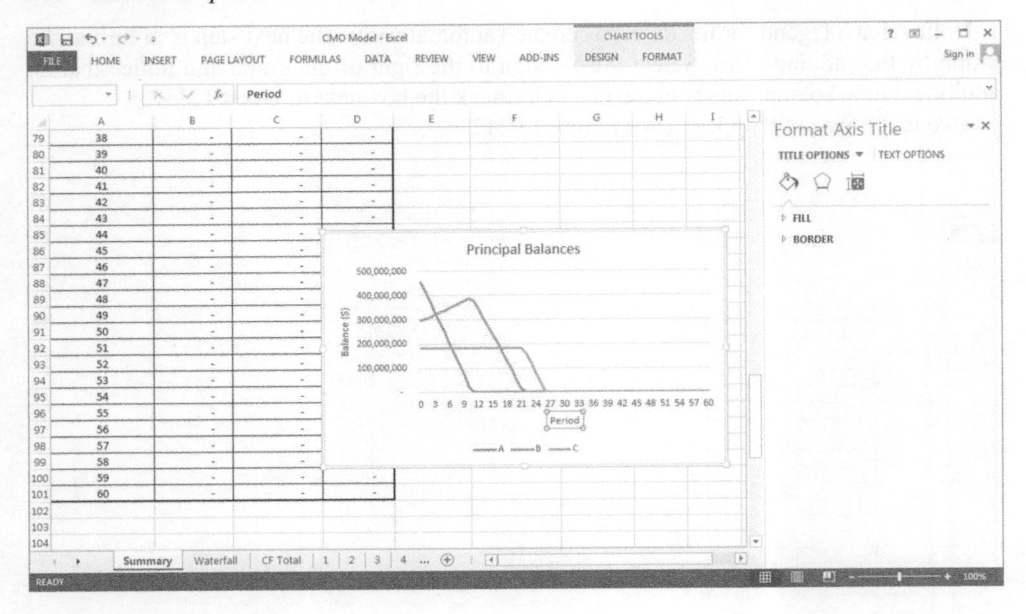

Figure 13.113 Axis Labels

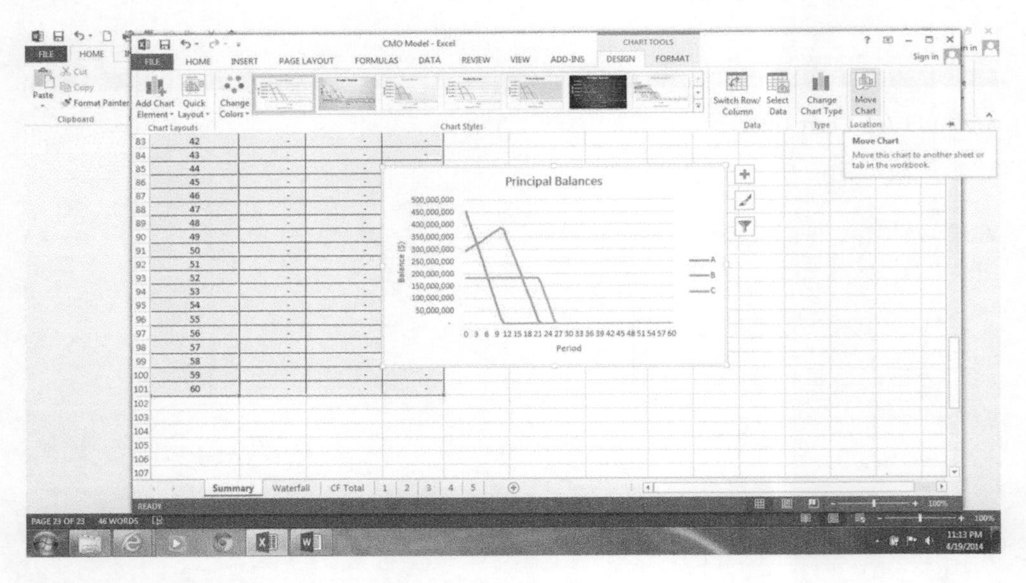

Figure 13.114 Move Chart

Change the labels as desired by clicking in the text boxes and typing over the words 'Axis Title' (Figure 13.113). The chart title may be changed the same way.

Finally, the analyst can move the graph to another page or create a new page for the graph. Select the chart and within the 'Design' tab select 'Move Chart' (Figure 13.114).

The box shown in Figure 13.115 will appear.

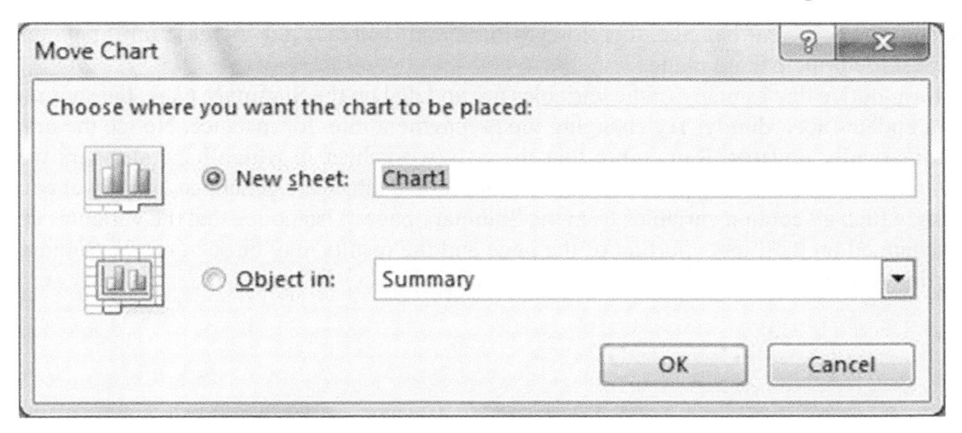

Figure 13.115 Create a New Page for the Chart

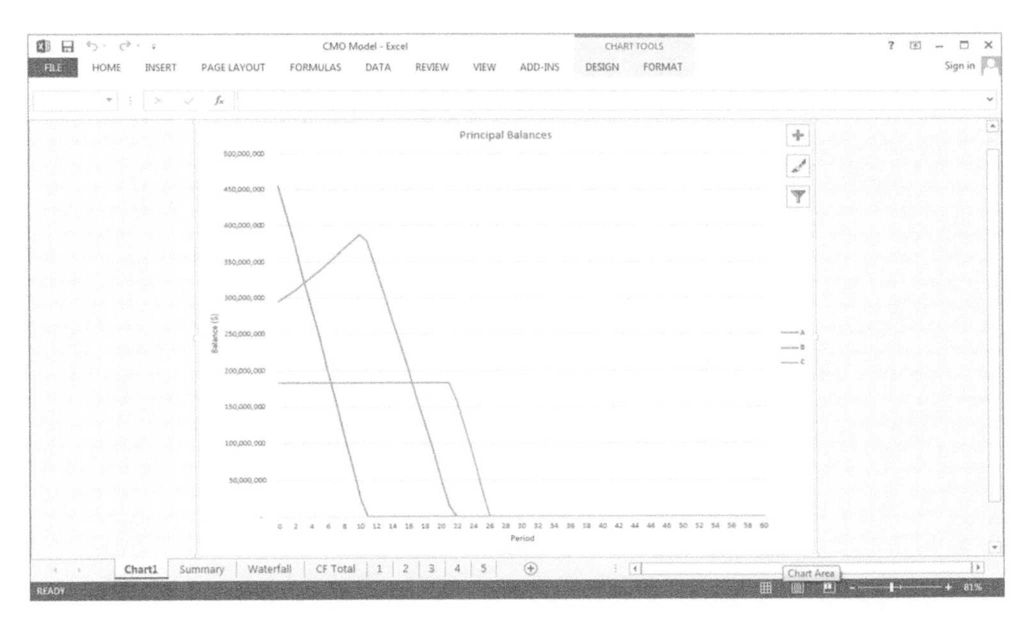

Figure 13.116 Final Output Chart

Select the desired location and then 'OK' (Figure 13.116). In this instance a new sheet was created titled 'Chart1'.

The chart produced indicates that the waterfall was created properly. Class A, as the senior tranche, will have a declining principal balance from period 0. This is because the principal in Class A gets paid before principal in any other tranche. Class B, which accrues interest, will have an increasing principal balance until Class A is fully retired. Once Class A is retired, the principal in Class B begins to pay off. Class C will have a steady principal balance until Class A and B are both fully retired. Remember that interest on Class C is paid off each period; the principal balance, therefore, will neither decrease nor increase until payments against the principal are made.

Last, notice that as many of the variables are updated on the Summary page, the summary data updates accordingly. Try changing the prepayment rate, for instance. Notice the graph automatically updates. Remember that the only worksheet in which hard numbers were entered was the Summary page. All data on subsequent tabs were generated entirely of equations, which all contain variables from the Summary page. Also notice that the variables may be changed on the Inputs portion of the page and the results may be seen on the Summary portion of the same sheet.

Index

Notes are indicated by 'n.' Page numbers in *italic* indicate figures.